Forensic Investigation of Clandestine Laboratories

Forensic Investigation of Clandestine Laboratories

Second Edition

Donnell R. Christian, Jr.

CRC Press
Taylor & Francis Group
Boca Raton London New York

CRC Press is an imprint of the
Taylor & Francis Group, an **informa** business

Second edition published 2022
by CRC Press
6000 Broken Sound Parkway NW, Suite 300, Boca Raton, FL 33487-2742

and by CRC Press
4 Park Square, Milton Park, Abingdon, Oxon, OX14 4RN

CRC Press is an imprint of Taylor & Francis Group, LLC

© 2022 Donnell R. Christian, Jr.

First edition published by CRC Press 2003

Library of Congress Cataloging-in-Publication Data

Names: Christian, Donnell R., author.
Title: Forensic investigation of clandestine laboratories / Donnell R. Christian, Jr..
Identifiers: LCCN 2021059347 (print) | LCCN 2021059348 (ebook) | ISBN 9780367629908 (hardback) | ISBN 9781032272849 (paperback) | ISBN 9781003111771 (ebook)
Subjects: LCSH: Forensic sciences. | Chemical laboratories.
Classification: LCC HV8073 .C53 2022 (print) | LCC HV8073 (ebook) | DDC 363.25--dc23/eng/20220405
LC record available at https://lccn.loc.gov/2021059347
LC ebook record available at https://lccn.loc.gov/2021059348

ISBN: 9780367629908 (hbk)
ISBN: 9781032272849 (pbk)
ISBN: 9781003111771 (ebk)

DOI: 10.4324/9781003111771

Typeset in Minion
by Deanta Global Publishing, Services, Chennai, India

Contents

4 Scene Processing   101

Disclaimer 2nd Edition

"The views expressed in this publication are those of the author and do not necessarily reflect the official policy or position of the Department of Defense or the U.S. government."

Author Biography

Donnell R. Christian, Jr. has over 30 years of forensic experience, split between two areas of expertise. He spent 15 years with the Arizona Department of Public Safety specializing in the clandestine manufacture of drugs and explosives. He is currently an explosives chemist for the Forensic Exploitation Directorate of the Defense Forensic Science Center. In between these engagements, he assisted in establishing forensic science programs in the developing democracies as the Director of International Training for Professional Business Solutions and as the Forensic Science Development Coordinator for the U.S. Department of Justice's International Criminal Investigative Training Assistance Program. With the International Criminal Investigative Training Assistance Program, he has assisted in establishing forensic science programs in the developing democracies of Armenia, Azerbaijan, Bulgaria, Georgia, Kazakhstan, Kyrgyzstan, Senegal, Turkmenistan and Uzbekistan as well as the post-conflict countries of Bosnia, Haiti, Kosovo and Iraq. Mr. Christian has utilized his experience as the basis for three books and numerous articles, to include: *Forensic Investigation of Clandestine Laboratories* (CRC Press 2003), *Field Guide to Clandestine Laboratory Identification and Investigation* (CRC Press 2004) and *Clandestine Lab Investigation Pocket Guide* (PBSI 2006) as well as chapters in *Forensic Science, An Introduction to Forensic Investigative Techniques* (CRC Press 2003, 2005, 2009, 2014) and *The Forensic Science Handbook* (Humana Press 2006.) Donnell has a Bachelor of Science degree in Chemistry and Police Administration from Northern Arizona University and a Masters' Degree in Criminal Justice from American Military University.

Introduction
First Edition

Drugs of abuse in the United States traditionally have come from a variety of foreign sources. Heroin and cocaine are produced in foreign countries. A vast amount of marijuana is cultivated and smuggled in from sources outside the U.S. However, law enforcement authorities must look inward to identify the source of clandestinely produced synthetic drugs that are increasing in popularity.

The clandestine production of drugs of abuse is not the only controlled substance that affects the public order. The clandestine production of explosives and explosive mixtures placed into destructive devices and used with criminal intent has greatly impacted the feeling of safety experienced by law-abiding citizens.

The manufacturing of controlled substances in clandestine labs is an ever-increasing problem within the U.S. Identifying and shutting down these operations has the greatest impact in stemming the flow of contraband substances. The effect of eliminating the ultimate source of the controlled substance being manufactured reaches far beyond the jailing of individuals arrested at the site. Everyone who would potentially have come in contact with the finished product, from the mid-level distributors to the end user, feels the ramifications of putting the manufacturer out of business.

The investigation of clandestine labs is one of the most challenging of law enforcement. It is a roller coaster ride of activity that requires every tool at its disposal. Traditional investigative techniques are used to develop information concerning the location of the clandestine lab and the identity of the operators. Forensic experts are used to corroborate information by establishing the identity of the final products as well as the manufacturing methods used to produce them.

No other law enforcement activity relies on forensic experts as heavily as does the investigation of clandestine labs. The forensic expert's involvement commences with the drafting of the affidavit used to obtain the search warrant. Their expertise is imperative to effectively process the crime scene. Experts analyze the samples from the crime scene in a forensic laboratory. Finally, they render opinions in a written report or in courtroom testimony. Occasionally the forensic expert may be called upon further to testify on auxiliary issues concerning the clandestine lab investigation that occur even after the criminal case has been adjudicated.

The identification, investigation and prosecution of a clandestine lab is a team effort. It is a collaboration of the efforts of law enforcement, forensic experts, scientists and criminal prosecutors to present a case that definitively demonstrates how a group of items with legitimate uses are being used to manufacture an illegal controlled substance. *Forensic Investigation of Clandestine Labs* was written to provide these groups the general information needed to understand how the different pieces of the clandestine lab puzzle fit together.

Individuals outside of law enforcement can benefit from the information in the first three chapters of *Forensic Investigation of Clandestine Labs*. Emergency responders such as police patrol officers, fire fighters, emergency medical technicians (EMT) and representatives from certain social service agencies routinely encounter clandestine labs during the course of their duties. Landlords, storage locker managers, and the public at large stumbled upon these operations with realizing they are looking at either an operational clandestine lab or its components. This knowledge will allow these groups to be able to recognize a potentially dangerous situation so they can report it to the appropriate authorities.

The goal of this book is to provide anyone involved in the investigation or prosecution of clandestine lab activity the information to guide him or her through the process of establishing the existence of a clandestine lab beyond a reasonable doubt. This book will not make the reader an expert in the clandestine manufacture of controlled substances. That can only be accomplished through training and experience.

The information in this book will provide an overview of clandestine labs. This will be accomplished by dividing the process into five sections that correspond to the various phases of the investigation and prosecution. The first section describes how to recognize a clandestine lab and the physical characteristics they all have in common. The second section deals with processing the site of a clandestine lab. The third section covers the analytical techniques that can be used in the laboratory to analyze evidence from a clandestine laboratory. The fourth section deals with the opinions that can be rendered from the physical evidence. The fifth and final section covers presenting the evidence in court.

Recognition of clandestine lab activity is the first step in the process. Section one, chapters one through three, describes what a clandestine lab is and the common elements to expect. The chapters will provide a profile of a clandestine lab operator; identify the chemical and equipment requirements, as well as the basic manufacturing techniques utilized. This section will dispel the commonly held, yet faulty, notion that the manufacture of controlled substances requires higher education, sophisticated equipment and exotic chemicals. The knowledge gleaned from this section should enable an individual to recognize a clandestine lab. An investigator should be able to articulate why a clandestine lab exists and subsequently secure a search warrant to proceed to the next phase of the process.

Some of the explanations of the manufacturing process may seem oversimplified to a forensic chemist. Yet investigators, attorneys, and jurors involved in the various segments of the investigation and prosecution of clandestine labs cannot be expected to have acquired the scientific knowledge necessary to understand the chemical processes involved. Using nontechnical terms with examples common to everyone should remove the scientific mystique. The understanding of the process of clandestinely manufacturing a controlled substance by a broad audience is more easily achieved using laymen's terms.

Forensic Investigation of Clandestine Labs does not describe in detail how to clandestinely make controlled substances. Unfortunately, there are already numerous sources of such information available to the general public. What this book does address is how controlled substances are made and how investigators, forensic experts, scientists and attorneys can identify the existence of a clandestine lab, compile the information necessary to establish what was being manufactured, how it was being manufactured; and finally, how to present the information to a jury for adjudication.

Knowing what a clandestine lab is and proving one exists are separate issues. Sections two and three (Chapters Four and Five) deal with the steps necessary to collect and identify all of the pieces of the clandestine lab puzzle. The information gathered from investigators must be evaluated. These chapters outline the steps required to process clandestine lab sites for physical evidence and analytical approaches that can be taken during the subsequent laboratory analysis.

Processing the clandestine lab scene is addressed in Chapter Four. It is more complicated than the traditional crime scene search normally associated with a narcotics investigation. Because of the chemicals involved, the site of a clandestine lab is by definition a "hazardous materials incident" and necessitates invoking different protocols for crime scene processing. Agencies such as the fire department, emergency medical personnel, local health and environmental quality personnel should be involved. The equipment requirements for processing clandestine lab scenes are more extensive because of the potential chemical exposures. Finally, there are a number of preliminary opinions that should be made at by evaluating the physical evidence observed at the scene, which necessitates an on scene expert.

Chapter Five addresses the options available to the forensic chemist who analyzes the evidentiary samples. Complete forensic laboratory analysis is a critical element of a clandestine lab investigation. The analysis of a reaction mixture is more complex than identifying the controlled substance it contains. Identification of precursor and reagent chemical as well as reaction by-products is necessary to establish the manufacturing method used. Identifying unique chemical components can be used as an investigative tool to connect the clandestine lab under investigation to other illegal activity.

Opinions, or "What does it all mean?" is the next section. A large amount of information is collected during a clandestine lab investigation. This section deals with collating information from various sources and creating a profile of the clandestine lab under investigation. What type of operation existed? What it was making? How was it being made? And, "How much could it make?" are some of the questions that will be addressed in Chapter Six.

All the work to this point may be useless if the expert's opinions cannot be relayed effectively to a jury. The final section deals with expert testimony. Chapter Seven discusses how to effectively educate the prosecutor, deal with defense attorneys, and presenting technical information to non-technical jurors.

The main focus of clandestine lab investigations in the U.S. is the manufacture of illicit drugs because the manufacturing of explosives is not illegal per se. However, placing the explosive final product of a clandestine lab into a destructive device is illegal. All of the techniques used to investigate clandestine drug labs can also be applied to the manufacture of explosive chemicals, compounds and mixtures. Chapter eight specifically addresses issues involving the clandestine manufacture of explosives.

The use of forensic evidence is essential to the successful investigation and prosecution of a clandestine lab, whether the final product is a drug or an explosive. The proper collection and preservation of the physical evidence followed by the complete analysis of the evidentiary samples are key elements. Their information is the cornerstone on which the forensic expert's opinion is based. If forensic evidence is properly handled the Court will have all of the information it needs to make a fully informed decision.

Introduction
Second Edition

The initial drafts for the first edition of *Forensic Investigations of Clandestine Laboratories* began in 1995. The intent was to address issues associated with the clandestine manufacture of contraband drugs in the United States. The first edition was released in 2004. At that point in history, clandestine lab activity was a domestic issue in which law enforcement could stem the flow of contraband drugs at its ultimate source. The main focus was on the manufacture of methamphetamine. Phencyclidine (PCP) operations were sporadically encountered. The production of crack cocaine was not considered a manufacturing operation. The rampant creation of analogs to modify the pharmacological effects of a controlled substance and avoid prosecution due to the prescriptive nature of the relevant criminal statutes was not an issue. Encountering operations that manufactured explosives or explosive mixtures was rare and with the exception of two major domestic incidents, limited to a small number of fringe ideologues or chemistry students with too much time on their hands.

The world is a different place today. The large-scale clandestine manufacture of contraband drugs, once the purview of domestic criminal organizations, has been commandeered by international drug cartels who have transferred the large-scale operations outside the United States. Domestic manufacturing methods have been simplified to reflect small user-based operations, creating a different set of hazards. The prevalence of analog drugs, particularly associated with fentanyl and synthetic cannabinoids, has skyrocketed. The use of homemade explosives (HME) to further radical ideologies has taken center stage in the global war on terror.

As a result, it became apparent that an updated version of *Forensic Investigations of Clandestine Laboratories* was needed. The principles associated with the clandestine manufacture of contraband substances, whether they be contraband drugs, explosives or chemical weapons, have not changed since publication of the first edition. What has changed is the mindset associated with the application of these principles. The perspective of the original treaties was that clandestine laboratories were a U.S. issue associated with the production of a limited number of contraband substances, specifically methamphetamine. The principles were broad enough that they could be incorporated into the manufacture of other contraband drugs as well as ancillary applications for the occasional situation in which the manufacture of explosives or explosives mixtures was encountered.

As previously stated, the world is a different place today. The compounds that are being clandestinely manufactured today may have changed, and the list has expanded. The countries in which these operations are located may be outside the United States, and the issue has become global. However, the basic principles associated with forensic investigation of clandestine laboratories remain the same. The generic process used to manufacture methamphetamine is the same generic process that is used to manufacture analogues of fentanyl and synthetic cannabinoids. The generic process used to manufacture a monomolecular explosive is the same generic process that is used to manufacture a fuel/oxidizer explosive. More importantly, the same generic process used to manufacture contraband drugs is the same generic process used to manufacture explosives.

The wonder of science is its continuity. This continuity can be utilized as a tool that will allow the clandestine laboratory investigator to transition from the investigation of contraband drug manufacturing to the manufacture of homemade explosives when the physical evidence dictates the change of direction. Once a person understands the process, they can take the chemicals, equipment and conditions located at the scene of a suspected clandestine laboratory operation and establish the manufacturing method and the probable end product.

The first edition was written with a law enforcement perspective. The collection, preservation, examination and analysis of physical evidence was performed to meet the standards of the U.S. criminal justice system. This perspective provided for the ability to debate the facts of the case in a court of law, allowing the trier of the fact (judge or jury) to decide the fate of the subject of the investigation.

The perspective of this edition has expanded to be incorporated as a tool for military and intelligence applications. The use of this information by these communities may differ from that of law enforcement and criminal justice applications. However, the standards used for the collection, preservation, examination and analysis of physical evidence during these forensic investigations must be the same or done with more resolve. This is due to the fact that the end results of these forensic investigations can, and do, result in prosecutions with extreme prejudice in which the subjects of these prosecutions cannot provide an alternate theory of the facts, as in a United States-based criminal prosecution.

The identification, investigation and prosecution of a clandestine lab is a team effort. It is a collaboration of the efforts of law enforcement, military personnel, forensic experts, scientists, intelligence analysts and criminal prosecutors to present a case that definitively demonstrates how a group of items with legitimate uses are being used to manufacture a substance with a nefarious use.

Forensic Investigation of Clandestine Laboratories was written to provide these groups with the general information needed to understand how the different pieces of the clandestine lab puzzle fit together. Individuals outside of law enforcement, military and intelligence communities can benefit from the information in the first three chapters of *Forensic Investigation of Clandestine Laboratories*. Domestic emergency responders such as police patrol officers, fire fighters and emergency medical technicians (EMT) as well as military personnel involved in combat operations may come into contact with clandestine labs as part of their routine activities. This knowledge will allow these groups to be able to recognize a potentially dangerous situation so that they can take the appropriate actions.

The goal of this book is to provide anyone involved in the investigation or prosecution of clandestine lab activity with the information to guide him or her through the process of establishing the existence of a clandestine lab beyond a reasonable doubt. This book will not make the reader an expert in the clandestine manufacture of contraband substances. That can only be accomplished through training and experience. This information is a stepping stone in that journey.

Basics of Clandestine Manufacture

1

Abstract

A forensic investigation in its most fundamental form can be equated to the assembly of a puzzle whose pieces have been mixed with pieces of other puzzles. The successful restoration of the puzzle hinges on the recognition of relevant pieces. Once they have been identified and separated from the mix, a meaningful attempt at reconstruction can commence.

Clandestine manufacturing operations, regardless of the end product, utilize ordinary items to produce extraordinary things. These operations are located in environments that are not associated with the manufacture of drugs or explosives. In many cases, the operators are living in or in close proximity to the operation, using everyday living items as the tools in the manufacturing process. The key to a successful forensic clandestine manufacturing investigation is to be able to discern what is being manufactured and the process being used in a situation that at first glance may appear to be nothing more than bad housekeeping.

This chapter will address lab operators, the general manufacturing process and the needs triangle. The characteristics of the four types of clandestine lab operators (small-scale, commercial, educated, terrorist) will be discussed. The generic manufacturing processes (extraction, conversion, synthesis and compounding) encountered in clandestine labs will be presented. The three sides of the "Needs Triangle" (equipment, chemicals and knowledge) will be reviewed. Practical examples of these concepts will be presented in an effort to establish a link between theory and practice.

The ultimate goal of the chapter is to provide the reader with the information required to recognize the pieces of a clandestine lab puzzle in a box of random puzzle pieces. Once the relevant pieces have been separated, the knowledge from this chapter should allow the investigator to begin the reconstruction process. If sufficient pieces are identified and correctly assembled, a clear picture of what was being manufactured and the process used to meet that end can be established.

DOI: 10.4324/9781003111771-1

1.1 Introduction

Clandestine manufacturing operations, regardless of the end product, utilize ordinary items to produce extraordinary things. Chemicals with legitimate residential, commercial and agricultural uses are the building blocks of contraband drugs and explosives. Variations of items found in homes around the world are repurposed to facilitate the manufacture of the illicit substances. Simply changing the lenses used to observe a group of chemicals and equipment can change their perceived use. Through rose-colored glasses, a combination of items appears to be a harmless part of the environment in which they are encountered. However, through gray lenses, those same items in that same environment take on an illicit meaning, with the shade of gray determining the darkness of the intent.

Before a clandestine manufacturing investigation can begin, the investigator must be able to recognize that such an operation exists. To do this, they must be familiar with the basic techniques used to produce contraband substances. This chapter will provide information concerning basic manufacturing techniques used by clandestine operators. Legitimate sources of chemicals and equipment will be used in the explanations, followed by examples of underground alternatives.

Clandestine labs come in a variety of shapes and sizes; their sophistication is limited only by the education and imagination of the operator. Complicated equipment and exotic chemicals are not required to manufacture drugs of abuse or explosives (contraband substances). Most of the equipment and chemicals found in a lab have legitimate uses and can be obtained from a variety of legitimate retail outlets. Therefore, the forensic clandestine lab investigator must be able to recognize the combinations of equipment and chemicals that are used to manufacture controlled substances and to determine whether the combination is coincidental or intentional.

For simplicity, the illicit production of drugs and explosives will be collectively referred to as a clandestine lab. The word "laboratory" or "lab" conjures up certain images and perceptions of a scenario. Using the words "manufacturing operations" to describe the same scenario summons a different set of images and perceptions. "Clandestine labs" is a generic term traditionally used to describe the illicit manufacture of contraband drugs. Slightly different terminology has been used to describe the illicit manufacture of explosives. The term "clandestine lab" will be used, regardless of the end product, to reinforce the fact that the same general processes are used to produce both drugs and explosives.

A clandestine laboratory is literally a secret room or building equipped for scientific research or manufacture. Clandestine labs may not in themselves be illegal. The substances they produce and the act of manufacturing them is what may be controlled.

Webster's Dictionary defines manufacturing as "the act of making goods, by hand or machinery." The legal definition of manufacturing is different. The Code of Federal Regulations 21CFR1300.01 defines manufacture as "the producing, preparing, propagating, compounding or processing of a drug or other substance or the packaging or repackaging of such substance or labeling or relabeling of the commercial container of such." This definition relates more directly to the production of drugs of abuse and is applicable in one form or another throughout the United States. Investigators should compare the federal definition with their local statutes and consult with district attorneys responsible for prosecuting clandestine labs for any differences.

The legal definition includes acts of packaging and labeling that are not normally associated with the "making of goods." The acts of preparing the final product for distribution or sale expand the perception of laboratory operations from just the traditional mixing and extracting of chemicals. Under this definition, the mirror act of making little ones out of big ones is the same as making the big ones to begin with.

Many criminal statutes dealing with the manufacture of controlled substances state that the possession of chemicals and equipment for the purpose of manufacturing a controlled substance is illegal. The statute may not specify that all of the equipment and chemicals must be present, but only that a combination exists sufficient for a reasonable person to believe that a manufacturing operation exists. Other statutes may require all of the components of the operation to be present. Either situation places the burden of demonstrating how the different components can be utilized to manufacture a controlled substance on the government. Understanding the different manufacturing processes will allow the investigator or prosecutor to differentiate and articulate how the presence of cold medications, rubbing alcohol, coffee filters and glass jars can be legitimately present in one situation and yet be used in a clandestine lab that manufactures a controlled substance in another.

The manufacture of explosives is a slightly different issue. The simple possession of explosives is not regulated to the same extent as drugs. Explosives have a wide range of legitimate applications. However, for public safety reasons, the manufacture, distribution and storage of explosives are regulated. In the interest of public safety, the U.S. government regulates the importation, manufacture, distribution and storage of explosive material through 27 CRF 55 and chapter 40 of Title 18 of the United States Code (18 USC 40).

The federal criminal laws associated with explosives do not define manufacture per se, deferring to the generally accepted definition. However, 18 USC 841 defines a "manufacturer" as "any person engaged in the business of manufacturing explosive materials for purposes of sale or distribution or for his own use." Additionally, 18 USC 842.p delineates the prohibition of "Distribution of Information Relating to Explosives, Destructive Devices, and Weapons of Mass Destruction" with the intent to commit a crime. (Intent being the operative word, which will be discussed later in this chapter.) As with drug statutes, the criminal laws within individual states have their own definitions and sanctions associated with the possession and manufacture of explosives.

1.2 Lab Operators

Traditionally, there are three distinct categories of clandestine lab operators: small-scale, commercial and educated. The size of the lab may vary among operator categories, but the principles that demonstrate an operation exists remain the same. World events have expanded the number of groups to include terrorists. The original three are generally associated with the production of contraband drugs, occasionally dabbling in explosives. Terrorist operations, by contrast, focus on the manufacture of explosives.

The *small-scale operator* is the one whose lab is most commonly encountered. He is generally a drug user himself, using his own product as well as selling a portion to support his habit. Financial gain in itself is therefore often not the only objective of this operation. All levels of such an operation usually take place at one location. Generally, these operators are uneducated. They obtain their chemicals through retail purchases at grocery and drug stores, local chemical supply houses or mail order suppliers. At times they will shoplift over-the-counter preparations that contain regulated precursor chemicals. Operations are conducted in single-family homes and apartments as well as hotel and motel rooms. These operations are found in the poorest part of a city or the most affluent. The types of motels and hotels used vary from skid row to major luxury chains. The distribution of the final product is usually in the same area as the lab, and the operator is usually the dealer. Small-scale drug manufacturers have no political agenda other than maintaining their lifestyle.

The *commercial operator* manufactures for financial gain. These operators may or may not be users themselves. Different portions of a commercial operation may take place at separate locations in an effort to avoid detection by law enforcement. The commercial operation may have one "cook" who holds the knowledge concerning the manufacturing process. He may or may not have chemistry training. The "cook" is only at the lab site during critical portions of the operation, and for the balance of time, so-called "lab rats" are present to monitor the operation and secure it from theft or detection. The commercial operator has an established network to obtain large quantities of the necessary chemicals and equipment required to manufacture his product. A separate distribution network for the final product is usually established away from the lab to avoid detection. The commercial operator does not generally participate in street-level sales.

During the 1960s, outlaw motorcycle gangs began producing their own methamphetamine in these labs and dominated the distribution of the drug within the United States. Today, there are two major forces fueling the methamphetamine trade within the United States: these include well-organized manufacturing and trafficking groups based in Mexico and a widely scattered series of local methamphetamine producers, predominantly based in rural areas around the country. They operate the well-organized, high-volume "super lab" defined by the Drug Enforcement Administration (DEA) as a clandestine lab operation that is capable of producing 10 pounds or more of methamphetamine in a single production cycle. During the height of domestic methamphetamine production, the DEA estimates that fewer than 5% of clandestine labs seized are classified as super labs. With increased enforcement efforts in the United States and changes in the law limiting access to precursor chemicals, large-scale production of methamphetamine has shifted to operations outside the continental United States (OCONUS).

Large-scale operators manufacture more than methamphetamine and have been operating OCONUS for years. Cocaine and heroin have been produced in large-scale operations for decades. The main source of the methylenedioxyamphetamine (MDA) family of "party" drugs is synthesized and tableted outside the United States. Fentanyl, its analogs and other opioids are also produced in OCONUS operations and imported into the United States. These operations may go undetected because the precursor chemicals utilized, which have legitimate uses, are not those used in the manufacture of methamphetamine, cocaine or heroin, which are the focus of most enforcement efforts.

The *educated operator* is the least often encountered type of clandestine lab operator. He usually has formal training in chemistry, which was obtained through traditional education or from on-the-job training. He therefore has located legitimate access to sources of regulated chemicals. He may even be using his job site as a manufacturing location without the knowledge of his employer. The educated operator may be a hybrid between the small-scale and the commercial operator. In some instances, he acts as the "cook" in operations that require chemical expertise, such as the synthesis of lysergic acid diethylamide (LSD) or fentanyl and its analogs. At other times, he can be found in a small-scale operation working as both manufacturer and distributor. Graduate students, engineers and government chemists have all been arrested for manufacturing controlled substances while using their employer's facility or for purchasing chemicals and equipment through their employer without his knowledge. The educated operator may or may not have a distribution network established. He may or may not be a drug user. Profit is much more the driving force for the educated operator than for his uneducated small-scale counterpart.

The *terrorist* has been added to the categories of clandestine manufacturing operations. Terrorists were considered a subset of the previously discussed groups. However, the prevalence of their activity and their focus on the manufacture and use of explosives deem them worthy of a unique designation.

The terrorist is a person or group who uses unlawful violence and intimidation, especially against civilians, in the pursuit of political aims. They can have one or more characteristics of the aforementioned groups. The brothers associated with the Boston Marathon bombing are characteristic of the small-scale operator. The Unabomber, a reclusive mathematics prodigy, is a classic educated operator. Any of the 68 groups designated as foreign terrorist organizations by the U.S. Department of State (DoS) or the group responsible for the 1993 World Trade Center bombing could be considered commercial operations. The individuals responsible for the bombing of the Alfred P. Murrah Federal Building in Oklahoma City, Oklahoma are small-scale operators with commercial-scale results. The common thread in all these is the manufacture and use of explosives in pursuit of a political agenda.

Clandestine lab activity associated with terrorists is not limited to the manufacture and distribution of explosives. Terrorist organizations, like those designated by the U.S. DoS, require operating capital. The manufacture and distribution of contraband drugs is a lucrative source of operating funds. As such, it is not uncommon to find a link between terrorist organizations and operations that manufacture and distribute drugs of abuse (Table 1.1).

1.3 Manufacturing Processes

There are a number of different processes that can be used to manufacture a controlled substance. The one employed will depend on the starting

Table 1.1 Operator Characteristics

Characteristic	Small-Scale	Commercial	Educated	Terrorist
Chemical Education	No	Yes	Yes	Maybe
Drug User	Yes	Maybe	Maybe	Maybe
For Profit	No	Yes	Maybe	Maybe
Political Agenda	No	Maybe	Maybe	Yes
Legitimate Chemical Supply	No	Maybe	Yes	Maybe
Single Location	Yes	No	Yes	Maybe
Local Distribution	Yes	No	Maybe	Maybe

materials used and the end product desired. Each process may be encountered alone or in combination with one or more of the others. One clandestine operation may incorporate multiple manufacturing processes to obtain the end product. The four basic manufacturing processes used in clandestine labs are extraction, conversion, synthesis and tableting/compounding.

The following is a generic example of a multi-method process used to create a controlled substance.

1. The necessary precursor chemical is **extracted** from a bulk substance.
2. The precursor is then **converted** into substance "B."
3. Substance B is combined with three other chemicals to **synthesize** compound "C."
4. Compound C is **extracted** from the reaction mixture using the appropriate solvent.
5. Compound C is **converted** from the free base into its salt form.
6. The salt form of Compound C is then **extracted** from the liquid.
7. The resulting product is packaged (**tableted**) for distribution.

This single operation thus actually contains seven processing steps. All four of the basic manufacturing processes were utilized at some point in the operation. The simplest operation will generally require at least two processing steps.

It is not uncommon for large-scale clandestine lab operators to perform individual manufacturing processes at separate locations. This is once again done in an effort to avoid detection. Precursor chemical extraction may take place at location one. Synthesis and purification extractions then occur at location two. Conversion into the salt form may be done at a third location, and packaging for sale at the point of distribution.

The presence of only one process of the sequence does not make the operation any less a clandestine lab. It is the investigator's responsibility to recognize the process or part thereof and to articulate how it fits into the manufacturing method the operators are using.

1.3.1 Extraction Process

Extraction labs remove raw materials from a mixture. This is accomplished by using the desired component's physical and chemical properties to separate it from the mixture. No chemical change in the raw material occurs during the process. Examples of extraction labs include hashish production, coca paste productions and extractions from pharmaceutical preparations (Figure 1.1).

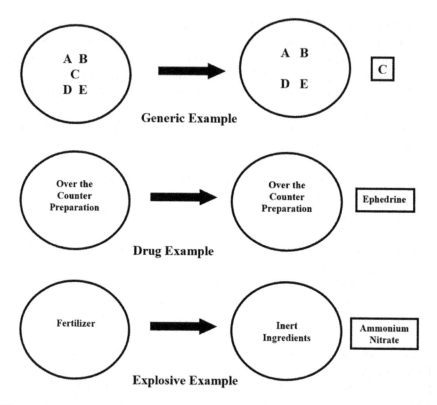

Figure 1.1 Extraction lab chemistry.

Separating the resin containing the cannabinoids and tetrahydrocannabinol (THC) from the marijuana leaves produces hashish and hash oil. Hashish is made by physically removing the resin from the leaves. The resin obtained is then collected and compressed into brick form. Hash oil is obtained by removing the resin from the leaves through the use of solvent extraction. In either case, the chemical structure of the extracted cannabinoids remains intact.

The production of heroin begins with the removal of the raw opium from the poppy. This physical extraction does not change the chemical structure of the morphine or codeine contained in the opium. It only removes it from the plant so that the morphine can be processed into heroin. Some operations take the additional step of chemically extracting the morphine from the opium prior to the conversion into heroin.

A number of over-the-counter pharmaceutical preparations contain the precursor chemicals used for the manufacture of controlled substances. These preparations are placed into a solvent, and the desired component is allowed to dissolve into the liquid. The liquid containing the component is removed from the solids and evaporated, leaving the component of interest.

Some pharmaceutical preparations contain a controlled substance in an aqueous solution. The item of interest is extracted from the solution by simply evaporating the water. This process by itself may not be illegal. However, the resulting product may be. The act of extracting chemicals that are regulated from an uncontrolled preparation may demonstrate the intent to perform an illegal act.

The extraction process can be found in explosive labs as well. Using the appropriate solvent, ammonium nitrate can be extracted from the inert ingredients added to fertilizers in an effort to deter their use in explosive mixtures. Capitalizing on trinitrotoluene's physical properties, it can be extracted from military munitions and repurposed into improvised explosive devices (IEDs).

As demonstrated in the example, the extraction process can be an important part of both the initial and the final phases of a manufacturing route. The purity of the starting material is as important as that of the finished product. Therefore, the isolation of these compounds through the extraction process is equally important in both situations.

1.3.2 Conversion Process

The conversion process takes a raw material and changes it into the desired product. This involves minor structural changes within the molecule of the compound or of the chemical's *salt form*. Functional groups may be added or removed from the molecule, somewhat like pieces on a tinker toy. The drug of interest can be changed from its salt form to the free base form or from the free base form to the salt form.

Examples of the conversion process in drug chemistry include the conversion of cocaine hydrochloride into free base or "crack" cocaine and the conversion of ephedrine or pseudoephedrine into methamphetamine by converting a hydroxyl group on ephedrine to a single hydrogen. Explosive examples include converting the inert free urea into the explosive urea nitrate.

The simplest conversion process is the conversion of a free base drug to its salt form, or from the salt form to the free base (see Figure 1.2a). The simple act of adding a strong acid or a base to a water solution containing a drug will convert it to its salt or free base form. This act changes the physical and chemical properties of the drug, allowing it to be extracted from liquid solutions. Depending on whether an acid or a base was added to the water, the drug either dissolves into or precipitates out of the water. Structurally, the drug remains unchanged.

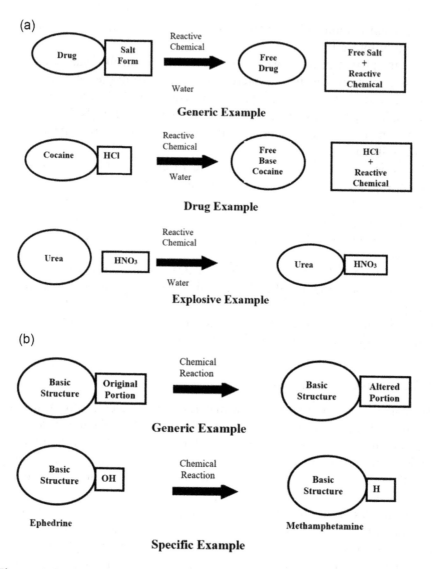

Figure 1.2 1.2a: Salt form conversion lab. 1.2b: Molecular structure conversion lab.

This conversion process functions in both directions. The creation of the salt form of a compound from the freebase is the conjugate of the creation of the freebasing process. The creation of urea nitrate is an example of this reverse reaction.

The addition or removal of a functional group(s) to or from a molecule is another form of a conversion process. A chemical reaction adds to, or takes away a portion of, the original compound, leaving the skeleton of the compound unchanged (see Figure 1.2b). The resulting molecule will have different physical and chemical properties. In the case of drugs, the original and final compounds will have different physiological effects on the body.

1.3.3 Synthesis Process

The synthesis process is a chemical reaction or series of chemical reactions in which molecules or parts of molecules are combined to create a new molecule. This process can be equated to a chemical erector set. It differs from the conversion process in that the skeleton of the resulting molecule is a sum of the molecules or significant parts of the molecules involved in the reaction. Lysergic acid diethylamide (LSD), phencyclidine (PCP), phenylacetone (P2P) and certain methamphetamine reactions are examples of drugs produced using the synthesis process.

In recent years, educated operators have been using the synthesis process and the specificity of the drug statutes to manipulate the structure of a controlled substance to create an analog with similar effects whose structure is different enough not to be subject to regulation. Analogs of fentanyl and the synthetic cannabinoids commonly referred to as "Spice" fall into this group of drugs.

Most monomolecular explosives are created using a synthesis process. One or more nitro ($-NO_2$) groups are added to the chemical serving as the explosive's structural foundation, using a series of chemical reactions facilitated by the appropriate reagent chemicals. The explosive nature of the resulting product increases with each addition (Figure 1.3).

Even though the synthesis process sounds complicated, many reactions do not require exotic equipment or lengthy reaction times. Some reactions can take place in plastic buckets in a matter of minutes or hours. Other reactions do require sophisticated equipment or extended reaction times to achieve the desired results. The only way to differentiate between a conversion process and a synthesis process is to compare the structures of the precursor chemicals with that of the final product.

1.3.4 Tableting/Compounding

The terms "tableting" and "compounding" have origins associated with the production of pharmaceuticals and have taken on meanings within the investigation of clandestine labs. Drug compounding is often regarded as the process of combining, mixing or altering ingredients to create a medication tailored to the needs of an individual patient. Compounding includes the combining of two or more drugs (or components of an explosive mixture).

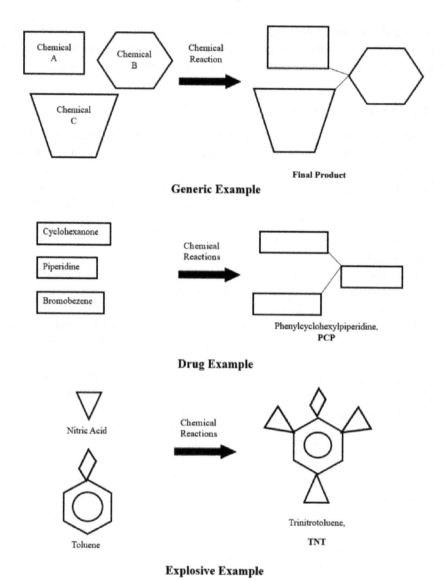

Figure 1.3 Synthesis lab chemistry.

Tableting adds a step in the preparation of the final product. The compounded mixture is placed into a press and compressed into tablets. Originally, the tablets were designed to mimic legitimate pharmaceutical preparations. This is true today. However, the impressions within tablet designs of today are used as a marketing tool by clandestine operators to identify the active ingredients. The tableting process can also apply to bulk quantities of contraband drugs. Hydraulic presses are used to compress cocaine, hash and heroin into solid bricks, some complete with a distinct logo, in preparation for distribution.

Clandestine labs that are involved in the tableting process are placing the finished product into dosage forms or into smaller, more salable units for distribution. Statutes in some jurisdictions include the act of packaging and repackaging in the definition of manufacturing. This language may interpret the act of placing the final product into a container for distribution (packaging) or dividing the container of final product into smaller containers (repackaging) as a manufacturing process. Thus, the act of making little ones out of big ones could also legally be equated to the actual production of the material itself.

This is a very subjective area. It is essential that the investigator consult with the local prosecutor to ascertain that the act of packaging items for sale or distribution can be statutorily considered manufacturing in that jurisdiction.

Compounding is an essential element of the manufacture of many homemade explosives (HMEs) utilized by domestic and international acts of terror. Many HME preparations are a mixture of specific oxidizing chemicals with an appropriate fuel combined in a specific ratio. How the components are mixed and placed into the explosive device is more critical than how similar drugs are packaged for distribution. The homogeneity of the mixture impacts the explosive potential of the mixture, whereas the uniformity of the amount of active ingredient in a drug is not of major concern (Figure 1.4).

1.3.5 Combination Labs

A combination of processes is used to manufacture a controlled substance. It is not uncommon for more than one process to be observed at any given clandestine lab site. The size and scope of the operation will often determine exactly how many processes are seen at the site.

A small-scale operator may extract the precursors needed from over-the-counter medication. He then converts the precursors into the controlled substance. Finally, he packages (tablets) the final product into dosage units or into smaller quantities for sale or distribution. All this can occur in a single hotel room, camping trailer, kitchen or garage.

Commercial operators, in an effort to avoid detection, may choose to perform different phases of their operation at separate locations. The extraction of precursor chemicals from legitimate sources may occur at one location, and the synthesis may be done at another. The oily free base compound is often transported to a third location and then converted to the solid salt form. The powder may be then transported to a final location, where it is prepared for distribution.

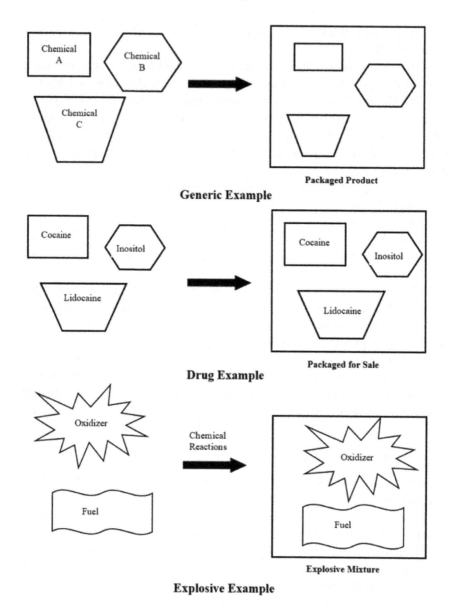

Figure 1.4 Tableting and compounding.

In both scenarios, the same processes existed. Precursor chemicals were extracted; the precursor was converted or synthesized into the controlled substance. The controlled substance was converted into a usable form; and finally, the final product was packaged for distribution. The only difference is that in the first scenario, everything took place at the same location, while in the second, these processes occurred in four separate locations but under the same umbrella.

This multi-phase manufacturing approach applies equally to the manufacture of explosives. Precursor chemicals are made or manufactured at one location and transported to a different location for conversion, synthesis or assembly. However, due to the hazardous nature of the components and their explosive end product, most HME operations perform all of the operations in the same general vicinity.

The production of fuel oxidizer explosives (FOX) is unique in that it is generally a single-step operation. The individual components have numerous legitimate commercial and agricultural uses and are generally nonreactive when properly packaged. However, their combination (compounding) creates a reactive mixture capable of explosive results when properly initiated.

1.4 Needs Triangle

Clandestine labs need equipment, chemicals and knowledge to be complete. This "Needs Triangle" theme is recurrent in many areas of science and life. As in any triangle, if any one of the three elements is eliminated, the system will not be complete. The amount of each component may vary, but each must be present for the operation to exist (Figure 1.5).

1.4.1 Equipment Needs

The equipment needs of a clandestine lab vary with the manufacturing method employed. Independently of the manufacturing method, two items are universally encountered in drug manufacturing operations: a triple-beam balance (found in any high school science lab) and surprisingly, pornography. Research indicates that methamphetamine and related psychomotor stimulants can increase the libido in users; this is in contrast to opiates,

Figure 1.5 Needs triangle.

which actually decrease the libido. However, long-term methamphetamine use may be associated with decreased sexual functioning, at least in men. Pornographic materials and a triple-beam balance are seen so often in clandestine lab seizures that law enforcement inside humor often declares their mere presence to "officially" designate the operation a clandestine lab.

All joking aside, there do exist common threads among the equipment needs of clandestine labs. Each manufacturing method will, of course, have its own equipment requirements, which can usually be satisfied by using the scientific equipment that was designed to perform the process. However, to avoid detection, many clandestine lab operators have designed alternatives to the traditional equipment, opting to use items that can be obtained from the grocery or hardware store.

The manufacturing methods used by clandestine lab operators include reflux, distillation and extractions. Understanding the mechanics of the scientific equipment used in the various manufacturing processes will allow the investigator to recognize the alternative equipment that is frequently encountered in clandestine labs.

1.4.1.1 Reflux

Refluxing is one of the most common methods used in the synthesis and conversion processes. This is a controlled boiling process in which the evaporated liquid is condensed and then returned to the reaction mixture. A slang term for the reaction mixtures found in clandestine labs is "soup," with the lab operator called the "cook." These are actually very appropriate terms. In essence, the manufacturing of a controlled substance using the reflux method is very similar to cooking soup (Figure 1.6)

When making soup, the cook combines the ingredients and boils them for a period of time. The only equipment necessary is a stove and a pot with a lid. When making a controlled substance using the reflux method, the cook combines the ingredients and refluxes them for a period of time. The refluxing apparatus is a specialized stove and a pot with a lid.

A source of heat is needed to cook the soup. A heating mantle is used in a traditional reflux apparatus. This is the equivalent of the stove a chef uses to provide the heat required to boil the soup. The shape and size of the heating mantles vary to fit the reaction vessel used in the refluxing operation. (Appendix C depicts a variety of commercially available scientific equipment used in the various manufacturing methods.)

All stoves have a method of regulating the heat they provide. Add too little heat, and the soup will not boil; add too much heat, and the soup will boil over. A rheostat is used to regulate the heat produced by the heating mantle. It is plugged into any standard electrical outlet and is the equivalent of the control knobs on an electric stove. Regulating the amount of heat administered by the heating mantle controls the rate of the reaction.

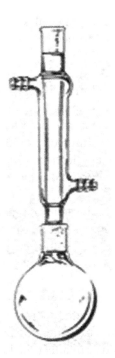 

Figure 1.6 Reflux apparatus examples.

Commercially available oil or water baths can be used as an alternative to a heating mantle. The reaction vessel is submersed partially or totally in the oil/water to provide a uniform source of heat around the reaction vessel when the reaction temperature is critical.

A chef needs a pot to cook his soup. A boiling or reaction flask acts as the pot. As a general rule, boiling flasks have a single opening, or neck. Reaction flasks have multiple necks. The multiple necks on a reaction flask do have legitimate purposes during the refluxing process. However, all that is necessary for a basic reflux apparatus is a single opening. The opening allows the ingredients to be added, pressure to be vented during the reaction and the product to be removed. Many clandestine recipes for controlled substances require a triple-neck reaction flask, while in reality a single-neck flask would suffice.

Boiling and reaction flasks are generally spherical with round bottoms that fit snugly into the appropriately sized heating mantle. The flasks encountered in clandestine labs vary in size from 25 ml (slightly less than 1 ounce) to 72 liters (approximately 18 gallons). There are also boiling/reaction flasks that have flat bottoms to allow the flask to sit on a hotplate or ring stand without support.

The necks on boiling/reaction flasks generally have ground glass fittings that provide a sealed connection for various auxiliary items such as condensers, addition flasks or thermometers. Smooth-necked flasks exist. Sealing connections with auxiliary equipment is accomplished by using cork or rubber stoppers.

A chef places a lid on his pot to keep the soup from boiling dry. The steam from the boiling soup condenses on the lid and drips back into the soup, replenishing the evaporating liquid. The steam escaping from around the sides of the lid relieves excess pressure. A reflux apparatus operates in much the same way. Various types of condensing columns or condensers act as the pot's lid. Each type has a specific scientific application. However, in the world of clandestine labs, all condensers act as the lid on the boiling soup.

A condenser has a compartment through which cool water is circulated. Steam from the boiling mixture condenses on the cool surface of the compartment. The condensed liquid then drips back into the boiling mixture. Vapors that are not condensed escape through an opening, relieving excess pressure on the system.

In some instances, the uncondensed vapors are vented into some sort of absorbent material, helping eliminate the odors associated with the production method. This also traps the toxic vapors that are generated by the reaction. Unvented fumes containing toxic compounds can affect the health of the lab operators or personnel tasked with the seizure of clandestine labs.

Not using a condenser has many hazards. Uncondensed vapors fill confined spaces that house clandestine labs and create a very toxic environment. A reflux apparatus without a condenser can potentially boil dry, creating a different set of toxic or hazardous compounds.

The venting process can itself be potentially hazardous. If the condenser becomes obstructed, pressure will build in the reaction vessel as a result of the boiling liquid. If the obstruction is not removed, one or more of several dangerous situations may occur. The pressure of the expanding gas could clear the obstruction violently. The expanding gas may compromise the connection between the condenser and the reaction flask, turning the condenser into a projectile. The structural integrity of the reaction flask or condenser may be compromised and result in a boiling liquid expanding vapor explosion (BLEVE).

Clandestine lab operators commonly create reflux apparatus utilizing ordinary household items. Hot plates have been used as a heating mantle. Countertop deep fryers have been used as an oil bath. Glass cookware has been used as reaction vessels. Condensers have been fabricated from copper or PVC pipes. The only limitation is the operator's imagination, so the clandestine lab investigator must also be thinking creatively in order to recognize things for what they really are.

1.4.1.2 Distillation

Distillation is the separation of a liquid from a solid or another liquid using evaporation followed by condensation. It is a modification of refluxing. It can be used as a technique to synthesize and separate compounds, or used solely as a separation technique.

Distillation uses the differences in the boiling points of the mixture's components to separate them. The component with the lowest boiling point will separate from a boiling mixture first. The mixture will maintain the temperature of the boiling point of that component until it has completely evaporated from the mixture. The mixture's temperature will then rise to the next lowest boiling point in the mixture.

Knowing the boiling point of the component(s) of interest allows the operator to isolate the component of interest. Even if the operator does not have a thermometer, he can isolate the desired component if he knows which boiling plateau the component of interest is in.

During the distillation process, a mixture of chemicals are heated to a boil. During this heating process, precursor chemicals can be converted into a desired end product or combined with other precursors to synthesize a new compound. At the same time, the unwanted by-products are separated from the mixture by evaporation (Figure 1.7).

The equipment used for distillation is the same as that used for refluxing. The individual components of the apparatus are rearranged to allow gravity to separate the condensing liquid from the boiling mixture rather than

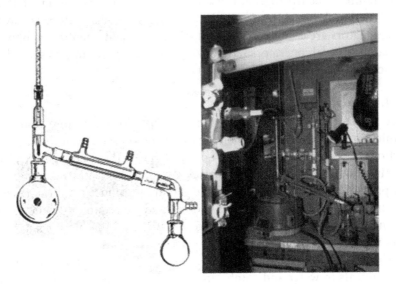

Figure 1.7 Distillation apparatus.

return it. As with a reflux apparatus, the distillation apparatus requires a heating mantle (stove), a boiling/reaction flask (pot) and a lid (condenser). The heating mantle provides the heat required to boil the ingredients in the reaction vessel. The vapors from the boiling ingredients then condense in the condenser. The orientation of the condenser is changed to allow gravity to separate the condensing liquid away from the boiling liquid instead of returning it to the mixture. It may then be collected in a reception flask.

In simple distillation, the temperature of the boiling liquid is monitored to determine when the component of interest is evaporating from the mixture. Fractional distillation allows the separation of components whose boiling points are similar by monitoring the temperature of the vapors that have traveled through a pre-cooling (fractionating) column.

A vacuum pump may be attached to either system to aid in the distillation process. The vacuum pump draws the vapors into the condensing column rather than allowing the pressure from the boiling liquid to force them into the column. This assists in expediting the process.

As with the reflux apparatus, clandestine lab operators often alter common equipment to create a distillation apparatus when the need arises.

1.4.1.3 Hydrogenation

Hydrogenation is a chemical reaction that adds hydrogen to a substance through the direct use of gaseous hydrogen. Under high pressure, in the presence of a catalyst and hydrogen, ephedrine can be converted into methamphetamine. The hydrogenator used in this method is commonly referred to as "the bomb" in scientific circles as well as the clandestine lab world. Hydrogenators are commercially available. As with all scientific equipment, the clandestine lab operators have developed alternatives to this specialized piece of equipment as well (Figure 1.8).

1.4.1.4 Bucket Chemistry

There are certain manufacturing methods that do not require traditional chemical apparatus. "Bucket" chemistry is an appropriate term because these reactions literally can take place in a plastic bucket. The chemicals are placed into the container and allowed to react. At some point in time, an extraction process is undertaken to separate the final product from the reaction mixture. No heat is necessary, but cooling may be required. Phencyclidine and methamphetamine can be produced using nothing more than plastic containers.

The primary explosive lead azide can be made by dissolving the precursor chemicals in water and combining the solutions in a separate container (bucket). The lead azide is essentially insoluble in water and can be easily isolated and placed into an explosive device.

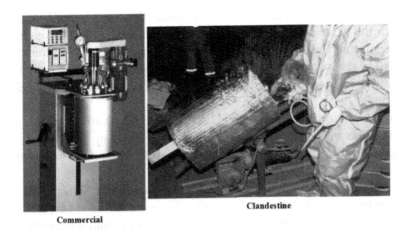

Clandestine

Commercial

Figure 1.8 Hydrogenators.

1.4.1.5 Extractions

Extraction is the act of separating a constituent from the whole. It may be performed a number of times during the manufacturing process. Clandestine lab operators rely on the component's physical and chemical properties to isolate it from the rest of the substances.

The two generic extraction mechanisms are physical and chemical. The label does not necessarily correspond to the use of the component's physical or chemical properties but is simply a means of explaining the process.

Chemical extractions use a component's ability to dissolve in a liquid (solubility) to separate it from the bulk substance. The component being extracted may be the compound of interest or some unwanted by-product(s). The process does not require sophisticated equipment. All that is required is a container to hold the original mixture and a liquid that the desired material will not dissolve in.

Insolubility can be used as well. The salt forms of many drugs are not soluble in most organic liquids. When a free drug is converted into the salt form, it will precipitate out of an organic solution. The act of converting the drug from one form to another changes its solubility properties. This allows its extraction from the solvent and any by-products that are soluble in the solvent.

The compound of interest does not have to be in a solid form to be extracted by a liquid. A liquid can extract other liquids from each other. The acidity (pH) of a liquid may or may not allow a substance to dissolve in it. For example, free base methamphetamine is an oily liquid at room temperature. It is soluble in acidic water solutions but insoluble in basic (pH > 7) solutions. Thus, adjusting the pH of the liquid can be used to chemically extract methamphetamine from a water solution.

Physical extractions mechanically separate the component(s) of interest from the balance of the material. In many instances, the act of chemically separating the desired component is only half the battle. In one form or another, the two chemically incompatible components still need to be physically separated. Specialized equipment has been developed to perform this task.

On the other hand, decanting and evaporation are two means of physical extraction that do not require specialized equipment. Both techniques use differences in their physical state to physically separate components or mixtures.

Decanting is a simple form of physical extraction that separates liquid/solid mixtures. The solid material is allowed to settle to the bottom of the container. The liquid is then carefully poured from the container, disturbing as little of the sediment as possible.

Gravity filtration is a simple physical extraction used to separate solids from liquids. All that is required is a funnel, a filtering material and a receptacle for the filtered liquid. The filter device is placed into the funnel, and the solid/liquid mixture is poured into the funnel. Gravity draws the liquid through the filtering material, leaving the solid trapped in the filter. This type of filtration can be slow and may not be conducive to removing solids of fine particle size.

The filter paper commercially made for gravity filtration applications is unnecessary. All that is required is a material that will allow a liquid to flow through and keep the solid on the other side. Coffee filters, sheets and women's silk underwear have all been found in clandestine labs being used to filter solids from liquids.

Vacuum filtration is a method of expediting the filtration process. A vacuum is used to draw the liquid through the filtering material. It is an efficient

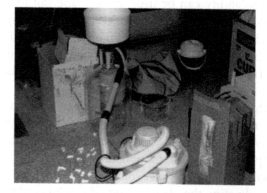

Figure 1.9 Vacuum filtration.

method of separating liquid from solids and is most efficient for solids of fine particle size (Figure 1.9).

Commercial vacuum pumps are commonly found at the scene of clandestine drug labs. Clandestine lab operators may use alternatives such as the compressor from a refrigerator or an air conditioner. Air compressors have also been replumbed to perform this function.

There are times during the manufacturing process when different types of liquids must be separated from each other. This is accomplished through the use of a separatory funnel (Figure 1.10). The liquid combination is placed into the separatory funnel, and the aqueous and organic liquids are allowed to form two distinct layers. The valve at the bottom of the funnel is opened, and the liquid is allowed to drain until the layer separation reaches the valve. The valve is then closed, and the desired liquid is saved for further processing.

There are alternatives to the use of the separatory funnel. The top liquid layer can be simply decanted off away from the top. Turkey basters have been used to separate one liquid layer from another. Sport water squirt bottles have also been utilized as a makeshift separatory funnel.

Evaporation can be used to separate a solid that is dissolved in a liquid and does not require specialized equipment. The liquid simply is allowed to evaporate, leaving the solid. An outside source of heat can be used to accelerate the process.

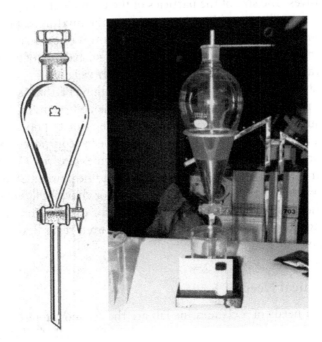

Figure 1.10 Separatory funnel.

Distillation is a form of evaporation used to extract one liquid from another. As previously discussed, it requires specialized equipment to evaporate, condense and capture the different liquid fractions. The temperature should be monitored. However, the compound of interest can be isolated if its position in the sequence of boiling plateaus is known.

There are some instances where chemical and physical extractions are used in tandem. In the final purification stage of production, the salt form of a drug is placed into a funnel containing a filtering material. The solvent is removed from the final product through a filtration process. Then, a different solvent, one in which the final product is not soluble, is poured onto the substance. This liquid chemically extracts the by-products from the final product and is physically extracted from the solid by filtration.

1.4.1.6 Compounding

The goal of the compounding process is the creation of an end product with a uniform consistency. Profit is the driving force behind the use of a compounding process in a drug manufacturing operation. From the drug manufacturing perspective, compounding is done to give the end user the illusion of purity regardless of the amount of diluents or adulterants that have been added.

Compounding plays a more prominent role in the manufacture of homemade explosives. The size of the particles of the individual components, the ratio of fuel to oxidizer, and how homogeneous the mixture, all impact the explosive performance of the mixture. Large clumps of oxidizer combined with fine particles of fuel impact the effective oxygen balance of the reaction, even if the ratio of fuel and oxidizer is balanced by weight.

Clandestine operators use any piece of equipment that is capable of grinding or mixing as a means of performing the compounding operation. There are numerous pieces of equipment with legitimate household, commercial and agricultural applications that have been repurposed from their intended purpose for the production of drugs and explosives. Coffee grinders and flour mills turn a clumpy substance into a fine powder. Stand mixers used in the kitchen or mortar/cement mixers stir or churn lumps of independent ingredients into a uniform mixture slated for distribution or destruction. The ingenuity of the clandestine manufacturer, regardless of the end product, knows no bounds.

1.4.2 Chemical Needs

The chemical needs of a clandestine lab are the second leg of the triangle. Most of the chemicals used in the manufacture of controlled substances and explosives have legitimate uses. Many can be obtained without restrictions through chemical suppliers or from grocery, drug or hardware stores. Some of the chemicals that do have restrictions have legitimate alternatives or sources that do not have restrictions on their distribution.

Some clandestine lab operators have a pragmatic solution: if they cannot legally buy a chemical, they will make it. A case demonstrating this philosophy occurred in Arizona, where a clandestine lab operator was utilizing phenylacetic acid as a starting material in a methamphetamine synthesis. Phenylacetic acid is a regulated chemical. To circumvent this problem, the operator manufactured it using mandelic acid as a starting material. He converted the mandelic acid into phenylacetic acid and continued the synthesis from there.

Nitric acid is an essential precursor for the manufacture of a wide range of monomolecular explosives. As a result, there are restrictions on its sale and distribution. However, nitric acid can be easily made by combining two unregulated chemicals. The hazards associated with the process seem to be a risk the HME operators are willing to take and need to be factored into the investigation and scene processing.

All of the chemicals used in the manufacture of drugs and explosives have legitimate household, industrial or agricultural uses. Some have legitimate home and hobby uses and may be found in anyone's kitchen, medicine cabinet, garage or workshop. Many can be found in bulk quantities on farms or manufacturing plants all over the world. The key to forensic clandestine lab investigations is the ability to recognize combinations of chemicals that together can potentially form a controlled substance.

Appendix D contains various lists of the chemicals utilized in the manufacturing of drugs and explosives. One table provides the legitimate home and hobby uses for commonly encounter chemicals. The table also notes the existence of a legitimate home or hobby use for each chemical listed. This information can be used by the investigator to establish whether a reasonable person would deduce that the chemical(s) has a legitimate reason for being at a particular location or in the possession of a given person.

Other tables within Appendix D list the range of clandestine uses that a variety of chemicals have in the manufacture of drugs and explosives. One set generically indicates what a particular chemical can potentially be used to manufacture. An associated table relates specific combinations of chemicals to the drug or explosive they can be used to manufacture.

When obtaining a search warrant, investigators should be able to articulate that the chemicals known to be associated with the clandestine lab do have legitimate uses but have no legitimate home or hobby use; an investigator should further be able to state which controlled substance this particular combination of chemicals may produce.

Certain combinations of chemicals with legitimate home or hobby use are commonly used to manufacture explosives or explosive mixtures. Large quantities of some of these chemicals are frequently procured for legitimate agricultural purposes and can be combined with additionally large quantities that are located at the same agricultural operation. In these instances, the totality of the circumstances must be taken into account when securing a search warrant for criminal purposes or initiating a sequence of events that culminates with an event that has extreme prejudice. The chemicals involved may be just being used for their intended purposes.

The three types of chemicals used in the manufacture of controlled substances are precursors, reagents and solvents. All three types are used at some point during the manufacturing process.

It must be noted that the illicit explosive manufacturing community refers to all chemicals involved in the manufacture of explosives or explosive mixtures as precursors. The reasoning for this is beyond the scope of this discussion. However, this discussion will refer to chemicals as either precursors, reagents or solvents for purposes of explaining a particular chemical's role in the manufacturing process.

A precursor chemical is a raw material that becomes a part of the finished product. It is the building block(s) with which the final product is constructed. In a conversion reaction, a precursor's chemical skeleton is altered to create the final product. In the synthesis process, precursors are chemically bonded together to produce the final product. In the compounding process, it is either the fuel or the oxidizer that produces an explosive mixture.

Reagent chemicals react chemically with one or more of the precursor chemicals but do not become part of the final product. During the process, a portion of the reagent may be part of an intermediate product but is removed prior to the formation of the final product.

The use of magnesium in the manufacture of PCP is an example. The magnesium reacts with bromobenzene to form the intermediate product phenylmagnesium bromide. This intermediate product reacts with another intermediate product, phenylcyclohexyl carbonitrile (PCC), to form phencyclidine (PCP). During the process, the magnesium is removed from the intermediate and returned to the reaction mixture solution.

A solvent does not chemically react with precursor or reagent chemicals. Solvents are used to dissolve solid precursors or reagents, to dilute reaction mixtures, and to separate or purify other chemicals.

Water is the most common solvent. It can provide the acidic, basic or neutral environment for precursor chemicals to react when influenced by a catalyzing reagent. Or, water can be used as a means to separate different chemicals based on their solubility preference. In either case, water does not become part of the intermediate or final product drug in the manufacturing process.

Organic liquids have the same solvent properties as water. They can act as a solvent during extractions or during a synthesis/conversion operation. The desired compound(s) dissolve in the organic liquid and can be separated from the bulk substance. No portion of the solvent becomes part of the compounds being extracted.

Some of the chemicals perform dual roles. For example, hydriodic acid (HI) acts as a reagent and as a solvent in the reduction of ephedrine to methamphetamine. This is because HI is not pure hydrogen iodide. It is technically a solution of hydrogen iodide and water. The hydrogen iodide acts as the reagent chemical. The water acts as the solvent in which the reaction takes place.

In the ephedrine/HI reaction, the water in the HI acts as the solvent. It does not become part of the final or intermediate products. It only provides the environment necessary to allow the iodide from the hydrogen iodide to attach to, and be removed from, the ephedrine molecule to form methamphetamine.

Identifying a chemical can be a challenge to those unfamiliar with the nuances of chemistry. Many chemicals have a variety of names. Many have a common name along with an official International Union of Pure and Applied Chemists (IUPAC) name. Common clandestine lab chemicals also have one or more slang terms (which may vary regionally) associated with them. For example, 2-propanone is commonly referred to as acetone. Therefore, investigators should rely on a forensic chemist to sort out the various chemical synonyms.

Pronunciation of chemical names and terms can also be a problem. An investigator verbally asked a chemist what "propozyfene" was. The chemist had never heard of the compound. It was not until the investigator spelled the compound that the chemist realized that the investigator was trying to pronounce "propoxyphene," a schedule II narcotic drug. For this reason, it is suggested that when non-technical personnel are attempting to describe or identify chemicals, they either write down the name(s) of the chemicals or select the chemical name from a prepared list of chemical names and synonyms.

1.4.3 Knowledge Needs

Knowledge is the final leg of the manufacturing triangle. Knowing how to combine the equipment and chemicals to produce a controlled substance is a necessary element. Knowledge is necessary to establish both capability and criminal intent. Knowledge may come from education (schooling or professional training), mentoring/apprenticeship, underground literature, and often even simply handwritten recipes that are bought and sold as property.

The Internet has unfortunately become a source for many "recipes" as well. Original methods have been taken from academic chemical literature and translated into simple recipes that can be followed by someone with no chemical training.

There are numerous ways to manufacture any given controlled substance. Each method has its roots in legitimate chemical or pharmaceutical literature. An example of this was found in a clandestine lab that converted mandelic acid to phenylacetic acid, which was then used as a precursor to methamphetamine. The origin of the recipe was tracked to a German pharmaceutical journal. The chemical proportions utilized by the recipe differed by a factor of 10 from the original article. The steps utilized in the reaction sequence were a translation from technical jargon into simple English (e.g. "reflux" in the original article was "boil" in the recipe, and "decant" was changed to 'pour').

Underground literature is a large source of the knowledge used by clandestine lab operators. Books like *The Secrets of Methamphetamine Manufacture* (for drugs) and *The Anarchist's Cookbook* (for explosives and booby traps) have long been staple sources of "how- to" information. The information in these books originated in legitimate chemical research, and in some cases, the underground authors even cite the original source of the information. The scary part of this plagiarism is that these are so often poor translations; lives are put in danger at every step of the process.

In the old days, when the Leukart reaction was the methamphetamine manufacturing method of choice, clandestine lab operators served apprenticeships under experienced "cooks" to learn the tricks of the trade. Recipes were guarded zealously and bought and sold as a commodity. Today, with the free flow of information over the Internet, these recipes can be obtained by anyone with a computer and a modem. There are websites dedicated to drug and explosive manufacture. There are newsgroups where an operator can shop for a new recipe or source of chemicals and equipment: a strange sort of support group for the underground chemist.

The downside to this information source for the underground chemist is that not all of the information is good. Some of the recipes do not even produce the desired product. Other recipes may explode when the operator follows the instructions provided.

One example of a recipe that does not work is the chickenfeed recipe for methamphetamine. There has long been a rumor that several manufacturers of chickenfeed place methamphetamine in their product to enhance the egg-laying capability of the chickens. If the clandestine lab operator follows a simple extraction process outlined in the recipe, so it goes, he should be able to isolate the methamphetamine. The only problem with the recipe is that the chickenfeed manufacturers do not place methamphetamine in their product.

Some states have a statute that makes the possession of chemicals and equipment for the purpose of manufacturing a controlled substance illegal. This can be problematic in that there are methods of manufacturing methamphetamine that utilize equipment and chemicals that can be commonly and legitimately found in any home in the United States. If this statute were interpreted literally, thousands of homes in the United States would be in violation of the statute. Just because there are over-the-counter cold or diet tablets, rubbing alcohol, iodine solution, swimming pool acid, glass jars, a turkey baster and coffee filters at a location, this does not make it the site of a clandestine lab. The knowledge must be proven to be present as well in order to prove intent. Did the person(s) who possessed the items know how to combine them in the proper sequence to make a controlled substance?

Knowledge in and of itself is not all that is required. If that were the case, half of the chemistry students at any given university would be in jail. The key to the statute is the words "for the purpose of." Did the person with the equipment, chemicals and knowledge intend to combine them to create a controlled substance? This is the question that needs to be asked when evaluating the knowledge requirement. In essence, did the person have the requisite criminal intent?

With regard to the manufacture of explosives, 18 U.S.C. 847 makes it unlawful:

> to teach or demonstrate the making or use of an explosive, a destructive device, or a weapon of mass destruction, or to distribute by any means information pertaining to, in whole or in part, the manufacture or use of an explosive, destructive device, or weapon of mass destruction, with the intent that the teaching, demonstration, or information be used for, or in furtherance of, an activity that constitutes a Federal crime of violence.

The operative words being: "with the intent that the ... information be used for, or in furtherance of, an activity that constitutes a Federal crime of violence." The purpose of this book is to provide to the relevant communities the information to be used to prevent crimes and acts of violence, and as such, is void of criminal intent.

1.5 Practical Examples

1.5.1 Extraction Lab Example 1

Extraction labs remove raw materials from a mixture. This is accomplished by using the desired component's physical and chemical properties to separate it from the mixture. No chemical change in the raw material occurs during the process. The process in itself may not be illegal. However, being able to recognize when the process is being used for illicit purposes provides the expert with the ability to fill in missing puzzle pieces.

Until the regulation of commonly used cold medicines, clandestine lab operators use over-the-counter medications as a source for the precursor chemicals needed for the production of amphetamine or methamphetamine. They grind them into a powder, which they place into a jar. A solvent is added to the powder, and the chemicals of interest are dissolved into the liquid, "extracting" them from the insoluble inert components of the tablet. The liquid is decanted into a glass pie pan, which is placed upon an electric hot plate to evaporate the solvent. The residue contains a relatively pure form of the desired precursor chemical.

The images in Figures 1.11 and 1.12 are an example of an extraction lab in which three extraction processes were taking place. First, the precursor was chemically removed from the medication with the solvent. Second, the liquid was physically separated from the solids. Finally, the solvent was physically removed from the precursor through evaporation.

Possession of the over-the-counter medications tablets is not illegal. The extraction of the precursor chemical components of the tablets may not be illegal. However, the combination of the quantity of tablets and the extraction process may be used to establish the intent to conduct an illegal activity.

1.5.2 Extraction Lab Example 2

Numerous vials of a veterinarian drug preparation containing ketamine were found at a clandestine lab located in a bungalow in a luxury resort, along with a variety of laboratory glassware. The operators were removing

Figure 1.11 Empty tablet bottles.

Figure 1.12 Tablet extraction evaporation.

the solution containing the drug from the injection vials, placing the liquid into beakers and evaporating the liquid. The resulting powder containing ketamine hydrochloride, a controlled substance under the local statutes, was then packaged in small zip lock bags. The operators were convicted of manufacturing a controlled substance in addition to possession of a controlled substance.

The local statute defined manufacture in the broadest sense:

> Manufacture means produce, prepare, propagate, compound, mix or process, directly or indirectly, by extraction from substances of natural origin or independently by means of chemical synthesis, or by a combination of extraction and chemical synthesis. Manufacture includes any packaging or repackaging or labeling or relabeling of containers.*

The operators "extracted" the drug from the original mixture. They also repackaged the substance in a manner commonly utilized by package contraband drugs for sale and distribution, meeting the "includes packaging or repackaging" clause of the definition.

1.5.3 Conversion Lab Example 1

Conversion labs perform one of the most commonly encountered processes associated with the manufacture of contraband drugs. The process takes a raw material and changes it into the desired product. This involves minor structural changes within the molecule of the compound or of the chemical's

---

* Arizona Revised Statute, 13-3401.17

salt form. Functional groups may be added to or removed from the molecule, somewhat like pieces on a tinker toy. The drug of interest can also be converted from its salt form to the free base form or from the free base form to the salt form.

Simply changing the pH of a water solution containing cocaine hydrochloride produces "crack" cocaine. Changing the pH with a basic reagent chemical removes the hydrochloride component from the cocaine molecule. This change in cocaine's salt form makes it insoluble in water. The free base form of cocaine has a lower boiling point, which allows it to be smoked.

This type of conversion lab can be encountered in a variety of situations. An individual can carry all of the components in his pocket. A vial of water and a small amount of baking soda are all that is necessary to convert cocaine hydrochloride into "crack." Larger-scale operations are slightly less mobile but utilize the same ingredients and can be encountered at or in the vicinity of the distribution point.

1.5.4 Conversion Lab Example 2

The Grignard reagent is a very reactive compound that is commonly used in chemical synthesis. The reaction between bromobenzene and magnesium produces phenylmagnesium bromide, a very reactive Grignard reagent. This compound is an essential intermediate component used in the synthesis of phencyclidine and its analogs. This reaction is self-driven and can be accomplished in a plastic bucket.

In response to information provided by a local street person, investigators found chemicals concealed behind a trash can in an alley (Figure 1.13). A plastic bucket containing a liquid with a strong ether odor with gray metallic particles at the bottom was found a short distance away (Figure 1.14). Laboratory analysis determined that the liquid contained bromobenzene, and the metallic particles were identified as magnesium turnings. The expert concluded that the Grignard reagent located in the plastic bucket was to be added to the phenylcyclohexylcarbonitrile (PCC) that was located in the vicinity to synthesize phencyclidine (PCP).

1.5.5 Conversion Lab Example 3

Ephedrine or pseudoephedrine is converted into methamphetamine using a simple reduction reaction. The chemical reaction substitutes a hydrogen atom for a hydroxyl group (-OH) to produce methamphetamine. The skeleton of the molecule is intact. However, the physiological effect of the drug on the body is dramatically different.

Figure 1.13 Grignard components.

Figure 1.14 Grignard mix.

Traditionally, the conversion of ephedrine into methamphetamine was accomplished by using a hydrogenator or a reflux apparatus. However, the ingenuity of the clandestine lab operator has created a situation in which this conversion can be accomplished using ordinary kitchen utensils. Understanding the basic principles involved in the conversion of the precursor chemical into the final product will allow the investigator to recognize ordinary items that have been adapted for use in a clandestine manufacturing operation.

A clandestine lab chemist was asked if a pressure cooker could be used to manufacture methamphetamine using the ephedrine/HI reduction process.

Using the cooking soup analogy, the chemist advised the investigator that the process described by the informant was viable and would produce the desired result. A pressure cook and mason jars were found during the subsequent search of the location. The operator used the mason jars as his reaction vessel and condenser. The pressure cooker was used as his heating mantle. A condenser was not required because the closed mason jars inside the pressure cooker produced a closed system, which contained the fumes that would normally have been condensed or vented away from the reaction.

The images in Figures 1.15 and 1.16 are examples of a pressure cooker that was used to produce methamphetamine in this fashion. Notice the corrosion resulting from the reaction between the acid fumes of the reaction mixture and the metal. This has caused the failure of some systems due to the corrosion clogging the pressure release valve. The overpressure produced a fatal mechanical explosion distributing hot fuming acid from the reaction mixture combined with glass and metal shrapnel from the mason jar reaction vessels and the fragments of the pressure cooker.

1.5.6 Synthesis and Extraction Example

The synthesis process is a chemical reaction or series of chemical reactions in which molecules or parts of molecules are combined to create a new molecule. This process can be equated to a chemical erector set. It differs from the conversion process in that the skeleton of the resulting molecule is a sum of the molecules or significant parts of the molecules involved in the reaction. Lysergic acid diethylamide (LSD), phencyclidine (phenylcyclohexyl piperidine, PCP), phenylacetone (P2P) and certain methamphetamine reactions

Figure 1.15 Pressure cooker reaction vessel.

Figure 1.16 Pressure cooker reaction vessel (interior).

are examples of drugs produced using the synthesis process. Monomolecular primary and secondary explosives such as triacetone triperoxide (TATP), nitroglycerin and trinitrotoluene (TNT) require one or more chemical reactions to fuse the requisite functional groups to the inert skeleton molecule, transforming it into an explosive.

Phencyclidine is produced in a multi-step reaction. During the process, bromobenzene, cyclohexanone and piperidine are chemically combined in a two-step process. The resulting molecule has the combined chemical skeletons of all three precursor chemicals.

The manufacturing of PCP is so simple that it is commonly described as "bucket chemistry." Figure 1.17 depicts the equipment required for this synthesis reaction. The image demonstrates that the required equipment can be very common and does not have to be sophisticated or exotic. In this example, it was estimated that over 100 pounds of PCP was produced in a rural operation in which all of the chemical reactions were conducted in 5-gallon plastic paint buckets.

1.5.7 Distillation Example 1

Nitric acid is an essential precursor chemical used in the production of nitrated explosives. Although it has numerous legitimate applications, its sale and distribution are controlled due to its hazardous nature and its illicit use by terrorists. It is commercially available at a concentration of 68% in water. Concentrated solutions containing more than 86% (fuming nitric acid) are more effective in explosive synthesis.

A distillation apparatus can be used to simultaneously synthesize, extract and concentrate nitric acid. The precursor and reagent chemicals are

Figure 1.17 Synthesis via bucket chemistry.

combined in the reaction flask and heated to a rolling boil. As the mixture boils, the chemicals react, producing nitric acid and its associated reaction by-products. The volatile nitric acid evaporates and is condensed and collected in a separate container. The reaction by-products are left in the reaction flask.

1.5.8 Distillation Example 2

Hydrogen peroxide is a strong oxidizer that is used as a precursor chemical in a number of homemade explosive mixtures. It is commonly available in water solutions of concentrations ranging from 3–6%, found in pharmacies, to laboratory use concentrations of 30%. As with nitric acid, the higher the concentration, the more effective hydrogen peroxide is in the manufacture of explosives.

Distillation can inversely be used to make concentrated hydrogen peroxide. In the previous example, distillation was used to separate the nitric acid from the impurities. In this example, distillation is used to separate the water "impurity" from the solution. It is a single process that simultaneously purifies and concentrates hydrogen peroxide.

1.5.9 Distillation Example 3

The operator in the situation illustrated in Figures 1.18 and 1.19 constructed a vacuum distillation apparatus to purify phenylacetone. The boiling container (Figure 1.18) was constructed from an 8" steel pipe with metal plates bolted onto the top and bottom. A kitchen hot plate was used to apply heat to

Figure 1.18 Distillation reaction vessel.

Figure 1.19 Distillation collection vessel.

the boiling container. The condenser was constructed of two sizes of copper tubing. Cool water was circulated through the makeshift condenser using a submersible water pump and a trashcan containing ice water. The reception flask was a vacuum flask connected to a beer keg that was connected to a commercial vacuum pump. This apparatus functioned as designed.

A note of caution when dealing with homemade equipment. Chemical compatibility significantly impacts the structural integrity of the component parts. The nonreactive nature of glass is one of the reasons it is used in the construction of distillation and reflux apparatus. A copper condenser may be a good choice when constructing a moonshine still. However, copper's incompatibility with nitric acid and hydrogen peroxide will have toxic, if not explosive, consequences when it is used to construct a distillation apparatus.

1.5.10 Extraction and Separation Example

During the final stages of a methamphetamine synthesis or conversion reaction, the reaction mixture is cooled to room temperature. The pH of the solution is changed, and an organic solvent is added to the mixture. The methamphetamine dissolves into the organic layer. The liquid combination is placed into a separatory funnel and allowed to form two distinct layers. The valve is opened, and the lower liquid is allowed to drain into a receptacle. The valve is then closed when the line defining the separation between the top and bottom layer reaches the valve (Figure 1.20).

Sport bottles can be used in a similar manner. The combination of liquids is placed into the bottle; the cap is put into place, the liquids are allowed to separate, and the bottle is then inverted. A vacuum is created in the air space above the liquids, keeping them from pouring out of the bottle. The lower liquid is removed by gently squeezing the bottle, forcing the liquid out of the restricting valve at the opening. This layer (the aqueous layer) is discarded. The remaining organic layer is transferred to a new container for further processing.

Novice laboratory operators may discard the wrong layer using this sports bottle adaptation. They may assume the top layer becomes the bottom layer when the sports bottle is inverted for use. They do not understand that layer sequence is independent of container orientation. The denser liquid will always settle to the bottom, regardless of how many times the container is inverted. This necessitates the sampling of containers thought to be waste material, as they may contain final product that was inadvertently discarded.

1.5.11 Filtration Example

Hydrogen chloride gas is added to the organic solution from the previous example. The free base drug and the gas react to produce the hydrochloride

Figure 1.20 Sport Bottle Separatory Funnel. 1.20a: (Left image) Commercial separatory funnel. 1.20b: (Right image) Clandestine lab alternative.

salt form of the drug, which is insoluble in the organic liquid. The solid–liquid mixture is poured into a Buchner funnel attached to a vacuum flask. The liquid is drawn into the vacuum flask when the vacuum is applied to the system, leaving the solid in the Buchner funnel. Acetone can be used to remove the reaction by-products from the solid. It is poured over the solid and drawn by vacuum into the flask. As the by-products are removed, the solid turns white (Figures 1.21a and b).

In this example, the drug was chemically extracted from the liquid by changing its salt from. The solid was then physically extracted from the liquid by vacuum filtration. Finally, the reaction by-products were chemically and physically extracted from the drug in a tandem operation. The acetone chemically extracted the by-products and was simultaneously physically extracted from the drug while being filtered under vacuum.

1.6 Summary

Clandestine labs come in all shapes and sizes. They range from the small-scale operator who produces just enough to sustain his personal habit to the large-scale operator who produces pounds at a time for commercial profit or wide-scale destruction. They are found in every segment of society and cross

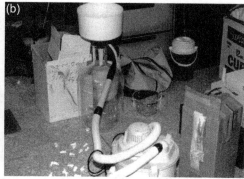

Figure 1.21 Vacuum source. 12a: Commercial vacuum pump. 12b: Shop vacuum cleaner.

all demographics. Race, religion, age, sex and economic status are neither indicators nor barriers.

All three legs of the clandestine lab triangle need to be present for a clandestine lab operation to exist. Equipment, chemicals and knowledge all must be present for the operator to produce the desired product. The equipment used ranges from technical and scientific to mere household kitchen utensils. The chemicals required vary from the very exotic that are only available through scientific supply houses to those that can be purchased over the counter at any drug, grocery or hardware store. Even though a science degree is helpful, all a clandestine lab operator needs to be able to do is follow directions.

Clandestine Lab Hazards

2

Abstract

This chapter deals with the different hazards associated with the seizure of clandestine manufacturing operations. It will provide a general overview of the different types of clandestine lab hazards. While some sections may be simplified and a refresher for experienced investigators, the goal is to provide the information required to give everyone involved in the investigations and prosecution of clandestine manufacturing operations a basic understanding of the scope of the hazards involved.

The three things clandestine labs have in common, regardless of their location or sophistication, are the simple facts that the operators usually have little chemical training, the operations are makeshift, and no two operations are alike. These three principal hazards exist whether the final product is methamphetamine or phencyclidine (PCP), flash powder or nitroglycerine.

There are numerous specific types of hazards associated with clandestine labs. These hazards have been grouped into categories and prioritized according to the immediate harm they can present to the personnel responding to the scene of a suspected clandestine lab. These hazard groups, in order of priority, are explosion, fire, firearms and exposure.

A clandestine lab is a mixture of physical hazards and chemicals with varying hazard levels and chemical compatibility. The hazards associated with some of the reaction by-products may be greater than those of any known chemicals. The generic rules that can be used when dealing with the hazards at a clandestine lab site will be discussed.

2.1 Introduction

Clandestine lab investigation is one of the most dangerous tasks law enforcement or the military has to deal with. The dangers go beyond the violence of the suspects involved in the illegal operation. There are seen and unseen hazards that must always be kept in the forefront of the minds of the personnel involved in investigation and seizure with these toxic operations. The effects of some of these hazards on a human being may not be seen until long after they were encountered.

DOI: 10.4324/9781003111771-2

This chapter deals with the different hazards associated with the seizure of clandestine manufacturing operations. It will provide a general overview of the different types of clandestine lab hazards. While some sections may be simplified and a refresher for experienced investigators, the goal is to provide the information required to give everyone involved in the investigations and prosecution of clandestine manufacturing operations a basic understanding of the scope of the hazards involved.

2.2 General Hazards

The three things clandestine labs have in common, regardless of their location or sophistication, are the simple facts that the operators usually have little chemical training, the operations are makeshift, and no two operations are alike. These three principal hazards exist whether the final product is methamphetamine or PCP, flash powder or nitroglycerine.

2.2.1 Little Training

With the exception of the educated operator mentioned in Chapter 1, clandestine lab operators have little chemical background or training. Their formal education is limited either in years or in content. The so-called "Mexican National" labs, operating in the western United States, are a good example. Illegal aliens are hired to tend methamphetamine production operations. They have little education and do only as they are told. They do not understand the physical and chemical principles of the reaction that they are tending, much less the hazards involved.

The insurgents compounding the ingredients of a fuel oxidizer explosive to place in a roadside bomb have no formal training or chemical education. The ratio of fuel to oxidizer is determined by information passed down through oral tradition, hand-written notes or a recipe obtained from the Internet. The concept of oxygen balance to maximize the explosive potential of their deadly mixture is outside their understanding.

The lack of chemical training can lead to hazards within the clandestine lab that endanger more than just the operators themselves and the enforcement personnel tasked with seizing the operation. Unsuspecting people in the area adjacent to the clandestine operation are exposed to many of the same hazards, often with disastrous consequences. There are numerous examples of lab operators, enforcement personnel and innocent by-standers being hurt as a result of actions taken by clandestine lab operators who do not understand the principles of chemistry and physics that govern the chemical reactions that occur within the operation. Houses and apartments have been destroyed by fire or explosions caused by operator error. Emergency responders and neighbors have been exposed to toxic fumes that are generated by the reaction of incompatible chemicals.

Clandestine lab operators neither understand nor practice common laboratory safety procedures. They are notorious for storing incompatible chemicals together; strong acids and strong bases are commonly found stored adjacent to each other. Organic acids are stored with oxidizing acids. Flammable liquids are stored near a source of ignition. Chemicals are routinely unlabeled or mislabeled to avoid detection. Waste material from reaction mixtures is often combined without regard to content or pH (Figure 2.1a and b).

Personal protection equipment (PPE) in these toxic environments is not used religiously, if at all. Respirators are often found hanging on the wall unused for what looks like an extended period of time. Chemically resistant gloves are conspicuously absent. The "cook" is initially identified by the chemical stains or acid burns on their clothes. The characteristic orange, white or silver gray stained fingers and hands are characteristic of handling nitric acid, peroxides or aluminum powder without the use of protective gloves, reinforcing the operator's lack of training and understanding of the hazardous nature of the chemicals they are working with.

The improper storage and handling of chemicals may lead to violent chemical reactions between incompatible chemicals, or the chemicals may react and create substances that are more toxic than the original chemicals. Emergency medical service (EMS) personnel often encounter operators at clandestine lab sites who have exposed themselves to the toxic waste or by-products of the combination of incompatible chemicals in reaction mixtures or waste material. The improper handling of the chemicals leads to human exposure to unknown chemical hazards and makes it difficult to treat the exposed operator without knowing what toxins he was exposed to.

An example of mixing chemicals to create a toxic atmosphere occurred when an operator mixed a waste solution containing acid with a solution

Figure 2.1 Improperly stored chemicals: (a) ether in a trunk and (b) abandoned explosive chemicals.

containing sodium cyanide. The resulting hydrogen cyanide placed the operator, one police officer and two EMS responders in the hospital. Fortunately, no deaths occurred in this scenario.

The concepts of vapor pressure, flash point, and flammable and explosive limits are usually not within the operator's knowledge base. Nationally, approximately 20% of clandestine drug labs are detected as a result of fire or explosion. Flammable vapors build up inside the lab space, reaching the flammable or explosive limit of the chemical involved. The operator lights a match, turns on a gas stove or turns on a light switch, thus igniting the fumes in the lab, resulting in a fire or explosion.

Unfortunately, statistics associated with the number of explosives manufacturing operations detected through fire or explosions are not readily available. The manufacture of explosives, by its very nature, presents a hazard of the end product prematurely performing the function it was designed for. Commercial explosive manufacturers go to great lengths to implement safety checks and balances to minimize risks. Untrained clandestine lab operators do not.

The images in Figure 2.2 are the result of an attempt to evaporate methanol from an extraction solution containing pseudoephedrine. The operator used methanol to extract the pseudoephedrine from an over-the-counter cold preparation. He placed the methanol solution on a gas stove to speed the evaporation process. The combination of the methanol vapors and the gas flame resulted in a fire that caused extensive damage to the two-bedroom bungalow. The operator was caught a week later when he once again caused a vapor explosion under similar circumstances in a motel room less than 2 miles from the location of the bungalow.

2.2.2 Makeshift Operations

The creativity of clandestine lab operators is truly amazing. Using the basic understanding of how scientific equipment operates, they have designed

Figure 2.2 Result of evaporating flammable solvents on a gas stove. (a) Fire's point of origin. (b) Container of the solvent involved.

equipment alternatives that allow them to avoid detection by not using scientific supply houses as an equipment source. Also, the homemade alternatives can be significantly cheaper than the actual item.

The problem with this situation is that even if the operator has taken the physics and mechanics of the equipment into account, he probably has not taken into consideration the interaction between the chemicals involved and the materials that the makeshift equipment is constructed with. Common glass kitchen utensils are not a substitute for Pyrex™ glassware, which is designed to operate at high temperatures. Rubber or cork stoppers are not a substitute for ground glass connections in reflux and distillation apparatus that uses organic solvents or strong acids. Reaction vessels made of steel are not designed to contain solutions of hydrochloric acid.

Figure 2.3 contains three examples of the ingenuity of clandestine lab operators. Figure 2.3a shows a countertop deep fryer being utilized as an oil bath heat source for a reflux operation. Figure 2.3b shows a hydrogenator that was constructed out of a beer keg and heat tape. Figure 2.3c shows a homemade condenser made from different diameters of copper tubing. Each one of these alternatives efficiently accomplished the task it was designed for.

2.2.3 No Two Labs Are Alike

There are hundreds of different methods of manufacturing controlled substances. However, only a small number of methods are actually encountered. Even when the same method is repeatedly encountered, there are still enough differences between clandestine labs to make each one unique.

It cannot be repeated emphatically enough that the personnel responding to the scene of a clandestine lab must be constantly vigilant for potential hazards. Even though the particular type of operation has been encountered numerous times before, it should always be treated as an unknown scenario. Complacency is the biggest danger to personnel investigating or processing

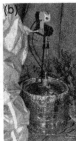

Figure 2.3 Makeshift operations: (a) oil bath reflux, (b) beer keg reaction flask, (c) copper tube condenser.

the scene. A less-than-aware attitude may lead to not recognizing booby traps, disregarding the presence of odd or unique chemicals, exposing one-self to toxic environments, or the improper handling of unsafe makeshift equipment setups. It can further lead to severe injury for the investigators.

The following is an example of complacency and over-familiarity that could have had a tragic end.

An investigator and a chemist were performing their initial walk-through of a suspected clandestine methamphetamine lab. The investigator was videotaping the event, describing how each piece of equipment could be used in the production of methamphetamine. The chemist's observations were not in sync with the investigator's comments. Upon observing homemade booby trap switches, the chemist deduced that the chemicals and equipment were consistent with the manufacturing of nitroglycerin, not methamphetamine. As a result, their approach to processing the scene shifted to address the change in the hazard potential associated with the suspected end product.

2.3 Hazard Priority

There are numerous specific types of hazards associated with clandestine labs. These hazards have been grouped into categories and prioritized according to the immediate harm they can present to the personnel responding to the scene of a suspected clandestine lab. These hazard groups, in order of priority, are explosion, fire, firearms and exposure.

2.3.1 Explosion

An explosion is a rapid chemical change that produces a large amount of heat and gas. It is the highest hazard priority because it can potentially do the greatest amount of damage to the responding personnel in the shortest amount of time. The source can be intentionally placed (i.e., a booby trap) or an unintentional result of the improper handling of chemicals.

Explosions have two effects on the body. Firstly, the resulting pressure wave can cause internal and external damage to the body by the blast directly or by the secondary effects that result from being struck by items that are thrown as a result of the explosion. The second effect is a result of the high temperatures that are created by the explosion. Both of these effects are reduced by distance; the farther away from the center of the explosion the exposed person is, the smaller the effect.

The three types of explosions that can be experienced in a clandestine lab are detonations, deflagrations and mechanical explosions. The only significant difference between the first two is the rate of the reaction. All have pressure waves and intense heat associated with them. The pressure wave of the explosion is proportional to the amount of energy released, which determines the effect on the body that is exposed to the explosion.

Detonations are chemical reactions that produce extraordinary amounts of heat and gas, resulting in a pressure wave that causes the damage. They are the result of known explosive materials or chemical mixtures that have a reaction rate of greater than 1000 meters per second. They also can be the result of the reaction between a mixture of incompatible chemicals or the undesirable outcome of the mishandling of unstable chemicals such as peroxides, white phosphorus or picric acid. Additionally, the end products of the operations that produce primary explosives used as improvised detonators are extremely unstable and subject to exploding if mishandled.

Deflagrations have a reaction rate of less than 1000 meters per second. They are explosions that result from the pressure created by a chemical reaction, usually combustion, which compromises the structural integrity of the container. They can result from the spontaneous ignition of an atmosphere that contains ignitable liquid vapors in an explosive concentration if the ignition takes place in a contained environment (i.e., a room). Confined deflagrations of combustible solids such as smokeless powder or fuel oxidizer mixtures can produce explosive results.

Mechanical explosions are the result of a pressure buildup in a container to the point at which it loses its structural integrity. A boiling liquid expanding vapor explosion, or BLEVE, is an example. The expanding vapors of a boiling liquid in a closed container create so much pressure within a closed container that the container explodes. Practical examples in this chapter demonstrate the effects of explosion in an actual setting.

The lack of chemical knowledge of the clandestine lab operator can create a situation that is prone to explosive results. Ethers are exposed to the atmosphere, creating unstable peroxide compounds. Picric acid stored in containers with metal lids and allowed to dry will form metal picrates on the threads of the lid that can explode when the lid is twisted. Metallic sodium or lithium stored improperly can explode when exposed to air or water. Flammable vapors may be allowed to collect in a confined space and when ignited will deflagrate, producing explosive results.

Figure 2.4 presents examples of improperly stored chemicals that pose an explosive hazard. Figure 2.4a is a jar of metallic sodium that is not covered with mineral spirits to protect it from exposure to air or water. The jar without a label in Figure 2.4b contains dry picric acid. The operator placed the label inside the jar so he would know what the contents were.

Other causes for explosions are intentional on the part of the operator. According to Drug Enforcement Administration statistics, approximately 10% of clandestine drug labs are booby-trapped. These booby traps may or

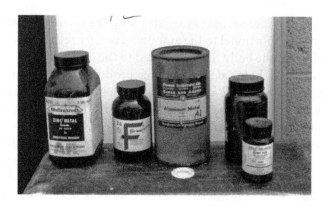

Figure 2.4 Improperly stored explosive chemicals.

may not be directed at law enforcement. No matter who the target, personnel processing clandestine lab scenes must be constantly aware of the potential existence of explosive devices.

The abatement of explosive hazards is accomplished by the use of explosive ordinance disposal (EOD) experts and chemists trained in the clandestine manufacture of controlled substances. They should jointly process all clandestine lab sites for the presence of explosive devices and dangerous chemical mixtures prior to the processing of the site for physical evidence. The EOD expert and the chemist should have complementary knowledge of each other's expertise. This team should be able to observe, recognize and be able to neutralize potentially explosive situations prior to the rest of the processing team entering the location.

It may be impractical to have a chemist involved in site exploitation operations of suspected explosive manufacturing operations conducted by military units. However, EOD personnel should consult with explosive chemists knowledgeable about the clandestine production of explosives and explosive mixtures to bolster their understanding of the hazards associated with the chemicals they could encounter during an exploitation operation and identify sample locations that would provide exploitable information upon laboratory testing.

Figure 2.5 is an example of a vapor explosion that occurred in a gamma hydroxy butyric acid (GHB) lab. Figure 2.5a demonstrates the pressure effect of the explosion. The wall at the center of the explosion was blown approximately 3 feet off center. Figure 2.5b shows the resulting fire damage. In this scenario, the operator was drying acetone from his finished product in a heated room (confined space). The acetone fumes had reached their explosive limit when the operator turned the light on in the room. The spark that occurred when the light switch was thrown ignited the vapors, causing the explosion.

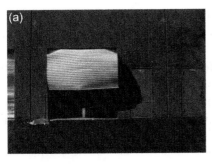

Figure 2.5 Vapor explosion damage: (a) vapor explosion damage, (b) fire damage.

2.3.2 Fire

Fire is second in the priority list of hazardous situations the clandestine lab investigator must consider. It has many of the same causes as explosive hazards. Booby traps, ignition of flammable atmospheres, incompatible chemical mixtures and the operator himself can contribute to the cause of a fire. A significant difference between fire and explosive hazards is lack of the pressure wave experienced from explosions. However, the heat and resulting combustion damage can be similar.

Flash fires and sustained fires are the two types of fire hazards. Flash fires result from the ignition of a flammable atmosphere. They are instantaneous and usually self-extinguish for lack of fuel. If contained, a flash fire may result in an explosion or produce minor pressure-related effects. A sustained fire, on the other hand, has a continuous source of fuel to feed the fire. The resulting heat and continuous flame allow the fire to spread beyond the initial area of the original ignition.

Tests done by the Phoenix, AZ Fire Department indicate that exposure to a flash fire may be survivable. If the victim were wearing fire-resistant clothing, significant damage would be limited to the unprotected areas. If the victim was not wearing fire-resistant clothing, the severity of the burns to the protected areas of skin would be determined by the type of clothing he was wearing. The test dummy wearing fire-resistant clothing did not fare as well when exposed to a sustained fire situation during the same series of tests. The clothing was totally consumed by fire when exposed to a sustained fire of flammable liquids commonly encountered in clandestine labs.

As with explosions, clandestine labs are commonly detected as a result of a fire. Figure 2.2 is an example that shows the results of a fire caused when the lab operator attempted to evaporate a flammable liquid (methanol in this case) on a gas stove. The fire could have been started just as easily if the flammable liquid had been on an electric stove. Any ignition source would have ignited the flammable vapors. Once the concentration in the room reaches the flammable or explosive limit, a spark from a light switch, a match struck for a cigarette or a muzzle flash from a discharging weapon could ignite the vapors.

Figure 2.6 is the result of a barricaded clandestine lab operator intentionally burning his lab in an effort to avoid detection. The operator could just as easily have waited until the investigators were inside the location before he ignited the flammable and combustible substances within the operation, causing incredible injury.

2.3.3 Firearms

As you can see, the clandestine lab investigator has more things to worry about than just getting shot, but that still ranks up there with the dangerous situations potentially encountered.

Individuals who abuse methamphetamine develop a paranoid psychosis. These individuals think their personal safety is continuously under attack. As a result, they can be unpredictable and violent and in general, are thus armed with some type of weapon. These people are irrational and subject to delusions and hallucinations.

The production and distribution of contraband drugs is a lucrative business worth protecting. The use of armed sentries in large-scale operations is

Figure 2.6 Three images of the progression of an intentionally set fire. (a) Initially set, (b) Fully engaged, (c) Scene that was processed.

just as much protection against the competition as from law enforcement. However, the individuals in charge of protecting the investment may not make the distinction and may use extreme measures regardless of the threat.

Explosives manufacturing operations are carried out by individuals with radical ideologies. The use of force and violence is engrained into their psyche. Firearms are a tool they use to protect and perpetuate their beliefs.

The threat of firearms is not over once the suspects have been arrested and removed from the scene. It is not uncommon for armed secondary suspects to appear at the scene once the processing has begun. This is a real hazard, since many of the people processing the scene are not authorized to carry weapons. Even those who are may not have them in their possession.

2.3.4 Exposure

Exposure to chemical and physical hazards is the lowest in the priority list of hazards. This does not mean that it is any less dangerous to the clandestine lab investigator. These are silent hazards. Their effects may not become apparent until after the exposure.

Exposures to chemical and physical hazards may have acute and chronic effects. Acute effects are experienced by exposure to hazards at high concentrations even for a short duration. The effects are felt immediately. The exposed person will generally recover if the exposure does not exceed the lethal limits, although there may be long-lasting side effects. Chronic effects are generally experienced as a result of numerous exposures to chemical hazards in a low concentration over a long period of time. The effects are cumulative, creating a toxic effect. These effects are usually different from the acute effects of the same hazard.

The difference between acute and chronic effects can be demonstrated in the example of a person who gets drunk and falls down every night for an extended period of time. The acute effects of the alcohol may cause the person to lose control of his motor functions, falling down and hitting his head. Hitting his head may have one acute effect, that of a fierce headache, but the effect is short-lived. The person eventually sobers up, and the headache goes away.

A chronic effect of drinking alcohol is different. Long-term exposure to alcohol may lead to cirrhosis of the liver, and brain damage can result from repeated blows to the head over an extended period of time. Neither of these effects will go away by simply removing the alcohol from the person after the fact.

2.3.5 Chemical Hazards

All chemicals are hazardous. The degree of hazard depends upon the chemical's properties, the proximity to other chemicals and the other chemicals' properties. While water is considered a benign substance, it does have an LD50 (lethal dose for 50% of the population) of approximately 90 grams per kilogram of body weight. Additionally, when this seemingly inert chemical is exposed to sodium metal, the combination becomes explosive.

Many common household products contain the hazardous chemicals that can be used to manufacture controlled substances and explosives. Acids can be found in swimming pool chemicals and batteries. Bases or caustic compounds are commonly used in drain cleaners. Flammable solvents found in paint thinners and camping fuels are used for extractions. A poisonous chemical used as automobile antifreeze is the precursor for an explosive. Just because these chemicals are found in the home, this does not make them any less hazardous.

The hazardous properties a chemical can possess are varied: explosive, flammable/combustible, corrosive, oxidizing, compressed (gases) and poisonous. The chemical's reactivity potential should also be taken into account. A single chemical may possess multiple hazardous properties. For example, nitric acid (HNO_3) is a corrosive, oxidizing poison. The tables in Appendix D list the hazards associated with the chemicals that are commonly encountered in clandestine labs.

The Global Harmonized System (GHS) was established in 2002, and adopted by the United States in 2006, as a means to merge the different methods of classifying chemical hazard into a single internationally accepted system. The overt labeling of chemicals using pictograms that depict the hazards associated with the chemicals is a key element in the GHS, as is the publishing of a Safety Data Sheet (SDS), formerly Material Safety Data Sheet (MSDS), for each compound. Figure 2.7 portrays the GHS pictograms and their associated hazards. Appendix F has consolidated some of the SDS information of chemicals associated with the manufacture of drugs and explosives into a single table.

Explosive chemicals undergo a rapid chemical change that releases a large amount of heat and gas. This rapid chemical change may result from a shock or friction, being exposed to a source of ignition, or sudden changes in temperature, water or air. Certain chemical combinations will explode or create a mixture that will explode at the least provocation.

Encountering explosive chemicals is rare in drug manufacturing operations, unless the operation's final product includes an explosive. However, there are numerous combinations of chemicals that will explode under the proper conditions. That is why it is imperative that properly trained EOD personnel and chemists evaluate clandestine lab sites prior to evidence collection.

By contrast, the main purpose of explosives manufacturing is to produce a product whose main purpose is to explode. As such, the precautions required when processing a suspected explosive manufacturing operation are

PHYSICAL HAZARD PICTOGRAM

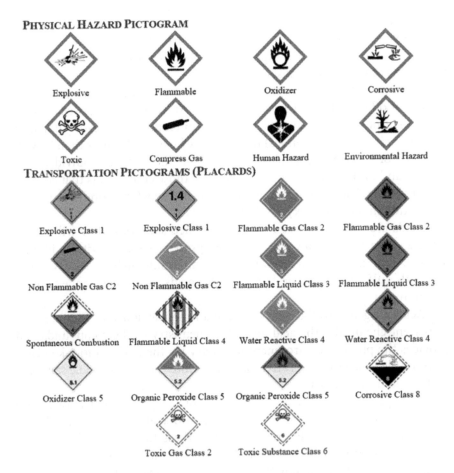

Figure 2.7 Chemical hazard pictograms.

orders of magnitude greater than those used during a suspected drug manufacturing operation. That is not to diminish the explosive hazard potential found within a drug manufacturing operation. The explosive hazard potential simply elevates from potential (drugs) to definite (explosives).

Most organic chemicals will sustain combustion if they are exposed to an ignition source or to enough heat. The chemicals with the greatest fire hazard are those that ignite easily at low temperatures. Flammable liquids are those that have a flash point below 100 degrees Fahrenheit (°F). Combustible liquids are defined as those having a flash point above 100 °F. Combustible chemicals such as phosphorus and magnesium are solids that can easily sustain combustion.

Many of the solvents used for extraction or purification purposes are extremely flammable. Solvents like methanol and acetone have low flash points and readily ignite; these chemicals also have the potential to create a flammable or explosive atmosphere. Solvents like xylene or mineral spirits do burn, but because of their higher flash point, they are less likely to form a flammable or explosive atmosphere.

Corrosive chemicals are those that can cause visible damage to metals, plastics or other materials (especially your skin). They are composed of acids and caustic or basic compounds. Acids are compounds that readily donate a proton (hydrogen) to a chemical reaction. They have a pH lower than 7 (with a pH of 2 or lower being considered corrosive). They can be subdivided into organic and mineral acids. Organic acids are those compounds that contain carbon. Mineral acids do not contain carbon but may (oxidizing) or may not (non-oxidizing) contain oxygen (Table 2.1).

Acids are soluble in water. In concentrated solutions, they will attack minerals and tissue. They can coagulate protein. Contact with oxidizing acids or reaction with organic material can result in fire. Metals reacting with sulfuric, nitric or hydrochloric acids can create an explosive environment.

Acids are generally found in synthesis and conversion labs. They are used as reagent chemicals. The acid alters the chemistry of the compounds and provides the physical environment necessary for the reaction to take place. They are also used to convert free base drugs into the water-soluble salt form that is sold to the end user.

Acids generally exist in a water solution of varying concentrations. Pure acids are rarely encountered. However, mineral acids such as hydrochloric acid and hydriodic acid do produce fumes. These fumes can fill the atmosphere in and around the lab area. As a rule of thumb, hydro_____ acids are very corrosive and will produce fumes (Table 2.2).

For identification purposes, if the label says acid in the chemical name, it should be considered corrosive. The strength and concentration may be unknown, but the compound will react like an acid. Therefore, treat all unknown acids as if they were extremely corrosive.

Table 2.1   Acid Cap Color Code

Cap Color	Acid
Blue	Hydrochloric acid (HCl)
Yellow	Sulfuric Acid (H_2SO_4)
Brown	Acetic acid (CH_3COOH)
Red	Nitric acid (HNO_3)
Clear/White	Phosphoric acid (H_3PO_4)
Black	Perchloric acid ($HClO_4$)
Black	Hydriodic acid (HI)

Table 2.2 Acid Relative Strength

Acid Name	Formula	Strongest
Perchloric Acid	$HClO_4$	
Sulfuric Acid	H_2SO_4	
Hydrochloric Acid	HCl	
Nitric Acid	HNO_3	
Phosphoric Acid	H_3PO_4	
Hydrofluoric Acid	HF	
Acetic Acid	CH_3COOH	
		Weakest

Caustics or bases are chemicals that readily accept a proton (hydrogen) in a chemical reaction. They have a pH greater than 7, with any pH greater than 12 considered corrosive. They can be subdivided into inorganic peroxides and organic amines. Inorganic peroxides are characterized by the presence of a hydroxyl ion (OH^-) group. The organic amines contain a characteristic amine ($-NH_2$) group.

Caustics generate heat when reacting with water, acids, organic material and some metals. They have a tendency to liquefy protein (e.g., tissue). Caustics are used to neutralize acids in the synthesis and conversion processes. They are used to adjust the pH of water solutions in the preparation of an extraction. They are used in conversion labs to convert the salt form of a drug into the free base form. In some instances, they can act as precursor chemical and become part of the final product.

Caustics can be found in solid, liquid and gas form. They are found in the pure form more often than their acidic counterparts. However, they can be just as easily found in a water solution of varying concentration. Words like caustic, hydroxide and amine in the chemical name indicate a compound with caustic or basic properties.

Oxidizers are compounds that provide oxygen to a reaction. These compounds may cause a fire if they come in contact with combustible material and can react violently when exposed to water or in a fire. Oxidizers contribute oxygen to chemical reactions, which increases the fire and explosion hazard because they provide a source of oxygen to sustain combustion in a normally oxygen-deficient atmosphere. An excess amount of available oxygen also can increase the reaction rate, making the combustion hotter and faster.

Oxidizers are used as reagent chemicals in the manufacture of drugs and explosives. More significantly, they are a major component of inorganic explosive mixtures. Oxidizers are generally found in the solid form. Some strong acids, such as nitric acid and sulfuric acid, act as oxidizers. Compound with names ending in "–ate" (i.e., chlorate, nitrate, permanganate) are compounds with strong oxidizing potential.

Hydrogen peroxide is a corrosive oxidizer. It is commercially available as an aqueous (i.e., water-based) solution at concentrations ranging from 3% to 35% for legitimate applications ranging from an antiseptic to a hair bleach or coloring product, and in wood bleach products, among other commercial applications. Concentrated solutions react violently with organic material, causing a fire or an explosion, and it is the precursor chemical to triacetone-triperoxide (TATP) and hydrogen peroxide organic material (HPOM) mixture, which are both considered extremely sensitive explosives.

As acids have bases as a conjugate, oxidizers have reducers. A reducer is a compound that can remove oxygen from or add hydrogen to a compound. They are used as reagent chemicals in conversion labs. Strong reducing agents react rapidly and violently. They are found in solid and liquid forms, and their labels will contain the words hydride or acetylide in the chemical name.

Compressed gases pose a dual hazard. Not only does the chemical inside the container have its own unique chemical hazards associated with it, but the container itself can pose a threat. The incompatibility of the contents with the container may cause the container to explode or discharge its contents unexpectedly. This often happens in situations where the operator has refilled a container with something other than what the container was designed for.

There is no accepted color code for compressed gas cylinders. The color of a compressed gas cylinder containing helium may vary depending upon the distributor. The only clue as to the contents of a legitimate gas cylinder is the connection on the top. For example, compressed gas cylinders that are to be used for compressed air have a unique fitting so that a flammable gas cannot be accidentally connected to the line.

In the world of clandestine labs, it is a given that the outside label of a container may not reflect what the contents are. Operators routinely replace the contents of compressed gas containers with chemicals other than those the container was designed for. These replacement chemicals often will corrode the brass fittings used to regulate the release of the compressed gases. This creates a hazard for anyone handling the container. Even the minor act of opening the valve may cause it to break, releasing the pressurized contents in a single violent rush. When the contents are unknown and may be under high pressure, and there is generally no safe way to sample the contents in the field, the field investigator may consider not even sampling the contents of the compressed gas cylinders and simply allowing the chemical waste disposal company to dispose of them properly.

Another compressed gas hazard is the homemade hydrogenator. These apparati are made of materials that were not designed to withstand the pressures or temperatures generated by the reaction. The minor act of touching the container may be enough to compromise the structural integrity of the container if it is under pressure, causing a BLEVE-like explosion that sprays the contents of the container all over the surrounding area (Figure 2.8).

A poison is a substance that in low concentrations will cause death or injury upon ingestion. They act on the internal systems of the body. Highly toxic poisons are usually gases or highly volatile liquids and have an oral LD50 of <5 mg/kg. Moderately toxic poisons may be solids or liquids and have an LD50 of > 5 mg/kg.

Chemical reactivity is the final hazard. How chemicals interact with each other can potentially pose a significant threat to clandestine lab responders. How the responders handle the chemicals at the scene is just as important as what lab operators did with them prior to the arrival of the good guys.

Chemicals are either compatible or incompatible. Compatible chemicals are those that can remain in close or permanent contact without a reaction. Incompatible chemicals react with undesirable results.

Incompatible chemical reactions generate heat that can cause a fire or explosion, form a toxic gas or vapor, or form a substance that is more toxic than the original compounds. The reaction can disperse a toxic mist or dust, produce a violent chemical reaction, or any combination thereof.

Water reactive and pyrophoric chemicals are examples of chemical groups that pose extreme reactivity problems. Water reactive chemicals hydrolyze with water, forming flammable, corrosive or toxic products. Pyrophoric chemicals react with the air and may spontaneously ignite.

Metallic sodium used in the Birch reduction is an example of a water reactive chemical. It reacts violently with water, producing sodium hydroxide and hydrogen gas. Sodium hydroxide is extremely caustic. Hydrogen is extremely flammable and under the right conditions, explosive. Under the right conditions, the water reacting with metallic sodium can produce an explosive hydrogen environment. If the hydrogen explodes, a caustic aerosol of sodium hydroxide will be dispersed. For this reason, metallic sodium is stored in mineral spirits or other non-aqueous solvent (Figure 2.9).

Figure 2.8 Compressed gas containers. (a) Alter freon containers, (b) HCl generator.

Figure 2.9 Magnesium/aluminum, a combustible solid mixture.

2.3.6 Physical Hazards

The final group of hazards associated with a clandestine lab is the physical hazards. These hazards include accidents, thermal exposure, electrical hazards and the dangers of confined spaces.

An accident is an unforeseen happening resulting in damage to people or property. Accidents are not the result of an unsafe act. For example, if a person opens a door without realizing there is someone on the other side, and the door hits the other person, causing them to spill red wine on their shirt, the person opening the door did not know there was anyone on the other side and could not foresee the collision; therefore, it was an accident.

On the other hand, if an investigator opens an unknown chemical container without the appropriate training or PPE, spilling acid on his shirt, damaging it and burning his skin, this was not an accident. The actions and damage were avoidable. The investigator should not have been handling the container without the proper training or minimally, without the proper PPE.

Processing the scene of a clandestine lab is the proverbial accident waiting to happen. However, knowing the causes of accidents can help prevent and eliminate them. The major causes of accidents are lack of preparedness, inattention, carelessness and fatigue.

There is no excuse for lack of preparedness. Investigators should have an idea of what manufacturing method the operator is utilizing prior to entry. With this knowledge, investigators should assemble the proper personnel and equipment to process the scene in a safe manner. The lab is not going anywhere; time is on the investigator's side. Therefore, the scene should not be entered or processed if the proper personnel and equipment are not present unless exigent circumstances exist.

Site exploitation of locations suspected of explosive manufacturing in a hostile environment is one of those exigent circumstances in which time on site may be limited. In these situations, the preparedness of the investigators is more critical. Their attention must be focused on the task at hand with an emphasis on properly implementing all of their knowledge skills and abilities associated with the manufacture of explosives and the associated hazards in an effort to identify and abate hazards. They need to utilize all of the tools at their disposal to efficiently process the scene in a safe manner.

Inattention and carelessness go hand in hand. Many seasoned clan lab investigators routinely see the same manufacturing process and become very lax in their handling of the situation. They begin to ignore common safety practices or fail to see obvious hazardous situations.

Fatigue can lead to such carelessness and inattention. Many clandestine labs are processed late at night, after a long protracted investigation or surveillance. Lack of sleep leads people to be not only tired but also frustrated. The lengthy procedure of safely processing a clandestine lab scene adds to the fatigue. Sleep deprivation also leads to many poor or hasty decisions, which in turn can lead to undesirable consequences.

Thermal hazards are another type of physical hazard. In this context, they relate to the environmental temperature rather than the heat of the reaction equipment in the lab itself, encompassing both extremes of hot and cold.

Heat stress occurs when the body is exposed to excess heat for an extended period of time. Such stress can affect the body's ability to regulate its temperature as well as other functions. Heat exhaustion can be debilitating but leaves no permanent effects. However, heat stroke can be fatal, and medical attention is required as soon as possible.

Heat effects can be minimized by acclimating the body to the temperature in the lab area prior to beginning work. It is not wise to go from an air-conditioned building or car to immediately working in an outdoor lab with an ambient temperature of 110 degrees. Allowing the body to get used to the temperature in the work area reduces the shock to the body.

Hydrating your body with fluids containing electrolytes is a good preventative measure. Avoid drinks containing diuretics such as caffeine (coffee, tea, soda). These cause the body to lose fluid faster than normal. These fluids could be used to cool the body down as perspiration if they were not being eliminated through the urinary tract so quickly.

The PPE used to process clandestine labs accentuates the problems related to heat. The characteristics of the equipment that protect the body from the environmental hazards in the lab area work against the body's natural ability to cool itself. Such clothing does not breathe or allow perspiration to evaporate and thus cool the body naturally (Figure 2.10).

Two common signs of heat stress are the loss of rational thought and the slowing of bodily functions. Inattention and carelessness lead to accidents. The reaction time of people suffering from heat stress slows down noticeably and is easily recognizable (by someone else) as changes in body language, speech pattern and perception of time.

It is imperative that the buddy system is used when working in situations involving elevated ambient temperatures. The buddy needs to be constantly aware of the condition of his partner. As soon as he sees signs of heat stress in his partner, they should both leave the area for a period of rehydration and cooling off.

The effects of heat can be minimized. Establishing a work-rest pattern will help reduce heat effects. Consistent breaks to cool the body and replenish fluid levels are good preventative measures in this battle. It is recommended that cotton underclothing be worn to wick the perspiration away from the skin. Also, during rest breaks, the PPE should be opened, allowing the body to breathe and cool.

Cold is the other side of the thermal hazard coin. As with heat, excess cold tends to distract a person's attention from the task at hand and lead to accidents. Cold stress occurs in cold, wet and windy environments, frostbite being the most common injury. Hypothermia is the extreme case, which can result in unconsciousness or death if not addressed immediately.

To reduce the effects of cold, clandestine lab responders should dress appropriately for the weather conditions. Layers of warm clothing are a wise choice, since layers can be added or removed as needed. Avoiding windy and wet conditions when possible also reduces the effect of the cold. Staying physically active generates heat and keeps the body warm. However, excessive

Figure 2.10 Assessment team in level B protection.

sweating also should be avoided. When the activity stops, the excess perspiration evaporates, increasing the effects of the cold.

As with hot conditions, an established work-rest pattern reduces the effects of cold. Frequent rest breaks remove personnel from the cold environment and allow them to warm up. This also keeps them from generating an excess of perspiration that will lead to excessive cooling and gives them the ability to adjust the number of warming layers to the current environmental conditions (Figure 2.11).

Electricity is taken for granted in modern America. We plug something into an outlet, turn the switch and it works. Electricity in the world of the clandestine lab operator is not that simple. The wiring of some of the equipment is makeshift, or the original design has been unsafely altered. Electrical boxes that were originally wired to the local building code have been altered to clandestinely divert current. A single outlet designed to draw 15 amps of current may have multiple extension cords attached to it, leading to equipment drawing 50 amps of current. Bare wires present a constant threat of electric shock or source of spark that can ignite a flammable atmosphere.

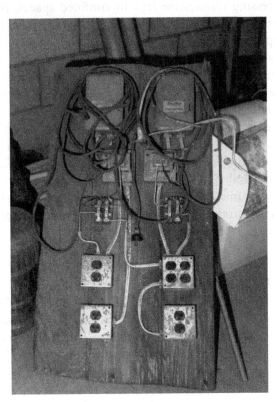

Figure 2.11 Clandestine lab wiring board.

To eliminate all electrical hazards prior to processing, turning off the electrical power to the site is simple but effective. However, prior to turning off the power, the assessment team needs to determine if there is some portion of the operation that would be adversely affected if the power were off. For example, it would not be wise to turn off the power if there was a cooling operation being powered by electricity. If the power were turned off, the cooling operation would stop and potentially allow the reaction to overheat, leading to an explosion or the like.

Confined spaces are those that have limited entry and exit openings, have unfavorable ventilation and are not intended for continuous occupancy. Almost by definition, a clandestine lab is a confined space. To avoid detection, they are squirreled away in bathrooms, attics, closets, a workroom off a garage, or in small outbuildings. The ventilation is usually minimal or non-existent. Even though it does not appear to be suitable for continuous occupation, some operators will spend hours or days at a time in the lab while the manufacturing process is under way. It is not uncommon for the lab area to contain so much stuff that there is hardly any room to move. This increases slip, trip and fall hazards.

When processing clandestine labs in confined spaces, the number of people in the area at one time should be kept to a minimum. Increasing the number of people in the area increases the potential for accidents and exposure to chemical hazards.

Personnel must be constantly aware of the limitations placed on them by their PPE. A person who normally has the agility of a rabbit is converted to a turtle when the PPE is worn. The PPE also reduces the person's dexterity and limits their mobility. The use of a self-contained breathing apparatus (SCBA) increases the operating size of a person by placing an air tank (adding 12 additional inches) behind them that is not usually there. Air purifying respirators (APRs) and SCBAs narrow peripheral vision and reduce the ability for verbal communication. These are unfortunately necessary evils when working in an environment containing numerous unknown and potentially lethal hazards.

2.4 Hazard Abatement

A clandestine lab is a mixture of physical hazards and chemicals with varying hazard levels and chemical compatibility. The hazards associated with some of the reaction by-products may be greater than for any known chemicals. The following generic rules can be used when dealing with the hazards at a clandestine lab site.

Do not process a clandestine lab without the proper training. Training provides the knowledge to recognize the potential hazards involved in any scenario. It also provides the knowledge of how to abate the hazards when they are encountered. Processing a clandestine lab without the proper training can lead to serious injury to the individual and the surrounding personnel.

Do not process unless all personnel are present. The processing of a clandestine lab scene involves a team of people and incorporates many specialized tasks. This includes the support personnel such as the fire department and emergency medical services. Safety may be compromised if the processing commences without an individual with a specific expertise. The operation is not going anywhere. Unless there are exigent circumstances, safety concerns overrule expedience.

Do not process without the appropriate safety equipment. The chemical and physical hazards involved in clandestine labs pose a variety of health hazards to the personnel processing the operation. Chemicals can enter the body in many ways, and specialized equipment is required to minimize such hazards. The use of the proper safety equipment will provide the responders with the protection necessary to minimize such effects of exposure. The operation should not be processed unless the proper safety equipment is present. As with the need for the proper personnel, the operation is not going anywhere. Therefore, for safety reasons, the operation should not be processed until the proper PPE is present.

Ventilate confined spaces. This reduces the concentration of explosive, flammable or toxic fumes in the area to below the hazardous limit. Proper ventilation also provides an atmosphere in which the need for respiratory protection is reduced or eliminated. This in turn reduces the potential for accidents that result from the use of bulky PPE.

Isolate from ignition source. All fires need fuel, oxygen and a source of ignition. Removing the source of ignition removes the capacity to burn. Turning on a light or lighting a cigarette at an inappropriate time can be an ignition source. Enforcement personnel should be reminded that the muzzle flash from a discharged weapon is sufficient to ignite a flammable atmosphere.

Remove heat from reaction mixtures. This will slow or stop the reaction that is taking place, reducing the amount of potentially explosive, flammable and toxic fumes that are being produced. Allowing the temperature of the reaction mixture to reduce to room temperature naturally eliminates the effect of drastic changes in temperature. This task should only be performed by a trained chemist who is able to identify the type of reaction being utilized.

Identify known chemicals from container label. Untrained personnel who encounter chemicals should relay any label information to trained personnel. Label information may provide trained personnel with insight as to what chemicals are potentially at the location. The information could help to establish the potential manufacturing methods and what additional chemicals may be at the location. It is important to remember that not all labels accurately indicate the contents of their containers; however, it is a start in the flow of information.

Do not handle chemicals unless trained. The handling of chemicals should be left to trained personnel. They can recognize chemical names that have hazardous potential. They also can recognize chemical combinations or other situations with hazardous potential. Untrained personnel should seal the area of the clandestine lab and wait for trained personnel to begin the abatement procedures.

Do not handle containers unless absolutely necessary. Safety outweighs the need to identify the contents of the container. When practical, law enforcement or military personnel should take evidentiary samples of containers where they are located and allow hazardous waste disposal specialists to move such containers at the time of disposal. Every time a container is handled, there is a potential for an accident. Therefore, reducing the number of times hazardous materials are handled reduces the potential for an accident.

Only trained personnel should do field testing. Field testing involves conducting chemical reactions with unknown chemicals. Trained chemists and hazardous materials responders have the training and experience to determine what field tests are appropriate for a given unknown. Applying an inappropriate field test may lead to an incompatible chemical reaction that creates a more hazardous situation than existed prior to the field test.

Incompatibility of unknowns with chemicals in test reagents is minimized with the use of modern instrumentation designed for use in austere environments. Although these tests are reliable, training is required to properly interpret the results. Misinterpretation of the results can lead to more severe problems than no results. This is equally true with chemical color testing.

Powdered sodium cyanide can easily be mistaken for a controlled substance. Administering controlled substance field tests that are commonly available to law enforcement officers can prove to be a lethal decision. Many of these tests contain an acid whose reaction with sodium cyanide produces hydrogen cyanide. Exposure to hydrogen cyanide is extremely hazardous if not deadly.

Chlorates combined with sugar are a common mixture encountered in fuel oxidizer explosives. When combined, the mixture has the appearance of a white granular powder. Acid is used as a time delay to initiate the explosion in certain booby traps. This is the same acid that is used in the reagents of commonly used field test kits. Although the test sample is small, the violent reaction and the associated heat and flame can injure the person performing the test.

Beyond establishing the pH of a liquid and whether it is organic or aqueous, without field instrumentation, there are few field tests that will establish probative information for the clandestine lab investigator. Evidentiary testing should be restricted to laboratory conditions. However, field tests to establish potentially hazardous characteristics of chemicals may be done by trained personnel for the purpose of separating and segregating chemicals before the chemical disposal company arrives.

Separate and segregate incompatible chemicals. Trained personnel should separate and segregate chemicals with known incompatibility. This reduces the potential for these chemicals to accidentally combine and create a more hazardous situation. Trained personnel use known label information and their training and experience to approximate what the contents of unlabeled containers are and separate them accordingly. They then segregate chemicals by placing those with similar properties in distinct groups. Placing a physical barrier between chemicals with known incompatibilities reduces the likelihood of chemical interaction if a spill should occur. Acids and caustics are differentiated by pH and segregated. Acids are also separated by type, because not all acids are created equal. Oxidizers are separated from organic material and reducers. Appendix F contains groups of incompatible chemicals that can be used to assist in the segregation process.

Seal containers. Sealing containers reduces the likelihood of incompatible chemicals combining. Some mineral acids produce fumes that can contaminate the air or mix with uncovered solutions containing incompatible mixtures. However, hot reaction vessels should not be sealed until they reach room temperature. If they are sealed prior to cooling, the act of cooling will create a vacuum seal, making the removal of the lid difficult and hazardous.

Do not handle compressed gas containers unless absolutely necessary. The contents of compressed gas containers are unknown. There is no way of safely determining such contents or the structural integrity of the container itself. The valves of compressed gas containers should not be manipulated unless absolutely necessary because they could break off from corrosion. They may release an unknown toxic gas into the atmosphere. Close valves of containers only if they are connected to reaction apparatus. Compressed gas containers should be secured and left for the chemical disposal company to handle (Table 2.3).

There is a number of further steps clandestine lab responders can take to minimize the effects of physical hazards.

Be in good physical shape. The human body has a remarkable ability to compensate for a wide variety of physical stresses that are placed upon it. Proper exercise and diet maximize the body's defenses against physical stress. It is recommended that clandestine lab responders exercise regularly and eat properly. It is also wise to have yearly physicals to help determine

Table 2.3   Hazardous Materials Handling Guidelines

Hazard Type	Abatement
Explosive chemicals	**Do not handle chemicals** unless trained or absolutely necessary. The handling of chemicals should be left to trained personnel. Trained personnel can recognize chemical names that have explosive potential. They also can recognize chemical combinations or other situations with explosive potential. Untrained personnel should seal the area of the clandestine lab and wait for trained personnel to effect the abatement procedures. **Separate incompatible chemicals.** Trained personnel should separate chemicals with known incompatibility. This will reduce the potential for these chemicals to accidentally combine and create an explosive situation. **Remove heat** from reaction mixtures. This will slow or stop the reaction that is taking place. This will reduce the amount of potentially explosive and toxic fumes that are being produced. Allowing the reaction mixture to reduce to room temperature naturally eliminates the effect of drastic changes in temperature. **Ventilate** confined spaces. This will reduce the concentration of explosive fumes in the area to below the explosive limit. **Do not open valves** **Close valves** of containers connected to reaction apparatus **Secure container** for chemical disposal company disposal.
Flammable/ Combustible	**Do not handle chemicals** unless trained or absolutely necessary. **Separate incompatible chemicals.** **Remove heat** from reaction mixtures. **Ventilate** confined spaces. **Isolate from ignition source.** All fire needs fuel, oxygen and a source of ignition. Removing the source of ignition removes the capacity to burn. (The muzzle flash from a discharged weapon is sufficient to ignite a flammable atmosphere.)
Acids and Caustics	**Do not handle chemicals** unless trained or absolutely necessary. **Identify the pH** and/or acid type of all unknown liquids. Knowing the basic characteristics of the liquids will provide insight into how they should be segregated. (See Table 2.1 for bottle cap identification guide.) **Seal all containers.** This reduces the likelihood of the mixture of liquids if the containers are spilled. Also, some mineral acids produce fumes that can contaminate the air or mix with uncovered solutions containing incompatible mixtures. **Separate acids and caustics.** Segregating liquids by pH will prevent violent reactions if a chemical spill does occur. **Separate acids by type.** All acids are not created equal. Oxidizing mineral acids react violently when they come into contact with non-oxidizing mineral acids and organic acids. Segregating acids by type will prevent violent reactions if a chemical spill does occur.

(Continued)

Table 2.3 (Continued)   Hazardous Materials Handling Guidelines

Hazard Type	Abatement
Oxidizers and Reducers	**Do not handle chemicals** unless trained or absolutely necessary. **Identify** known chemicals from container label. Unfortunately, there is no screening test for the rapid identification of oxidizers or reducers commonly used in the field by chemists investigating clandestine labs. Unlabeled containers should be treated as unknowns, leaving the field identification to the chemical waste disposal company. **Seal all containers.** This reduces the likelihood of the mixture of incompatible chemicals if the containers are spilled. **Separate oxidizers and reducers.** Segregating chemicals by class will prevent violent reactions if a chemical spill does occur. Segregate oxidizers from organic material and other combustible material to prevent fire in the case of spill.
Compressed Gas Containers	**Do not handle containers** unless trained or absolutely necessary. **Do not handle containers unless absolutely necessary.** Safety outweighs the need to identify the contents of the container.

any medical conditions that may have an adverse effect upon the body when placed under the physical stress involved in clandestine lab response.

Be rested. Being well rested also reduces the effects of mental and physical fatigue that leads to inattention and carelessness. This may not be realistic under the circumstances involved in the seizure of a clandestine lab. However, every effort should be made to minimize the effects of fatigue. Frequent rest breaks are necessary to reduce the effects of heat and cold stress on the body. These breaks also provide a mental break to reduce the effects of mental fatigue.

Drink liquids. Proper hydration is necessary to allow the body to regulate its internal temperature. It is wise for a responder to hydrate his body with liquids high in electrolytes prior to arrival at a clandestine lab scene. Replacing fluids during regularly scheduled rest breaks is essential to maintaining the fluid levels the body requires for optimum efficiency. This one suggestion cannot be emphasized too often, especially in hot climates. Avoid caffeinated drinks because of their diuretic effects.

Minimize number of people exposed. Keeping the number of people to a minimum in the lab area at any given time helps to reduce the potential for accident. A clandestine lab has a limited amount of space in which to operate. Increasing the number of people in this small space, coupled with the reduction in agility and dexterity brought on by the limitations inflicted by the PPE, increases the potential for an accident. Also, by limiting the number of people in the lab area itself, it is possible to limit the number of people exposed to the maximum amount of hazard. Bottom line is: if you do not need to be in the lab area, don't go in.

Utilize a buddy system. The buddy system is an essential safety consideration. The effects of heat and cold stress and chemical exposure may not be apparent to the person exposed. Exposure may lead to clouded thinking, irrational responses, distortion of time and compromising of motor skills. A buddy is needed to monitor the physical well-being of his partner and determine whether he is functioning normally. At the first sign of abnormal behavior, it is the responsibility of the buddy to have the pair leave the area for rehabilitation. Also, if one of the partners is involved in an accident, the buddy is there to render aid.

2.5 Practical Application

2.5.1 Mechanical Explosions

A mechanical explosion occurs when the structural integrity of a container is compromised as the result of excessive pressure inside the container. This situation can occur by design, as in the case of a pipe bomb. However, when associated with clandestine labs, it is more often than not a result of an equipment modification performed by the operator. The following examples are the result of operator equipment modifications.

The operator attempts to vent the fumes emanating from the top of a reflux condenser into a makeshift filtering device. The opening of the condensing column can became obstructed as a result. The structural integrity of the boiling flask can be compromised by stress cracks that were the result of improper handling. During the reflux process, pressure inside the reaction vessel increases to such a point that one of three things could happen. First, the pressure could clear the obstruction at the top of the condensing column. The excess pressure would be vented into the filtering device. This may or may not have an adverse effect, depending upon the construction of the filter.

Second, the connection between the boiling flask and the condensing column could fail. The pressure in the system would then propel the condenser like a missile into the ceiling. The pressurized contents of the reaction flask would spew from the opening like a fountain, bathing everything in the area with boiling hazardous chemicals.

The other possibility is that the structural integrity of the reaction flask would fail. The weak point in the vessel would succumb to the internal pressure, propelling broken pieces of glass like shrapnel from a grenade and coating the immediate area with the boiling reaction mixture. This option presents the dual hazard of chemical exposure and being impaled by the reaction flask fragments.

2.5.2 Mechanical Explosions

A clandestine lab operator repeatedly used a kitchen pressure cooker as a reaction vessel. Over time, the pressure relief valve corroded to the extent that that the vessel would not pressurize. To rectify the situation, the operator soldered the opening closed. This particular operator placed the reaction mixture directly into the altered pressure cooker, placed it on the electric kitchen stove and turned the stove on. The acids in the reaction mixture weakened the metal during repeated uses of the pressure cooker. The pressure from the boiling reaction mixture reached the point at which the pressure cooker lost its structural integrity and exploded. The metal lips on the pot portion, which held the lid on, sheared off. The pressure propelled the lid into the stove vent, spewing hot reaction mixture over the immediate area. Fortunately, no one was in the vicinity of the kitchen at the time of the explosion.

2.5.3 Vapor Explosions

The operator of a large-scale GHB operation constructed a drying room to evaporate the residual acetone from his final product. During the drying process, the concentration of the acetone vapors inside the drying room reached the explosive range. A spark was generated inside the drying room when the operator turned on an interior light. This resulted in a vapor explosion, whose effects are demonstrated in the images of Figures 2.5a, b, 2.12 and 2.13.

Vapor explosions have two demonstrable effects. The initial flash fire shows thermal damage, as opposed to deep charring as seen in a sustained fire (Figures 2.5b and 2.12). The rapid combustion in a confined space creates an overpressure situation, resulting in structural damage. This phenomenon can be observed in the garage door damage seen in Figure 2.5a and Figure 2.13, demonstrating the impact of moving a wall at the center of the explosion, which was blown approximately 3 feet off center.

2.5.4 Compressed Gas Hazards (Example 1)

Compressed gases are utilized in a variety of clandestine manufacture operations and pose multiple hazards. First, the chemical within the container may have a hazardous component. Second, the container itself may be unstable and pose a physical hazard to anyone attempting to handle it. Third, because of the second condition, there may be no safe way to determine the status of the first. Simply put, the investigator does not know what is in the container, and the container's condition may be too hazardous to determine what is inside. The following examples are used to provide some

Figure 2.12 Vapor explosion fire damage.

Figure 2.13 Vapor explosion blast damage.

insight as to the hazard potential of compressed gas containers encountered in clandestine labs.

Hydrogen chloride gas ($HCl_{(g)}$) is used to convert free base drugs into the hydrochloride salt form. One method bubbles commercially available HCl_g into a mixture of extraction solvent and free base drug. The free base drug reacts with the HCl_g, creating a solid, which precipitates out of solution. Clandestine lab operators commonly generate their own HCl_g using household chemicals. They place the chemical mixture into containers such as compressed gas cylinders, plastic gas containers or other containers that can be sealed, and the pressurized gas is vented in some manner. The result is a pressurized container of HCl_g (Figure 2.14).

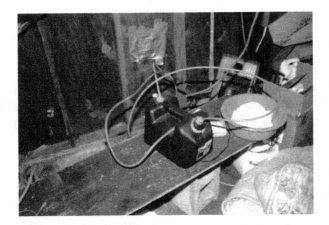

Figure 2.14 Hydrogen chloride gas generator.

The metal that compressed gas cylinders are constructed of is incompatible with the HCl_g. The HCl_g corrodes the brass valves or reacts with the metal of the container itself. The corroded valve can break during use, or the container eventually loses its structural integrity. Both situations lead to a discharge of pressurized HCl_g.

The other situations are not much safer. The containers used by the operators may be resistant to the corrosive nature of the chemicals involved, but they are not designed to withstand significant pressures. As a result, HCl_g is continually discharged into the atmosphere until the chemical reaction between the ingredients has finished.

2.5.5 Compressed Gas Hazards (Example 2)

Clandestine lab operators utilizing the Birch reduction (commonly referred to as the "Nazi" method) obtain the liquid ammonia required from agricultural areas that use it as a fertilizer. The operators use the propane tank from a gas bar-b-que to transport and store the stolen ammonia. Over time, the ammonia corrodes the valve on the tank to the point that the valve does not function. It may break when operated, resulting in the release of pressurized ammonia into the atmosphere.

2.5.6 Compressed Gas Hazards (Example 3)

Under pressure, ephedrine can be reduced to methamphetamine in the presence of a catalyst, acid and hydrogen. Clandestine lab operators have designed an apparatus to facilitate the hydrogenation process that uses a 2-liter plastic soda bottle. Under normal conditions, plastic soda bottles can maintain pressures in excess of 500 pounds per square inch. However, the conditions in a hydrogenation reaction expose the bottle to temperatures and chemicals the container was not designed for. The constant increasing and decreasing of pressure during the hydrogenation process, coupled with the heat generated by the chemical reaction, compromises the structural integrity of the bottle to the point where any shock will cause the plastic skin of the bottle to rupture and peel open like an over-ripe watermelon, releasing pressurized hydrogen and corrosive chemicals.

2.6 Summary

There are numerous hazards associated with clandestine laboratories, all of which can be minimized with proper training and education. The key is for the responders to use the knowledge, skills and abilities provided in the training to recognize potential hazards and take the preventative measures necessary to minimize the potential for undesirable effects.

All clandestine labs have three things in common. The operators have little training, they are generally makeshift operations, and no two operations are exactly alike. Keeping these three principles in mind allows the scene investigator to have a greater appreciation for the number of things that can potentially go wrong while processing the scene of the operation.

While on location, all personnel should continually be evaluating the scene for potential hazards. What can explode? Is there anything or any situation that could cause a fire? Are firearms available to undesirable people? Are there unseen things or situations that could potentially present a hazardous situation, either now or in the future? Finally, what can be done to reduce or eliminate these hazardous situations?

There are numerous things that can go and do go wrong at clandestine lab scenes. The post-scene debriefings of incidents in which something did go wrong reveal a common thread. The undesirable effect was a direct result of someone not following the established safety procedures. The lesson is that using the established safety procedures and equipment reduces the effects of the hazards involved with clandestine labs to the degree that will allow anyone processing the scene to leave as healthy as they arrived.

Basic Toxicology

3

Abstract

Exposure to the materials involved in the illicit manufacture of drugs and explosives is a topic that does not receive the level of attention required to instill the significance of the short- and long-term effects repeated, or in some cases ancillary, encounters with these substances can have on the individuals charged with investigating and dismantling these operations. As in many situations, knowledge is power. Understanding the adverse health effects of exposure to a chemical substance, toxicology, provides the knowledge required to protect against their toxic effects.

This chapter will provide generic information concerning how toxic materials enter the body, what happens when the toxin reaches its target (mode of action), and methods to minimize the entry of toxic materials into the body. Factors that influence the toxicity of a material and its short- and long-term effects on the body will also be discussed. The goal is to provide a basic understanding of how the materials used to manufacture drugs and explosives can adversely impact individuals who encounter them as part of a clandestine laboratory investigation. This chapter is not to act as a substitute for professional training as required by the U.S. Occupational Safety and Health Administration (OSHA) or the international equivalent. A secondary purpose is to encourage individuals actively involved in clandestine laboratory investigation to seek out and embrace training associated with hazardous materials, because the hidden hazards of these may be more dangerous than the physical hazards associated with the seizure and disposal of the laboratory items themselves.

3.1 Introduction

Exposure to hazardous chemicals is one of the dangers of processing clandestine lab scenes. Exposure to the materials involved in the illicit manufacture of drugs and explosives is a topic that does not receive the level of attention required to instill the significance of the short- and long-term effects repeated, or in some cases ancillary, encounters with these substances can have on the

DOI: 10.4324/9781003111771-3

individuals charged with investigating and dismantling these operations. As seen in the Basic Hazards chapter, the hazardous effects vary widely among chemicals. The short-term effects of a chemical exposure can differ from the long-term effects. In some instances, a single exposure can be lethal. In other situations, the effects are seen in the exposed person's offspring. That is why chemical exposure is the silent hazard.

Toxicology is the study of the adverse health effects of exposure to a chemical substance. As in many situations, knowledge is power. To have the power to protect one's self from the effects of chemical exposure, a person needs to know where exposures can occur, how toxins enter the body, and the physiological effects of the various chemical classes. Knowing how a chemical affects the body provides the knowledge required to protect against its toxic effects.

The goal of this chapter is to provide basic knowledge and understanding concerning the toxic effects of the hazardous chemicals encountered at clandestine lab scenes. This chapter will address the routes chemicals use to enter the body, the conditions that affect the absorption of chemicals, chemical toxicity ratings, types of toxin and their systemic effects. This chapter is not a scientific treatise concerning the toxicology of hazardous material. It provides basic explanations of how toxic materials can enter the body and what happens when they do.

3.2 Entry Routes

Hazardous chemicals must have contact with the body to have a toxic effect. The point of contact determines the entry route and can affect the body's response to a particular toxin. For example, the body's response to a dermal exposure to sodium cyanide differs from ingesting the same compound. Cyanide is a chemical asphyxiant that inhibits the blood's ability to carry oxygen. If the cyanide does not enter the circulatory system, it cannot affect the blood's oxygen-carrying ability. Dermal exposure (i.e., skin contact) prevents the cyanide from entering the circulatory system unless the skin has been compromised through abrasion or laceration. However, if the sodium cyanide is ingested or inhaled, it can be absorbed into the circulatory system, resulting in toxic effects.

The three entry routes through which toxic materials enter the body are inhalation, dermal adsorption, oral ingestion and direct contact. As chemicals can have multiple hazardous effects (e.g., nitric acid is a corrosive, an oxidizer and a poison), they can also have multiple toxic effects. The entry route of a toxin plays a role in the toxic effect the substance has on the body.

3.2.1 Inhalation

Inhalation is the most common and efficient entry route for toxins to enter the body. It is the most common entry route because people have to constantly breathe to survive. People can survive for extended periods without eating or touching things. Thus, the respiratory system is constantly exposed to the outside environment, which may contain toxic substances (Figure 3.1). It is the most efficient entry route because the respiratory system provides a direct conduit from the environment to the body's circulatory system. Once the toxin is in the circulatory system, it is a simple bus ride on the blood express to the toxin's target organ of choice.

The respiratory tract is comprised of the upper airway, the lower airway and the alveoli. Each portion has a specific function. Some portions filter out toxic material that would inhibit the oxygen/carbon dioxide exchange in other portions. Other portions exchange toxic materials between the blood and the air that is breathed.

The upper airway is comprised of the nose and larynx. Its purpose is to increase the relative humidity of the incoming air. It is lined with hairs called ciliated epithelium, whose purpose is to keep large particles in the incoming air from getting into the lower airway and alveoli. The upper airway's moist

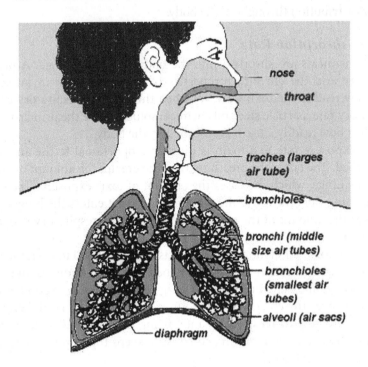

Figure 3.1 Respiratory system.

environment provides an atmosphere for water-soluble materials to dissolve and potentially enter the circulatory system.

The lower airway is comprised of the trachea, bronchi and bronchioles. They are lined with mucus-coated ciliated epithelium. The purpose of the lower airway is to prevent particles that pass through the upper airway from getting into the alveoli. The ciliated epithelium moves the trapped particles up the lower airway to the oral cavity for elimination. Smoking and the use of cough suppressants can in effect paralyze the ciliated epithelium, affecting its ability to filter and remove particles from the incoming air. Water-soluble toxins can be adsorbed in the mucus, also affecting the function of the ciliated epithelium.

The alveoli are the tiny sacs that are attached to the bronchioles. They contain the blood vessels that exchange oxygen and carbon dioxide. Another function of the alveoli is to exchange volatile toxins with the incoming air and eliminate them in the expired air.

The alveoli provide the largest surface area of the body that is constantly exposed to the environment. They provide a direct connection between the environment and the body's circulatory system. Toxins can enter the circulatory system directly at this point. The act of breathing constantly refreshes the supply of toxic substances that can be introduced to the circulatory system for distribution throughout the body.

3.2.1.1 Absorption Rate

Not all chemicals are absorbed into the body at the same rate. A person's physiology and the chemical properties of the toxin work in tandem to affect how readily the toxins are absorbed through the respiratory system. Respiratory rate, particle size and chemical solubility are the principal issues that affect how readily a toxin is absorbed by the body.

The exposure rate to a toxin is directly proportional to the amount of air breathed. The faster the breathing rate, the greater the amount of air that enters the lungs, which increases the amount of toxic exposure. The slower the breathing rate, the lower the amount of air that enters the lungs, which decreases the amount of the toxin that can enter the respiratory system for absorption.

To limit toxic inhalation exposure, it is recommended to keep the respiratory rate to a minimum by being in good physical condition and maintaining one's composure during processing activities. Being in good physical shape influences the efficiency of the lungs' oxygen exchange ability, which assists in reducing the respiratory rate. Being in an excited state increases the respiratory rate. Maintaining composure keeps the respiratory rate low, minimizing the exposure.

The substance's particle size affects whether or where the toxic effect will result. Solid particles ranging in size from 5 to 30 microns are filtered out in the upper airway. Particles in the 1–5-micron range are filtered out in the lower airway. They collect on the mucus-covered epithelium and are transported to the oral cavity for elimination. These particles can also embed themselves into a portion of the respiratory system, causing infection or disease. Asbestos is an example of an insoluble particle that embeds itself in this way, producing adverse effects on the respiratory system.

When working in toxic environments, it is imperative that responders do nothing to affect the ciliated epithelium's ability to remove particles from the incoming air. Smoking and the use of cough suppressants could inhibit the cleansing movement of the cilia, allowing particles to embed themselves into some portion of the respiratory system or be introduced into the circulatory system.

A substance's water solubility will determine its toxic effects through inhalation. Water-soluble substances will dissolve in the mucus of the upper and lower airways. This may produce a localized toxic effect or allow the compound to be absorbed into the circulatory system. Water-insoluble substances can travel past the upper and lower airways into the alveoli, where they can enter the circulatory system.

3.2.2 Dermal Absorption

Dermal absorption is the second route of entry of toxins into the body. Many people have the misconception that the skin is impervious to everything. If the contact with the substance does not produce some sensation of pain or discomfort, there cannot be a toxic or hazardous effect. This is not the case when considering the skin as an entry route for toxic materials.

The is skin comprised of the dermis and epidermis (Figure 3.2). By weight, it is the largest organ of the body, and it has two main functions. The first function is to protect the internal organs of the body from adverse environmental conditions. The second is to regulate the body's temperature. Both of these functions affect how toxic substances are absorbed into the body.

The epidermis is the top layer of skin cells, which provides the first line of defense from environmental toxins. It can intrinsically repel the toxic effects of a number of substances. The condition of the epidermis plays a critical role in its ability to repel toxins. Dryness or skin damage as a result of cuts or abrasions allows toxins to bypass the protective traits of the epidermis. Corrosives can damage the skin and provide an opening for other toxins to enter.

The dermis comprises the lower layers of the skin. It contains the sweat glands and ducts, oil glands, fatty cells, connective tissue and blood vessels.

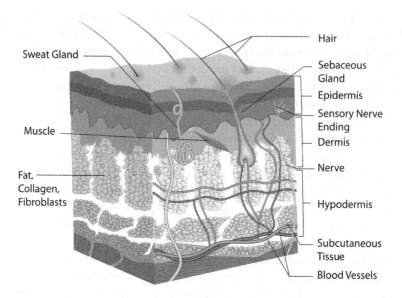

Figure 3.2 Skin cross section. Courtesy Shutterstock.com.

It provides a nonselective diffusion environment for toxins to travel into the circulatory system.

The rate at which toxins are absorbed through the skin is affected by a number of factors. In combination, the relative effect of one may affect the relative effect of one or more of the others. The factors that affect the rate of dermal absorption include skin damage, hydration state, temperature, concentration and carriers. Each of these affects whether a substance is repelled or absorbed by the skin.

Skin damage is the simplest to understand. The skin can be equated to the plastic coating that protects a package's contents from water damage. A cut or an abrasion in the skin can be equated to a break in the package's plastic coating. Breaking the plastic coating provides an entrance for the water to enter the package. Breaking the protective coating provided by the epidermis provides toxins with a direct route into the circulatory system.

The hydration state of the epidermis can affect its ability to absorb or repel toxic substances. The epidermis has an optimum hydration state. If the skin is too dry, it may more readily adsorb liquid toxins or let them pass through to the dermis. If the epidermis is over-hydrated, it may provide an environment that will allow the toxins to diffuse through it into the blood vessels in the dermis.

The ambient temperature affects the dermal defenses in a couple of ways. It can affect the hydration state of the skin. Skin perspires in elevated temperatures. The perspiration increases the hydration state of the epidermis, leading to the effects described earlier. The sweat ducts also provide a conduit for toxins to travel past the epidermis into the dermis and potentially, into the circulatory system.

A second effect concerns the blood vessels in the dermis. In hot environments, the blood vessels expand to increase the blood flow in the dermis in an effort to reduce the body's internal temperature. In cold environments, the blood vessels in the dermis contract to restrict the blood flow in the dermis in an effort to keep the body warm. In elevated temperatures, the surface area of the blood vessels increases, increasing the ability for a diffusion transfer through the dermis into the circulatory system.

The concentration of a substance plays a role in its effect on the skin. The epidermis is very resilient. It can compensate for the toxic effects of low concentrations of a given substance. Exposures to these low concentrations produce little or no toxic effect. However, higher concentrations of the same substance can produce devastating effects.

Chemical carriers can provide a ride for a toxin through the skin's natural defenses. A solid toxin that the epidermis would normally repel may pass through it if it is dissolved into a solvent carrier that permeates the epidermis. The solvent shuttles the dissolved toxic substance past the epidermis into the dermis and then into the circulatory system.

3.2.3 Ingestion

Ingestion is the final mode by which toxins enter the body and is the least effective method. The physical state of the substances introduced through this mode and the nature of the digestive tract keep the amount and the toxic effects to a minimum (Figure 3.3).

Solids and liquid substances enter the body through this mode, which makes intentional ingestion difficult. The mouth is a relatively guarded entry point. People have the common sense not to intentionally eat or drink something that has a hazardous potential, as opposed to dermal contact, for which unintentional direct contact with the substance is not uncommon.

Inadvertent ingestions of toxic substances are not unusual. It is not uncommon for people in toxic environments to handle items that are placed into the mouth. Smoking and eating in an area where toxic substances are located provides the opportunity for toxins to be placed into the mouth if a contaminated hand handles the cigarette or food. For example, people taking notes at a clandestine lab scene will place the pen they have been writing with in their mouth. They have also been handling toxic chemicals, which have been transferred to the pen. The toxins are transferred to the gastrointestinal tract when the pen is inserted into the mouth. The only saving grace in this

Digestive System

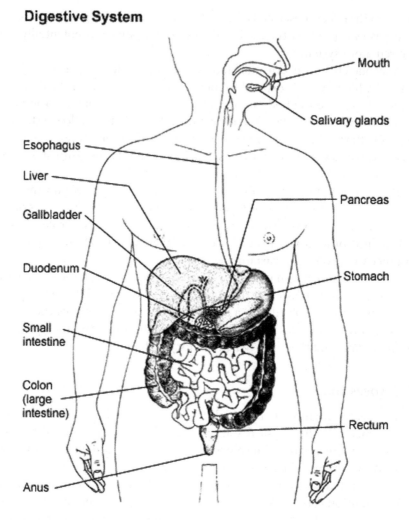

Figure 3.3 Gastrointestinal tract diagram.

incident is that the concentration of the substance may be low enough that it does not produce a toxic effect.

A third consideration is the environment within the digestive system. The highly acidic and alkaline environments in the various portions of the gastrointestinal tract have a tendency to neutralize the toxic effects of many substances before they can be introduced into the blood stream. Even if a toxin is introduced, interaction with the digestive juices may alter the toxic properties of the substance before it has a chance to enter the circulatory system.

3.3 Modes of Action

Every substance has a different effect on the body. The mode of action refers to the physiological system that is affected by the exposure. The three general modes of action consist of the physical, the chemical and the enzymatic.

The physical mode of action refers to how the substance interacts with all tissues of the body. The substance's hazardous characteristics can be used to characterize its physical mode of action. For example, a substance with corrosive properties will produce the same effect no matter what tissue it comes in contact with; i.e., corrosive substances are non-selective in the type of tissue they react with.

The chemical mode of action refers to how a substance interacts with specific tissues of the body. It can be characterized by generic toxic properties of the substance. For example, chemical asphyxiants interfere with the blood's ability to carry oxygen. The reactions are tissue specific and deal with the chemistry between the substance and the tissue involved.

The enzymatic mode is also referred to as the physiological mode. It refers to how the substance interacts with specific enzymes in the body. The substance can enhance or inhibit the processes of the enzyme it affects.

3.4 Influences on Toxicity

A number of variables will affect the body's response to a given toxin. The length and the degree of the exposure play a determining role. The chemical's physical properties, its concentration and the duration of exposure have their own influences on the body's reaction. Even environmental factors such as the temperature play a role in how a substance will react with the body. The variables that affect the toxicity of a substance are the length of the exposure, the degree of exposure, toxicity factors of the substance, factors concerning the exposure, environmental factors, and factors concerning the person exposed.

3.4.1 Length of Exposure

As discussed in Chapter 2, the length of exposure to a hazardous substance will determine the effects the substance has on the system. Also, the short-term effects can and do differ from the long-term effects.

Exposures to toxic hazards have acute and chronic effects. Acute effects are experienced on exposure to hazards of high concentrations and of short duration. The effects are felt immediately. The exposed person will recover if the exposure does not exceed the lethal limits. Chronic effects are generally experienced on exposure to hazards of low concentration over a long period of time. The effects are cumulative, creating a toxic effect. These effects are usually different from the acute effects of the same hazard.

The duration of an exposure may be too short to produce an acute effect. The body may have enough natural defenses to ward off any adverse effects. However, because of the depletion of these natural defenses, subsequent short-term exposures may have a compounded acute effect if the body has not had an opportunity to regenerate its defense mechanisms.

In some instances, the body stores the toxin in the fat cells, liver or other target organ. The concentration increases until it reaches a toxic level. Numerous insignificant short-term exposures may not produce any acute effects. However, the cumulative exposure may produce a chronic toxic effect.

3.4.2 Degree of Exposure

The degree of exposure differs from the length of exposure. Exposure length is strictly a measure of the time a person is exposed to the substance. The degree of exposure deals with the substance itself. It relates to the substance's chemical properties. The concentration and the specific duration of the exposure are all interrelated and affect how the substance will react with the body. Degree and length of exposure are interrelated in that changing one of the properties of the degree of exposure will alter the length of time required to produce a toxic effect. The group of exposure properties interacts in different ways. The effect produced results from the relationships between the different factors surrounding the exposure. The factors concerning the compound itself, the circumstances surrounding the exposure, the exposed person, and the environment of the exposure will combine and establish the total hazardous potential of the exposure.

3.4.3 Compound Factors

The substance itself has properties that establish its inherent toxicity. Its chemical properties, the concentration, the duration of the exposure and interaction with other chemicals will interact in different ways to produce different effects. The ratio of these is different in each exposure. Thus, the effects of the same substance will vary depending on this ratio.

Each chemical has an inherent set of physical and chemical properties. Some have a hazardous potential. Some physical properties prevent the substance from interacting with the body. For example, under normal circumstances, solids and liquids cannot be inhaled. However, changing the solid into an airborne dust or a liquid into an aerosol allows these physical states to be inhaled.

The substance's chemical properties will determine what effect it will have on the body when an exposure occurs. For example, under normal circumstances and without outside influences, a corrosive chemical will not burn or explode. However, contact with an incompatible substance may create an explosive situation.

The concentration of the substance during the exposure will affect the body's response. The body has the ability to compensate for an exposure to low concentrations of a wide variety of chemicals. Its internal defenses can neutralize or compensate for the toxic effects of a given substance. However, high concentrations of the same substance during the same exposure time period may overload the body's ability to defend against or compensate for the toxic effects.

The duration of the exposure will affect how the body reacts. The effects of instantaneous contacts differ from those of prolonged contact. Instantaneous contacts may not allow time for the substance to interact with the body, or the body may have sufficient defenses to compensate for that exposure duration.

A dramatic example of this is exposure to liquid nitrogen. Liquid nitrogen has a boiling point of −196 degrees C (−321 degrees F). The human body can tolerate a direct instantaneous exposure to liquid nitrogen. Beyond that, the temperature of the liquid nitrogen overcomes the body's ability to compensate, and severe damage results.

The interaction with other substances can potentially affect the substance's toxicity. Contact with other chemicals may alter the substance's physical state, which may affect the entry routes available to the substance. For example, the reaction between cyanide salts and acids converts a cyanide salt from a solid (with low inhalation potential) to hydrogen cyanide gas (with high inhalation potential). Dissolving a substance into a liquid may alter its ability to be absorbed through the skin.

3.4.4 Exposure Factors

The circumstances surrounding the exposure will affect the hazard potential. The entry route, the duration and the number of exposures will determine what effects the body will experience. The combination and ratio of these factors will affect the body's toxic response.

The entry route is the primary factor in the exposure scenario. If the substance cannot get into or have contact with the body, it cannot produce a toxic effect. As discussed earlier, different entry routes provide access to different physiological systems. A toxin's effect may differ depending on the entry route. For example, under normal conditions, cyanide salts have negligible potential for dermal absorption. However, they are readily adsorbed by ingestion. A person can handle these highly poisonous substances with minimal effect from dermal absorption. However, when the person's contaminated hands first come into contact with food that is then ingested, that insignificant exposure just became lethal.

The number of exposures influences the toxic effects of a substance. There are certain substances known as sensitizing agents. The initial encounter with the substance may not produce an effect. However, all subsequent exposures produce a toxic reaction. Other substances can have a cumulative effect. The body may store a substance until the concentration builds up to the point where it reaches a toxic level.

3.4.5 Personal Factors

The personal factors are the factors that can be directly attributed to the individual who is exposed. The exposed person's age, sex, health and genetics are directly related to how their body will react to an exposure to certain toxic substances.

A person's metabolism changes over time. The metabolism of an infant is different from that of a teenager entering puberty. A young adult body's ability to bounce back from injury and disease is greater than that of someone who has reached retirement age. This difference in metabolism between the ages dramatically affects the body's ability to fight off the effects of toxic substances.

A person's sex may determine whether or how a person is affected by an exposure to a particular toxic substance. In some instances, the metabolism of a male differs from that of a female. In other instances, a toxin may target a specific organ. If the toxin specifically targets the reproductive organs of a particular sex, it will not have a detrimental effect on the opposite sex. This can be of special concern to women of reproductive age. There is also a group of toxins that can lead to birth defects (teratogens) or fetal death (embryonic toxins).

The health of the person exposed to toxic substances will affect the body's response to the exposure. The body's ability to counter the effects of exposure to toxic substances is diminished if the immune system is fighting off disease or infection. Poor health extends beyond illness. Fatigue can be included in this category. If the body is run down due to lack of sleep or other fatigue factors, its ability to ward off toxic effects is reduced because the metabolism is not functioning at its optimum level for the person's age and normal physical health.

A person's genetic makeup will play a role in how certain toxic substances will affect the body. Some people are genetically predisposed to have a toxic reaction to a substance another person is not affected by. Just as some people are allergic to dogs, pollen or a variety of other allergens, some people will demonstrate a toxic reaction on exposure to a particular chemical, while others will exhibit no toxic symptoms on the same exposure.

Factors related to the environment affect whether or how a toxin enters the body. The carrier, ambient conditions and chemical interactions determine whether a particular exposure will produce toxic effects.

The carrier with which a toxin is associated can provide the toxin with an entry route that would not normally be available to it. Cyanide salts are an example of a solid poisonous substance that cannot easily enter the body under normal circumstances. However, accidental contact with ingested food provides the carrier needed for the entry route. The same cyanide salt that cannot normally be absorbed dermally can enter the system through the skin if it is dissolved in the appropriate carrier solvent under the appropriate ambient conditions.

The ambient conditions during the exposure have their effect. Cold slows the body's metabolism to one extent or another. Cold affects dermal exposure by closing pores and restricting the blood flow along the surface of the skin. Conversely, heat increases the potential for toxic exposures. Heat opens pores and increases the hydration state of skin, which increases the potential for solvent-soluble toxins to permeate the epidermis into the dermal layer that contains the blood vessels. Also, the blood flow in the surface of the skin is increased in an effort to cool the body. The dilated capillaries provide a greater surface area and potential for toxins to enter the blood stream.

3.5 Distribution and Elimination

Many toxins do not stay at the initial contact site. They enter the body through one of the exposure routes and travel through the body to a specific target organ. The movement of the toxins through the body is facilitated by the circulatory system, which consists of the blood and lymphatic systems. These systems also provide a method to remove toxins from the body.

Figure 3.4 graphically depicts the flow of toxins. Toxins enter the circulatory system via inhalation, dermal absorption or ingestion. Liquid and water-soluble solids are filtered out in the liver and kidneys and excreted from the body in the urine. Solid toxins are excreted in the feces. Volatile toxins are exchanged with the incoming air and expelled into the expired air. The toxins that are not otherwise eliminated from the body are transferred into the cellular fluid, where they are stored in the organs, soft tissue or fat cells.

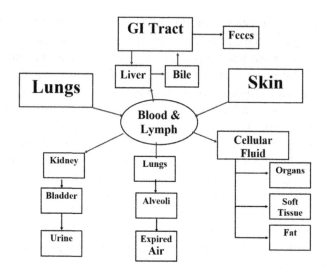

Figure 3.4 Toxin distribution chart.

3.6 Toxicity Measurements

There are a number of ways to measure the toxicity of a substance. Some of the measurements relate to the lethal dose. Others have to do with the amount a person can be exposed to before they feel any effects. Other measurements relate to the instantaneous exposure or the amount of exposure a person can experience over a period of time. All of these measurements identify the dose threshold of a substance a person can experience before suffering adverse effects.

Most toxicity measurements are derived from epidemiological and animal test data. The test values are usually derived from exposures given to rats and extrapolated to human ratios. Values are reported as a ratio of weight of substance per weight of subject. For example, a toxic value of 5 mg/kg rat means that the 5 mg of substance will produce the toxic effect in a 1-kg rat. This can be extrapolated to mean that 500 mg of substance would be needed to produce the same effect in a 100-kg human.

Many substances have an established LD_{50} value. The LD_{50} is the concentration of a substance that produces a lethal response in 50% of the test population. In other words, it is the concentration of the substance that will kill 50 of 100 of the test subjects.

Not all substances react in the same way with the body. Some have an immediate effect. Others can come into contact with the body at low concentrations without demonstrating symptoms of exposure.

Low–dose response substances produce an almost immediate effect on the body. The percentage of the population affected by exposure to a given substance increases with the concentration of the substance. The rate of increase of the affected population will vary from substance to substance. At some point, the concentration will reach the LD_{50} for the substance. Eventually, a concentration will be reached at which everyone in the population will experience the toxic effects of the substance (Figure 3.5).

Substances with a high dose response will not produce a toxic response until a certain concentration is reached (Figure 3.6). Up to that point, the body's natural defenses can counteract the toxic effects of the substance. Once the concentration of the substance reaches a threshold value, the population begins to exhibit the toxic effects. From that point, the percentage of the population experiencing toxic effects increases with the dosage concentration. As with substances with a low dose response, a concentration will be reached at which everyone in the population will experience the toxic effects.

The relative toxicity of a substance is a function of the substance. Each compound has its own relative toxicity. Table 3.1 contains the terms commonly used to describe the toxicity of a substance. Each term has an associated concentration range. Extremely toxic compounds can cause death upon a minute exposure, whereas you can literally bathe in relatively harmless substances before a toxic effect is experienced.

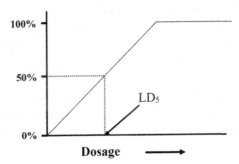

Figure 3.5 Low dose response.

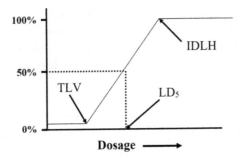

Figure 3.6 High dose response.

Table 3.1 Relative Toxicity

Rating	LD$_{50}$ (oral rat)	Example
Extremely Toxic	<1 mg/kg	Hydrogen cyanide (HCN)
Highly Toxic	1–50 mg/kg	Mercuric chloride (HgCl$_2$)
Moderately Toxic	50–500 mg/kg	Sodium hydroxide (NaOH)
Slightly Toxic	0.5–5 g/kg	Cyclohexanone
Practically Nontoxic	5–15 g/kg	Methanol
Relatively Harmless	>15 g/kg	Water

From Drug Enforcement Administration, *Clandestine Laboratory Training Guide*, Vol. 1, p. 22.

3.6.1 Exposure Guidelines

A number of agencies have established guidelines used to determine the toxic levels of various substances. The National Institute of Occupational Safety and Health (NIOSH), the OSHA and the American Congress of Government Industrial Hygienists (ACGIH) each have their own rating systems. Each measures a specific effect and is used to establish safety parameters to create a safe work environment.

The threshold limit value (TLV) is an exposure guideline established by the ACGIH. It corresponds to the concentration required to produce toxic symptoms. Low TLV values indicate a substance with a low dose response. High TLV values indicate a substance with a high dose response. The exposure concentration values are reported as either a time weight average (TWA) or as a ceiling (C). The TWA value is the average concentration a person can be exposed to over a period of time (either 8 or 40 hours). The C value is the maximum concentration for an instantaneous exposure.

The relative exposure limit (REL) is the exposure limit established by the NIOSH. These values are similar to the TLV values and are directed toward industrial applications.

The most relevant exposure level to clandestine lab seizures is the IDLH. The IDLH is the concentration that NIOSH has determined to be Immediately Dangerous to Life and Health. It is the concentration that will cause death or serious injury upon a single unprotected exposure.

The OSHA has established permissible exposure limits (PEL) for numerous substances. These values are similar to the NIOSH REL values and can be found in chapter 29 of the Code of Federal Regulations (CRF) section 1910.1000 Table Z-1.

The significance of the PEL values is that they hold legal weight. An employee cannot be exposed to concentrations exceeding the PEL of a substance without the appropriate personal protective equipment (PPE). Law enforcement and emergency responders are not exempt from these regulations. Even under exigent circumstances, law enforcement and emergency responders are required to adhere to the PPE requirements established by 29 CRF 1910.120 when concentrations exceed the PEL.

3.7 Toxin Properties

Now that we know how toxins enter and affect the body, we can talk about toxins themselves. A toxic material is a substance that in a relatively small quantity is capable of producing localized or systemic damage. They come in a variety of physical states and can affect the body in a number of different ways.

3.7.1 Physical States

Toxins come in all physical states. They can be in the basic physical states of solid, liquid or gas form. They could be in some combination of these, such as fumes, smoke, aerosols, mists, vapors or dust. Knowing the various physical states in which a toxic substance can be encountered provides a means to determine what type of personal protection will be required.

All substances have a normal physical state of solid, liquid or gas. The normal physical state is related to the standard temperature and pressure (uncontained at room temperature). Changing the temperature or the pressure conditions can affect the physical state of the substance.

Water is the simplest substance that can be used to demonstrate the effects of temperature. Water is a liquid in an open container at room temperature (25 °C). When the water's temperature is reduced to 0 °C by placing it into a freezer or exposing it to liquid nitrogen, the water turns into a solid. When the temperature is elevated to 100 °C by adding heat from an external source, the water turns into a gas.

Changing the pressure conditions under which the substance is stored will affect the substance's physical state. Increasing the pressure in a closed container will condense a gas into a liquid, and the container will experience an increase in temperature. (That is why a bicycle tire's temperature increases as air is added.) Reducing the pressure over a liquid will convert the liquid into a gas and reduce the temperature of the container. (That is why frost forms on a container of evaporating liquid nitrogen.)

The conditions encountered in clandestine labs are dynamic. Ambient temperatures can be at extremes of hot and cold. Boiling reaction mixtures can vaporize substances that are normally liquids at room temperature. The excessive heat generated in an automobile trunk in 110-degree weather may pressurize chemical containers. Rapidly cooling a boiling liquid that has been closed may create a vacuum, which may make opening the container difficult. All of these environmental factors will affect the physical state of the substance.

The dynamics of the environment in which the substance is encountered leads to hybrid physical states. They are transitory or a combination of the basic physical states. These hybrid physical states include vapors, fumes, smoke, aerosols, mists and dust.

A vapor is the gas phase of a substance that is a solid or a liquid at standard temperature and pressure (room temperature). Vapors are produced when the temperature of the substance is elevated through natural (high ambient temperatures) or artificial means (heat being added to a reaction mixture). The substance returns to its natural state when its temperature is reduced.

Fumes are vapors from substances that are solid at room temperature. They are the result of heating the substance, which produces airborne particles less than 0.1 micron in diameter. They can aggregate into fine clumps and eventually settle out of the air. Lead and iodine are examples of substances that can produce fumes.

Smoke is the result of incomplete combustion. The particle size is greater than 0.5 microns. These particles do not generally settle out of the air.

An aerosol is a stable suspension of solid or liquid particles of various sizes in the air that will eventually settle out of the air. Aerosols are usually the result of a mechanical distribution of the substance that atomizes the substance and disperses it into the air. The size and weight of the particles do not allow them to remain airborne.

Mists are liquid aerosols formed by liquid vapor condensing on airborne solid particles, which may or may not be visible.

3.7.2 Toxic Properties

Toxic materials produce localized or systemic damage to the body. That is, the toxic effect is experienced either at the point of contact or when the toxin enters the body, travels to the target organ and disrupts its function in some manner. Localized effects are experienced immediately. Systemic effects may be seen immediately, they may not manifest for years or they may occur in the exposed person's offspring.

Chapter 2 discussed the hazardous properties of the chemicals found in clandestine labs. This section will discuss how the hazardous properties correlate with toxic effects on the body. According to their effects, toxins can be divided into corrosive, asphyxiant, irritant, respiratory, systemic, external and special toxins.

Corrosives are localized toxins that can cause visible damage or irreversible alteration to human tissue at the point of contact. They include acids and caustics. The effects are generally considered to be localized. However, the damage can lead to systemic effects. For example, exposure to hydrochloric acid fumes can produce localized damage to the alveoli. This localized damage produces a systemic effect on the body's respiratory system.

Asphyxiants affect the supply of oxygen to the body. Even small reductions in the oxygen supply to the body can potentially produce a variety of effects on the body. Asphyxiants are divided into simple and chemical (Table 3.2).

Simple asphyxiants displace the oxygen in the atmosphere. There is no direct interaction with the body. They simply reduce the amount of an essential element available to the body. Table 3.3 demonstrates the effects that can be felt as the result of lack of oxygen in the air. These chemicals may be inert when they come into direct contact with the body. Freon, which is used legitimately as a refrigerant and clandestinely as an extraction solvent, is an example of a simple asphyxiant.

Chemical asphyxiants are substances that affect the blood's ability to carry oxygen. This involves a chemical reaction with the hemoglobin. The results will be similar to the oxygen concentration in the air being reduced. Carbon monoxide (CO) and hydrogen cyanide (HCN) are examples of chemical asphyxiants.

Irritants are compounds that disrupt the function of the system at the point of contact. They are common in the respiratory system. Water-soluble irritants such as acids can be absorbed into the mucus of the upper respiratory system, disrupting the ciliated epithelia's ability to function properly. These effects are usually experienced immediately. Non-water-soluble

Table 3.2 Oxygen Deficiency Effects

% Oxygen	Effects
21	No abnormal effects
12–16	Increased breath volume
	Accelerated heartbeat
	Impaired attention, thinking and coordination
10–14	Faulty judgment and coordination
	Rapid fatigue
	Intermittent respiration
	Permanent heart damage may occur
6–10	Nausea and vomiting
	Inability to perform vigorous movement or loss of all movement
	Unconsciousness and death
<6	Spastic breathing
	Convulsive movements
	Death in minutes

Table 3.3   EPA PPE Selection Guidelines

Respiratory Protection	Protective Clothing	Hand and Foot Protection	Optional Equipment
		LEVEL A	
• NIOSH approved positive pressure full face piece SCBA, or • NIOSH approved positive pressure full face piece supplied air respirator with an EEBA (5-minute duration minimum)	• Totally encapsulated chemical protective suit specifically designed to resist permeation by chemicals that are encountered	• Gloves: outer and inner chemical-resistant gloves • Boots: chemical-resistant with steel toe and shank	• Coveralls • Long underwear • Hard hat • Two-way radio communication system
		LEVEL B	
• NIOSH approved positive pressure full face piece SCBA, or • NIOSH approved positive pressure full face piece supplied air respirator with an EEBA (5-minute duration minimum)	• Hooded chemical-resistant clothing made of materials resistant to the chemicals encountered (overalls) and long-sleeved jacket: coveralls; one- or two-piece chemical splash suit; disposable chemical-resistant overalls	• Gloves: outer and inner chemical-resistant gloves • Boots: chemical-resistant with steel toe and shank	• Coveralls • Long underwear • Hard hat • Two-way radio communication system • Face shield • Boot covers (chemical resistant)
		LEVEL C	
• NIOSH approved full face piece or half mask air purifying respirator	• Hooded chemical-resistant clothing made of materials resistant to the chemicals encountered (overalls) and long-sleeved jacket: coveralls; one- or two-piece chemical splash suit; disposable chemical-resistant overalls	• Gloves: outer and inner chemical-resistant gloves • Boots: chemical-resistant with steel toe and shank	• Coveralls • Long underwear • Hard hat • Two-way radio communication system • Face shield • Boot covers (chemical resistant) • EEBA
		LEVEL D	
• No respiratory protection required	• Coveralls	• Boots: chemical-resistant with steel toe and shank	• All the above plus: • Eye/face protection • Gloves

irritants travel into the lower respiratory system. The irritating effects disrupt the lung's ability to perform its oxygen exchange function. Asbestos is the most notorious non-water-soluble irritant. However, phosphine and phosgene gas, which are by-products of methamphetamine manufacturing operations, are examples related to clandestine lab operations.

Allergens are substances that cause an immunological response upon exposure. The effect may be localized or systemic, and the point of contact determines the effects. For example, a dermal exposure to a substance may or may not produce a reaction. However, ingesting the substance may cause nausea, vomiting or some other allergic reaction.

Systemic toxins affect the function of a physiological system. They travel from the point of contact to the target organ before the toxic affect is experienced. The central nervous system is affected by asphyxiants, hydrocarbons such as hexane, and metals like lead. Aromatic solvents such as benzene and toluene can affect the circulatory system. The function of the liver can be affected by aromatic compounds, chlorinated solvents such as chloroform and carbon tetrachloride, and hydrocarbons. Halogenated compounds, such as Freon, and metals affect the kidney's function. The spleen is affected by halogenated aromatic compounds. Aromatics, pesticides and organic metal compounds affect the reproductive systems. Appendix F lists the target organs of the chemicals commonly associated with clandestine labs.

External toxins are those that target the body's external organs. Various solvents, oils, metals and corrosives target the skin and affect its ability to defend itself from toxic exposures. The eyes are the external target organs for corrosives, solvents, oils and lacrimators.

Special toxins are the silent toxins. Their effects are not seen immediately, and they may produce effects that are unrelated to the effects of the initial exposure. Carcinogens are substances that cause uncontrolled cell growth (cancer). Mutagens are substances that cause a change in the genetic code. Teratogens cause non-lethal congenital birth defects. The effects of mutagens and teratogens will be seen in the offspring of the exposed person. Embryonic toxics will cause fetal death. Reproductive toxins are sex specific and target the reproductive organs. Appendix F also lists the special toxic effects that are produced by exposure to the chemicals encountered in clandestine labs.

3.8 Personal Protective Equipment (PPE)

The use of PPE is a means to protect the body from toxic effects of the chemicals encountered while processing a clandestine lab. The results of the impact of these chemicals are readily apparent in the physical condition of the clandestine lab operators who are exposed to these chemicals without PPE protection.

In the United States, the use of PPE while working with hazardous materials as part of a person's employment is not a choice. Section 1910.132 of chapter 29 of the U.S. Code of Federal Regulations (29 CFR 1910-132) mandates the use of PPE when employees deal with hazardous materials as a job requirement. The same regulations require employers to provide the equipment and the training necessary to use the PPE effectively.

Each incident has a unique set of circumstances that require the use of a specific set of PPE. The PPE required for a specific situation depends: on the identified hazards, the level of contamination, the physical location of the scene and the work being performed. The PPE required to deal with corrosive liquids is different from that required for toxic gases. The PPE required to deal with a room full of factory sealed containers is different than that for a room of the same chemicals in open containers, some of which have been spilled. The PPE required to work in a confined space is different from that required to address the same hazards that are located in an open field with a blowing breeze.

In general, PPE is designed to provide a barrier to toxins via inhalation and dermal exposure. PPE is not designed to protect the user from ingestion exposures, which are completely avoidable using universal safety measures.

Most PPE is designed to protect against contact exposure. The most common exposure is by contact with hands because you are required to touch things. PPE places a barrier between your skin and hazardous materials, minimizing or eliminating the potential for dermal contact exposures.

Respiratory exposure has the greatest potential for injury because the lungs have the greatest surface area exposed to toxic environments. Fortunately, the overall number of materials that can enter the system is smaller than the number that can cause injury through contact with exposed skin. Additionally, the dilution of these airborne toxins in air reduces the amount of the hazardous substance that actually gets into the lungs. In general, the farther you are away from the source of the airborne hazard, the less concentrated it is, and the less effect it will have on your system.

3.8.1 Eye Protection

Regardless of the scenario, eye protection is a must. Eye sight is precious, easily lost and easy to protect. Eye protection associated with clandestine lab investigation can be divided into three general groups: spectacles (safety glasses), goggles and face shield.

Spectacles (safety glasses) are a protective device to shield the wearer's eyes from a variety of hazards. They do a great job of providing impact protection; however, they do have a few weaknesses. Safety glasses usually have small gaps around the lenses that can make your eyes vulnerable, especially to liquids and dust. Safety glasses provide protection from impact from flying debris and a degree of splash protection. When contending with splash hazards, airborne dust and flying debris, safety glasses are a minimal requirement, but safety goggles will prove to be a better option.

Goggles are a wraparound protective device that fits the face surrounding the eyes in order to shield them from impact, splash and vapor hazards. They provide 360-degree protection due to a tight, form-fitting facial seal. Examples where safety goggles are the better option include metal grinding, dusty conditions, chemical exposure and more. All of these situations have a higher-than-normal chance of a foreign object getting into your eyes from the side.

There are three types of goggles: direct vented, indirect vented and non-vented. Direct venting goggle offers impact protection only, with the vent holes allowing air flow around the eyes, helping to reduce lens fogging. They offer more protection from flying debris and splash hazards than safety glasses. However, the vent holes allow small particles such as dust to enter. Additionally, fumes and vapors can also flow through the vent holes unabated, leading to exposure of the eyes to the hazards associated with these chemicals.

Indirect venting goggles offer impact protection with some degree of splash and particle protection. The vents that allow air circulation are covered, reducing the ability for liquids and particulates to enter the vents. The vent covers do reduce air circulation, which in turn increases the potential for lens fogging.

Non-vented goggles have a lens frame construction without holes for air circulation. This design provides maximum protection, as there are no holes for liquids or particulates to enter. However, the lenses are subject to fogging.

A face shield is a supplementary protective device worn with safety glasses or goggles to shield the wearer's face from certain hazards. Safety glasses and goggles only protect the eyes from exposure to potential hazards. The rest of the face is exposed and subject to the effects of contact hazards or flying debris. A face shield does not protect from dust or fumes. However, its solid nature proves an effective barrier to splashes and flying debris.

3.8.2 Gloves

The use of appropriate hand protection is required when the hands are exposed to hazards such as those from skin absorption of harmful substances; severe cuts or lacerations; severe abrasions; punctures; chemical burns; thermal burns; and harmful temperature extremes. The selection of the appropriate hand protection shall be based on an evaluation of the performance characteristics of the hand protection relative to the task(s) to be performed, conditions present, duration of use, and the hazards and potential hazards identified. The focus of PPE selection related to hand protection is the resistance to a wide variety of chemicals that are or could be encountered.

The four primary chemical-resistant materials used to make gloves for hazardous materials applications are natural (latex) rubber, butyl rubber, neoprene and nitrile.

Natural (latex) rubber gloves are comfortable to wear, which makes them a popular general-purpose glove. They feature outstanding tensile strength, elasticity and temperature resistance. In addition to resisting abrasions caused by grinding and polishing, these gloves protect employees' hands from most water solutions of acids, alkalis, salts and ketones. Latex gloves have caused allergic reactions in some individuals and may not be appropriate for all employees. Hypoallergenic gloves, glove liners and powderless gloves are possible alternatives for people who are allergic to latex gloves.

Butyl gloves are made of a synthetic rubber. They protect against a wide variety of chemicals, such as peroxide, rocket fuels, highly corrosive acids (nitric acid, sulfuric acid, hydrofluoric acid and red fuming nitric acid), strong bases, alcohols, aldehydes, ketones, esters and nitro compounds. Butyl gloves also resist oxidation, ozone corrosion and abrasion, and remain flexible at low temperatures. Butyl rubber does not perform well with aliphatic and aromatic hydrocarbons and halogenated solvents.

Neoprene is a synthetic rubber material that provides excellent tensile strength and heat resistance. These gloves offer good pliability, finger dexterity, high density and tear resistance. They protect against hydraulic fluids, gasoline, alcohols, organic acids and alkalis. They are compatible with some acids and caustics. They generally have chemical and wear resistance properties superior to those made of natural rubber.

Nitrile is a synthetic rubber material that offers chemical and abrasion resistance and provides protection from chlorinated solvents. Nitrile gloves are a good choice for jobs requiring dexterity and sensitivity. They stand up to heavy use even after prolonged exposure to substances that cause other gloves to deteriorate. They offer protection when working with oils, greases, acids, caustics and alcohols but are generally not recommended for use with strong oxidizing agents, aromatic solvents, ketones and acetates. Nitrile gloves have superior puncture resistance. Nitrile gloves are often referred to as "medical grade" and are a good general duty glove for protection from a broad spectrum of hazardous material.

3.8.3 Clothing

Protective clothing provides a barrier between your skin and the hazardous chemicals. 29 CFR 1910.132 does not specifically address the construction of protective clothing. Therefore, you need to use your judgment when selecting the appropriate protective clothing to address the hazards specific to the scene under investigation. Depending on the material, the protective clothing will either impede the exposure by minimizing the skin contact or prevent exposure by repelling the hazards completely. Protective clothing is either disposable or reusable.

Disposable protective clothing is designed for single use and come in two basic forms. The first type is constructed with a paper-like material designed to repel dust, vapors and aerosol mists. Protection for liquids is not its primary function, as the liquid will permeate the material over time. However, this material will minimize exposures when smaller amounts of the hazard are splashed on the fabric. These materials do not provide protection against fire or extreme heat and can be flammable. They can also absorb certain amounts of flammable or combustible liquids, increasing the fire hazard.

Tyvex® and Micromax® are examples of paper-like disposable protective clothing. They are uncoated man-made material that provides protection from particulates and incidental liquid and vapor contact.

Coated fabric material is just as the name presents. A material has two distinct layers. The base layer is composed of natural or man-made fabric covered by a second layer of chemically resistant material such as rubber, neoprene or plastic. These materials repel liquids and can be used in situations where extended exposures to hazardous liquids are anticipated without worry about the hazards permeating the protective layers and exposing the skin to the hazardous substance.

Tychem® is an example of a material consisting of multiple layers of chemically resistant materials fabricated to provide maximum protection against chemical exposure. Like the paper-like fiber materials, coated fabrics do not generally provide protection against fire or extreme heat. There are some specialized coated fabrics that provide this protection, but they are not in common use.

Reusable clothes can be used repeatedly after appropriate decontamination and cleaning. Work clothes should be made of natural fibers such as cotton or wool and provide a barrier against particulates. They readily adsorb liquids and provide a minimal barrier against vapor exposures. However, they do provide a secondary barrier from liquid splashes when coupled with paper fiber disposables. Work clothes generally do not provide protection from heat and fire and will readily burn.

Work clothes should never be made from synthetic fibers like polyester, Rayon, Dacron, etc. These fibers melt before they burn. If they melt next to the skin, their removal requires removal of the top layers of skin, which is extremely painful.

Work clothes can provide heat and fire protection if made of the correct material. Nomex® is a man-made fiber with fire-resistant (not fireproof) qualities. The fabric's weave and the garment design will determine its resistance to liquid and vapor exposure. Clothing used by fire fighters is commonly made of Nomex®.

Kevlar® is a man-made fiber with fire-resistant qualities that exceed those of Nomex®. Like Nomex®, Kevlar®'s weave and the garment design will determine its resistance to liquid and vapor exposure. It is the same fiber that is used to make body armor. Fire fighter clothing is transitioning to Kevlar® because of increased fire resistance as well as increased durability.

3.8.4 Respiratory Protection

The respiratory protection requirements are outlined in a specific section of 29 CFR 1910.134. In brief, this section states that in order to control occupational diseases caused by breathing contaminated air, the primary objective shall be to prevent atmospheric contamination by accepted engineering control measures. When effective engineering controls are not feasible, or while they are being instituted, appropriate respirators shall be used. This means that if hazardous materials are suspected of being present, you must use the appropriate level of respiratory protection while assessing the environment. Also, if you cannot measure or reduce the concentration of airborne hazardous materials, you must wear respiratory protection. In the United States, employers are required to provide respirators that are applicable and suitable for the purpose intended when such equipment is necessary to protect the health of the employee.

The two types of respirators are air purifying respirators (APRs) and supplied air respirators.

An APR is a device used to filter contaminants from the air. They do not supply fresh air. They CANNOT be used in oxygen-deficient atmospheres that contain an oxygen content below 19.5%.

Types of APRs include dust masks, which are used to filter dust and other particulates (solid substances) from the air but provide no protection from vapors, and respirators, which use a single-use cartridge to filter out the hazardous substances.

Respirators can cover the full or half face. The half mask style only covers the nose and mouth, leaving the eyes and the rest of the face exposed to the hazard. The full mask covers the entire face, giving eye protection. Both masks use a negative pressure filtration mechanism, which means that air is only drawn through the filters when you breathe in.

Powered APRs use a blower to pass the contaminated air through the filtering cartridge. This creates a positive pressure situation, which does not require the person to breathe or use a sealed mask to get the benefits of filtered air. This allows the user to wear loosely fitting masks or hoods and does not require fit testing. These filtration systems are more appropriate for particulate environments as opposed to those that contain toxic fumes encountered in clandestine laboratories.

The following environmental conditions must be met for APR use:

- Oxygen content of between 19.5% and 23.5%
- Hydrocarbon content of 5 milligrams per cubic meter (mg/m^3) of air or less
- Carbon monoxide content of 10 parts per million (ppm) or less
- Carbon dioxide content of 1000 ppm or less
- Lack of noticeable odor

Supplied air respirators are devices that provide the highest level of protection against toxic materials and are required for work in oxygen-deficient atmospheres. They supply air that is known to be free of contamination so that the user can freely breathe a reliable supply of air with the proper concentration of oxygen. Air-line respirators have an air hose that is connected to a fresh air supply from a central source. The source can be compressed air cylinder or an air compressor that supplies Grade D breathing air.

Self-contained breathing apparatus (SCBA) has a limited air supply that is carried by the user, allowing greater mobility and fewer restrictions. Emergency escape breathing apparatus (EEBA) provides oxygen for durations shorter than those provided by an SCBA. These units are designed for emergency situations that call for immediate escape from IDLH environments.

The following provides a quick reference to the levels of respiratory protection offered by each type of respiratory PPE device as established by the NIOSH:

- No Protection (1)
- Dust Mask (10)
- Half Face Cartridge Respirator (10)
- Full Face Cartridge respirator (50)
- Powered Air Respirator (50)
- Positive Pressure Supplied Air (10,000)

Looking purely at the numbers, one would think that the SCBA at the bottom of the list would always be required or be the first choice. However, numbers can be misleading. The PPE used should be equal to the concentration of the actual or anticipated hazards. In a confined space with a totally unknown atmosphere, an SCBA is completely appropriate until the actual chemicals are identified and their concentration established. When sampling known substances in a well-ventilated area, a respirator of some type may be all that is required. Some scenarios may not require respiratory protection at all.

3.8.5 Levels of Protection

The U.S. Environmental Protection Agency (EPA) has established four levels used to describe the general type of PPE required to address scenes with varying degrees of hazards. To keep things simple, there is a single-letter designator for each level. Level A is used to designate the PPE requirements for the most hazardous situations. Level D describes situations that are essentially void of hazards. Levels B and C describe situations in between. Table 3.3 summarizes the PPE requirements for each level.

3.9 Summary

The chemicals encountered in clandestine labs can produce a variety of effects on the human body. Some are totally inert. Others generate an immediate lethal response. A great majority are somewhere in the middle.

The toxic effects of the chemicals involved in clandestine labs have the potential to affect the personnel processing beyond those who have direct contact with the clandestine lab scene. There are many unwilling people who can potentially experience the toxic effects of the chemicals in the lab. The operators; the people who are living in the lab area, including spouses, their significant others and children; the people who subsequently move into the apartment, house or motel room after the lab has been removed and not properly decontaminated; all have the potential to come into contact with the toxic materials. The responders may inadvertently bring toxic substances home or to the office, exposing families and coworkers to the hazards. The people involved in the criminal justice system, including people in the property room, court clerks and attorneys, may come into contact with the evidence throughout the adjudication process.

As stated at the beginning of this chapter, knowledge is power. Knowledge provides the power to prevent toxic exposure. However, with the knowledge of all of the potentially harmful things that can happen as a result of encountering any or all of the toxic substances, the clandestine lab investigator must ask the question: "Why am I investigating clandestine labs?"

Scene Processing

4

Abstract

The hazards indigenous to a clandestine lab scene make it impractical for even the best-prepared investigator to waltz through the scene, take a few photographs, throw the relevant evidence into a paper bag and go back to the office. Making sense out of what is found is far more difficult. The seizure of a clandestine lab often goes beyond the scope of a traditional crime scene. The dynamics of hazardous materials involved must be taken into account, and the safety of the personnel processing the scene must be paramount. However, the process cannot lose sight of the goal of preserving the physical evidence that indicates the existence of a clandestine laboratory. This balancing act can be accomplished through the use of specialized teams with specific functions. The number of people involved can give the process a circus-like atmosphere; however, with a documented set of policies and procedures delineating the responsibilities of each of the teams in place, order can be derived from what appears to be total chaos.

This chapter begins with a discussion concerning the sequence of events that encompass the seizure of a clandestine lab, starting with the planning and culminating with the disposal of the associated hazardous materials. This will be followed by a more detailed discussion concerning the process associated with the collection and preservation of relevant evidentiary items.

4.1 Introduction

To the uninitiated, the seizure of a clandestine lab may seem to be the culmination of the investigation. The long hours of surveillance, witness interviews, informant debriefings, information confirmation and search warrant preparation all do lead to the excitement of the actual seizure. Yet, what seems to be the end of a process is in reality only the beginning. In some aspects, finding the clandestine lab is the easiest part of the investigation. Making sense out of what is found is far more difficult. The hazards indigenous to a clandestine lab scene make it impractical for even the best-prepared investigator to waltz through the scene, take a few photographs, throw the relevant evidence into a paper bag and go back to the office.

DOI: 10.4324/9781003111771-4

101

The seizure of a clandestine lab should be a well-orchestrated event involving teamwork and timing. In any team sport, each player has a specific job to do if the team is to be successful. The seizure of a clandestine lab requires that same teamwork and coordination. A clandestine lab seizure is a scheduled event requiring a number of small teams to perform their specialty in a coordinated sequence. These teams can be grouped into support teams and seizure teams.

The support teams are comprised of mostly non-law enforcement or civilian personnel of military organizations and services that may be needed during the various portions of the seizure process. These support teams include members from fire departments, emergency medical services (EMS), hazardous waste disposal companies, local health and environmental protection agencies, child protective services and animal control. The circumstances surrounding the seizure can be used to predict exactly how many of these services may be needed.

The seizure teams are comprised of various law enforcement or military components that possess specialized training concerning clandestine labs. Each group has a specialized function. In the criminal justice realm, these groups are similar to those used in the processing of any major crime scene. The military equivalent would be the personnel responsible for site exploitation. In either situation, the hazards involved with clandestine labs require personnel and safety procedures to be incorporated into the processing procedures. Both explosive ordinance disposal (EOD) technicians and a forensic chemist who specializes in the manufacturing of controlled substances should be included in this group even if they are not part of a typical scene response team.

At this point, it is important to interject comments about the ultimate use of the results generated by a forensic clandestine lab investigation. The information is generically used for justice or intelligence purposes by the end users of the law enforcement or military communities, respectively. Although the goals and objectives for the use of the information may be different, the tools and techniques used to collect, process and analyze the information are the same. The ultimate goal of the forensic investigator is to provide objective opinion concerning evidence that was acquired and examined based upon sound forensic investigative techniques and good laboratory practices.

In some circles, military (expeditionary) forensic science is considered a poor relation to its law enforcement cousin. The esprit de corps of each group drives the belief that their approach is the most relevant. However, under close examination, both groups are using the same techniques to accomplish the same goal.

The most significant difference between military and law enforcement approaches to forensic investigations is the collection and preservation of evidence. The manner in which evidentiary samples are collected and preserved at the location of interest can differ between the military and criminal justice modi operandi. However, the individuals from each group process their location using the same sampling fundamentals, understanding the causes and effects of cross-contamination as well as how the Locard effect impacts the relevance of the samples they ultimately seize.

Crime scene processing and site exploitation are terms used by the law enforcement and military communities, respectively. They are essentially the same, with subtle differences. Crime scene processing is the collection and preservation of physical evidence for the purpose of investigating and prosecuting crime. Site exploitation is the systematic search and collection of information, material and persons from a designated location to facilitate subsequent operations or support criminal prosecution.

The fundamental difference between crime scene investigation and site exploitation is the operating environment and diverse objectives. Expeditionary (military) forensic science focuses on the analysis of material collected in the theatre of war. The materials collected differ from traditional forensic evidence. The primary objective is intelligence for force protection, targeting of insurgents, and network analysis in order to attack the network and to defeat the device.

Site exploitation can be deliberate or hasty depending on mission requirements and operating environment. Deliberate site exploitation involves coordination and comprehensive planning. Hasty site exploitation involves minimal planning and preparation as a result of the need for speedy execution.

This is opposed to the primary objective of traditional forensic science used in criminal justice, which is to protect the rights of the accused. This is accomplished by using a methodical approach to search, collect and preserve physical evidence in such a manner as to ensure its integrity from the crime scene to the courtroom. A conventional crime scene investigation is conducted, where the site is defined and secured for the required duration of the event.

Each community has its own terms to describe similar activities. The previous discussion concerning crime scene investigation and site exploitation is an example. This is the use of different terms by the military and civilian communities for the description of processes associated with the collection and preservation of physical evidence at a clandestine lab. Hopefully, the terminology used in the following discussions will be generic enough to apply to both communities.

4.2 Training

The hazardous nature of seizing a clandestine lab requires people with specialized training. The Occupational Safety and Health Administration (OSHA) is in charge of enforcing the regulations established by sections 1910.120 (HAZWOPER), 1910.134 (respiratory protection) and 1910.1200 (hazard communication) of chapter 29 of the Code of Federal Regulations (29 CFR…). Under 29 CFR 1910.120, an employer is responsible for providing training sufficient to educate employees concerning the dangers involved in working in a hazardous environment. As a result of such training, employees should be able to understand the hazards and risks associated with clandestine labs, understand the potential outcomes associated with an emergency response resulting from a clandestine lab, recognize and identify hazardous substances, and understand the need for additional resources when seizing a clandestine lab operation. The employer is also responsible for providing periodic/annual refresher training to keep employees abreast of the current information concerning the hazards in their work environment.

To address these requirements, in 1986, the Drug Enforcement Administration (DEA) began a training program designed to inform the personnel who investigate and respond to clandestine labs about the hazards involved in scene investigations. The objective of the program was to bring the DEA into compliance with 29 CFR 1910.120, 1910.134 and 1910.1200. The program initially covered DEA special agents, forensic chemists, diversion personnel and task force personnel responsible for clandestine lab enforcement. The program included an initial 40-hour safety school, annual 8-hour recertification courses, a 32-hour advanced course for site safety officers and a comprehensive medical surveillance program.

There are a variety of training resources available to agencies that require safety training concerning clandestine labs. The DEA has offered the initial 40-hour training course to law enforcement agencies that respond to clandestine lab scenes. Groups like the Clandestine Laboratory Investigators Association (CLIA) and the Clandestine Laboratory Investigating Chemists Association (CLIC) have sponsored 8-hour recertification courses. Private companies such as Network Environmental Systems, Inc. of Folsum, CA have been contracted to provide similar safety training to interested law enforcement agencies. State and local environmental quality agencies and fire departments can provide training on the handling of hazardous materials. The training information and resources an agency needs to educate its employees and comply with 29 CRF 1910.120 are available if the agency is willing to explore the options that are available outside the traditional law enforcement training network.

The U.S. Bureau of Alcohol, Tobacco, Firearms and Explosives (ATF) developed a similar training program to address the increase in the number of homemade explosives (HME) operations that have been encountered. The format and methods used to process an HME operation mirror those for a drug manufacturing operation. However, for a number of reasons, this program has not expanded to the degree of its DEA counterpart.

4.3 Seizure Stages

Regardless of the purpose (intelligence vs. criminal justice), the seizure and processing of a clandestine lab is an orchestrated event, one that involves securing potentially violent individuals, then processing the scene and dealing with an environment that often contains hazardous chemicals. The involvement of hazardous chemicals necessitates the seizure process to be broken into a particular sequence of steps. Each step requires a specific expertise. The steps of the seizure process include the pre-raid planning, the briefing, securing the site, the hazard evaluation and abatement, the search and site control, and finally, the disposal of the hazardous waste.

4.3.1 Planning

The planning of a clandestine lab seizure starts long before the affidavit for a search warrant is written. Hopefully, the responsible agency has identified the resources it will need to process the scene and will have established policies and procedures that delineate the notification mechanism necessary to coordinate their use during the seizure. Without proper planning, a domestic clandestine lab seizure can turn into a nightmarish multi-headed dragon that law enforcement is not prepared to slay. Although military site exploitation, outside the United States, does not involve the mixture of civilian agencies involved in a domestic crime scene investigation, planning is no less essential for the safe and successful completion of the mission.

The development of an action plan and the policies and procedures required to safely process a clandestine lab is a labor-intensive process. It is also expensive to provide and maintain the training and equipment required. This can be a major administrative hurdle that an agency must take into careful consideration before jumping into the clandestine lab seizure pool.

To address this situation, many jurisdictions rely on a larger agency (i.e., a state or large metropolitan law enforcement agency) or pool their resources with other agencies to create a task force that specializes in the investigation and seizure of clandestine labs. These alternatives provide the manpower and resources necessary to safely process the clandestine lab scenes. Other agencies simply provide a response team that will process the scene. In many cases, the paperwork and physical evidence are handed back to the investigating agency for prosecution once the seizure process is completed.

In an active military counter-insurgency engagement, it may be impractical to restrict site exploitation activities to a relatively small group. It is not uncommon for a military squad to encounter a clandestine explosive manufacturing operation during a routine patrol. The approach taken to identify, collect and preserve items of forensic value in a hostile environment cannot be compared to the sedate atmosphere of the civilian crime scene. The squad that initially encounters the manufacturing operation may very well be responsible for many, if not all, of the subsequent activities associated with exploiting the site. This increases the importance of the planning and training of personnel who would potentially encounter clandestine manufacturing operations under these conditions.

There are numerous criminal and civil statutes associated with clandestine labs that must be addressed during a domestic criminal investigation. Some of these issues are not directly or indirectly related to the manufacture of the contraband substance that is the focus of the investigation. However, if not addressed during the planning phase, these issues have the potential for assigning civil and criminal liability to the individuals who are associated with the seizure of the clandestine lab.

Many of the traditional civil and criminal issues associated with the domestic seizure of a clandestine manufacturing operation do not apply to site exploitation activities conducted during military operations. However, the policies and procedures of the military branch conducting the operation, as well as the internationally accepted Law of War and laws of the country in which the operation is conducted, are relevant and must be taken into consideration.

Most of the following discussion will focus on the collection and preservation of evidence from a U.S. criminal investigation perspective, i.e., a crime scene investigation. A crime scene investigation requires more actions than site exploitation. Most, if not all, of the actions required for successful site exploitation are contained within the crime scene investigation process. Understanding the crime scene process provides an understanding of the site exploitation process, which allows one to be adapted to the needs of the other by eliminating irrelevant processes or steps.

There are a number of resources that should be identified as part of a clandestine lab response policy. Other law enforcement agencies in the region that may have beneficial investigative expertise should be sought out, such as the DEA or state narcotics enforcement agencies. The local fire department and EMS should be on standby in case of explosion, fire or other medical emergency. Hazardous waste disposal companies need to be notified and arrangements for payment initiated. Local health and environmental quality departments must be notified for public health reasons. Child protective services may be required because children are frequently (albeit innocently) involved.

During the planning stage, in a criminal justice scenario, the primary investigator consults experts to ensure that the known information is consistent with a clandestine lab operation. The investigator identifies the location of the suspected operation, the information concerning why he believes there is a clandestine lab at that location, and the expert's opinion, and prepares an affidavit. The affidavit is then presented to a judge, who determines whether there is sufficient evidence to establish the probable cause necessary to grant a search warrant.

The affidavit should also describe the hazardous nature of the chemicals involved and a formal request that they be disposed of professionally after the hazardous items have been properly documented and the necessary samples taken. This statement preemptively notifies the court that there is a potentially hazardous situation and that the law enforcement agency does not have the facilities to safely store or dispose of the chemicals seized; and further, that they will make every effort to document and identify the items necessary to prove the state's case while protecting the rights of the accused and maintaining public health and safety. Quite a responsibility in itself! It is imperative, therefore, that the document be very carefully worded, since it must be so all-inclusive.

It should be stressed that the bulk of the substance discovered will be *disposed of* rather than simply destroyed. "Destroy" implies intent to deceive. "Dispose" indicates a plan to mitigate the hazards after the appropriate documentation and preservation steps have been taken. Even if the scene was properly documented and the necessary samples were taken, without the appropriate court authorization, the investigator may inadvertently cross the destroy/dispose line, negating all the evidence that is collected at the scene.

4.3.2 Briefing

All of the teams associated with the seizure of a clandestine lab should be brought together at the briefing. These teams include the entry team, the search team, EOD and hazardous materials personnel, forensic chemists, fire department and EMS representatives, and representatives from other law enforcement agencies that may be assisting in the seizure. Personnel from local health or environmental quality department and child protective services may be notified that their services may be needed, but their presence at the briefing is not required unless preliminary information exists indicating that it is necessary.

During the briefing stage, the lead investigator provides a history of the case, identifying the suspects and the location of the operation. Clandestine lab experts and chemists brief the group on the hazards associated with the type of operation that they can expect to encounter. The staging area and command post locations are disclosed. A separate entry team briefing is performed shortly after the main briefing. Although this is a short stage of the seizure process, it is necessary to coordinate all of the resources that will be needed to process the scene.

4.3.3 Entry/Arrest

The entry/arrest stage is definitely an "adrenaline-pumping" and exciting phase of the seizure process. It is also the most dangerous. The danger lies in the team's lack of control over the unknown scenario. An entry team trained in clandestine lab seizures should be able to enter and secure a location in less than 2 minutes. However, because of the unpredictability of the situation, during those 2 minutes anything can go wrong. The operator's mental state and level of drug-induced psychosis is a total unknown. The exact state of the manufacturing process is in question at the time of the entry. There may also be booby traps present in any part of the lab or its environs. These hazards are exponentially increased in an explosive manufacturing operation, as the end product is by its nature designed to explode.

The entry team has two functions: the first is to secure all the personnel within the immediate lab area, and the second is to be the eyes and ears of the evaluation/abatement team.

The entry team's primary function is to secure the lab location. This is done by securing, detaining and then removing any people from the lab area. Who these people are and their relationship to the lab is irrelevant. The seizure phase is neither the time nor the lab area the place to sort out who is involved in the operation and who is at the location for some other reason, however unrelated. This initial step is mandatory for their own safety and the safety of the personnel who subsequently will be processing the lab scene for physical evidence. Leaving detained personnel in the lab area is not wise because it needlessly exposes everyone to the hazardous materials in the lab. Leaving desperate suspects in the lab area also may further provide them with the opportunity to attempt to destroy evidence, thus dramatically increasing the potential for a hazardous materials exposure.

The second function of the entry team is to act as the eyes and ears of the evaluation and abatement team. What the entry team sees, hears, and in some cases smells or tastes is vital information that is used by the evaluation team to develop the scene's abatement plan. Using an entry team that has been trained in clandestine manufacturing techniques enables them to recognize the sights and odors of the chemicals and equipment that are commonly used in clandestine labs. The entry team is in the lab area such a short amount of time that their training must provide them with the ability to recognize significant items and the importance of relaying said information to the abatement team for use in the formulation of the abatement plan.

The entry team should be dressed in such a manner as to provide protection from the hazards they may encounter. In selecting the clothing worn during the entry phase, the team needs to consider inhalation and dermal exposures. All entry team equipment provides some level of dermal protection. However, there is a difference in philosophy when it comes to the use of respiratory protection. One extreme advocates not using any respiratory protection at all during the entry and arrest phase, reasoning that the physical mobility, increased peripheral vision and ability to give verbal commands outweighs the hazards of a short-term exposure the team will experience. The other extreme counters that the lab atmosphere may contain lethal concentrations of any of many substances. Training and practicing entries while using the protective equipment helps overcome the restrictions of the PPE. Wearing the equipment also adds the additional psychological shock value of a dynamic entry to any people in the lab at the time, possibly buying critical time for the raiders. Anything that can be done to discourage last-minute tampering at the clandestine lab is recommended.

There are times at which a clandestine lab is encountered by law enforcement, fire departments or EMS personnel during activities unrelated to clandestine lab investigations. The response to these situations is similar to those of an investigation-initiated seizure. An emergency on-the-scene briefing takes place and indeed, requires all the professional expertise accumulated from other raids. The initial responders act as the entry team and remove all personnel from the area. They report to the site safety officer what they know about the inside of the lab area. In essence, they act as the entry team to become the eyes and ears of the abatement team that they will then call.

4.3.4 Hazard Evaluation/Abatement

In clandestine lab investigations, the whole concept of crime scene processing changes once the site has been secured by the entry team. The main goal of the operation from this point on is to create and maintain a safe work environment. If this goal is not achieved, the other goals of the operation may not be attained either.

The site safety officer is responsible for the implementation and maintenance of safe work practices at the scene of a clandestine lab seizure. His tasks include determining the hazard potential of the lab scene, establishing work zones, and determining the appropriate levels of protection for the various stages of the balance of the seizure operation. The total dynamics of the responsibilities of the site safety officer is beyond the scope of this book. However, this section will touch on some of his basic responsibilities so as to provide an understanding of what safety measures should be put into place.

The entry team's observations provide useful information that the site safety officer can use to develop the site control plan. Having this firsthand information from an objective source allows the site safety officer to make decisions concerning what PPE will be used, in establishing work zones, in ensuring that safe work practices are followed and further, making other vital decisions concerning work place safety.

Interviewing the operator or other personnel who were removed from the lab area by the entry team (after arresting and reading them their Miranda rights if applicable) can supply useful information concerning what hazards may be inside. Some operators immediately invoke their right to remain silent and will not speak to law enforcement personnel. Other operators are perversely proud of their operation and are willing to tell someone who "admires the sophistication" of the operation all the details of what they manufacturing and how they are doing it. The interviewer and the site safety officer must always consider the source of this information and bear in mind that the operator may have an interest in providing misinformation. His understanding of the technical aspects of the operation may be limited, and he may not know the real names of the chemicals involved. For these reasons, an expert in clandestine manufacturing techniques should sit in on the interview if possible. The expert should not only know the technical aspects of manufacturing a wide variety of controlled substances but more importantly, should recognize the slang terms for the equipment and chemicals that are commonly used in the various manufacturing processes. The lab operator may only know his chemical components by these names himself.

4.3.5 Site Control

Site control is a primary responsibility of the site safety officer. He determines what tasks will be performed, and where, based upon an area's level of contamination. He also limits access to highly contaminated areas to personnel who have the required training and are wearing the appropriate PPE. The site safety officer will further divide the scene into areas based on their level of contamination: hot, warm and cold zones.

The hot zone consists of the area immediately surrounding the manufacturing operation, or areas with open chemical containers. This area may only encompass the area surrounding an ice chest containing closed chemical containers, or it may incorporate an entire house in which every room was used for some portion of the manufacturing operation. Who has access to the hot zone and what level of protection they should be wearing are the important issues. The hazardous nature of the hot zone mandates that access be limited and time inside minimized. Once the hazards have been abated or removed, the hot zone can be downgraded to warm.

The warm zone is the area immediately adjacent to the hot zone. Its access should also be limited. However, the level of protection required here might not be as great. Items from the hot zone are moved into the warm zone, where they can be processed under controlled conditions. Hot zone workers take rest breaks and are decontaminated in the warm zone. Emergency rescue personnel stage here in case an accident requiring a rescue occurs in the hot zone. Even though the hazards are not as great, access to the warm zone should be limited to trained personnel wearing the appropriate PPE. The warm zone can be downgraded to a cold zone once the sampling process is completed and the chemical containers have been sealed and segregated.

The cold zone is a hazard-free zone where the command post is located. Here the lead investigator and site safety officer coordinate the seizure activities. Any eating, drinking and smoking should only take place in the cold zone to reduce the possibility of ingesting hazardous materials. An access point to the warm and hot zone is established here to ensure that only authorized personnel enter the contained area and to document who had access to the crime scene for later court purposes.

4.3.6 Personal Protective Equipment (PPE)

In conjunction with site control, the site safety officer must establish the levels of PPE that will be required during the various stages of the processing operation. He relies on the information from the entry team to determine the level of protection the evaluation team will require during the evaluation and abatement phase. The evaluation team's information will be used to determine protection levels required for the subsequent stages.

Personal protection equipment levels range from A through D. Level A provides the greatest level of protection for all entry routes. Level D protection is not much more than work clothes.

Level A PPE is total encapsulation and provides the highest level of respiratory and skin protection. It is used in atmospheres that are immediately dangerous to life or health (IDLH) and require extreme dermal protection from compounds to which skin exposure in small concentrations will result in impairing or life-threatening health effects. The use of supplied air provides for operation in oxygen-deficient environments.

Level B PPE provides protection similar to level A, but the configuration of the equipment does not offer quite the same level of dermal protection. However, the respiratory system is totally protected. This level of protection is recommended for the assessment phase of operational clandestine labs. Some agencies utilize this level of protection for their entry team.

Level C PPE provides level B barriers to dermal exposure, but respiratory protection has been downgraded from supplied air to an air purification respirator. This level of protection is appropriate when the composition and concentration of the work atmosphere are known and continually monitored. Level C protection **is not** appropriate for oxygen-deficient atmospheres.

However, it is appropriate for work in ventilated areas in which the chemical composition of the hazardous materials involved is known.

Level D protection is used when the potential for chemical contact is minimal. It can be equated to industrial work clothes and is appropriate to protect the worker from incidental exposures.

The evaluation and abatement team has two primary functions: the first is to identify and neutralize any potential hazard within the hot zone, and the second is to create a less hazardous environment so that further scene processing can occur.

Using the appropriate PPE, the evaluation and abatement team will enter the hot zone to identify and evaluate the potential hazards. Team personnel should have a solid knowledge of improvised explosive devices (IEDs, i.e., booby traps), clandestine manufacturing techniques and hazardous materials chemistry. A forensic chemist and an EOD technician form a complementary team that can be used to perform the evaluation and abate the identified hazards. They should minimally have equipment that can monitor the atmosphere's oxygen content and the level of flammable/explosive vapors. They may also use equipment that can identify and quantify levels of a variety of specific hazardous materials. The site safety officer ultimately is responsible for selecting the equipment used to evaluate and monitor the hot zone's atmosphere.

Abatement can begin once the hazards have been identified. The EOD technician performs "render safe" operations to items with explosive potential. A forensic chemist trained in clandestine manufacturing techniques can shut down active chemical reactions (see Table 4.1). Open chemical containers are sealed, and confined spaces are safely ventilated. Once these abatement procedures have been conducted, and a safe work environment has been established, the actual processing of the scene can begin.

4.4 Scene Processing

The processing of a clandestine lab scene is a unique combination of processing a crime scene and a hazardous materials incident. Therefore, two considerations must be balanced during this phase. First, the integrity of the evidence must be maintained. Second, the exposure to the hazardous materials must be kept to a minimum. These tasks can be accomplished simultaneously by implementing three steps: planning, documenting and sampling.

The number of people used to process the lab area should be kept to a minimum to accomplish both of these tasks. Minimizing the people in the hot zone minimizes the number of people exposed to the highest contamination levels. Since the lab area is generally a confined space, increasing the number of people in the area increases the potential for an accident. A small

Table 4.1   Reaction Shutdown Guide

- Determine whether the reaction is being heated, cooled or both.
- Remove heat from the reaction vessel.
- Maintain cooling to the reaction until the reaction appears to have gone to completion and the vessel is cool to the touch.
- Remove obstructions and ventilation tubing from the top of condensing columns.
  - Cooling reactions can create a vacuum that could hamper the dismantling of the apparatus.
  - If the reaction is being vented into water, the vacuum created by the cooling reaction could draw the water into the hot reaction mixture, leading to a violent reaction.
- If necessary, compressed gas containers connected to a reaction vessel should be turned off at the source.
  - The pressure of the reaction vessel should be allowed to naturally reduce to atmospheric pressure.
  - Release of pressurized contents of the reaction vessel may create a toxic exposure.
- Systems under vacuum should be slowly brought back to atmospheric pressure.
  - Water or oxygen in the air may react violently with the chemicals in vessels under vacuum.
- Filtration processes should be allowed to naturally go to completion.

processing team can more efficiently perform the sequence of the tasks required to preserve the integrity of the evidence.

4.4.1 Planning

Planning is an essential step of any process. The first step of establishing the plan for processing the clandestine lab scene is to touch nothing but rather, to use the powers of observation. Every clandestine lab is different, and the person in charge of the scene processing must avoid the desire to start moving things from the hot zone before evaluating the totality of the scene. The team should walk through the scene, making mental notes and asking themselves questions like: What type of lab is this? What process(es) is apparent? Which chemicals are present? What equipment is present? What does the operator seem to be trying to make? Once a picture of what the operator was trying to make and how he was trying to make it has been developed, a plan of how to process the scene to prove the hypothesis can be developed. The examples in Section 4.5 demonstrate why a walk-through should be conducted before making assumptions or processing every lab the same way.

A word of caution is needed at this point. The objective of a forensic investigation is to allow the physical evidence to dictate the facts of the case. The scene should be looked at in a totally objective light. The physical evidence at the scene needs to drive how the search is conducted. The scene-processing team needs to guard against getting caught up in the "get the bad guys" mentality of the seizure and focus on identifying, collecting and preserving the physical evidence that indicates the presence of a clandestine manufacturing operation. This is not to say that they cannot focus on collecting evidence characteristic of a clandestine lab that also has common uses. However, investigators should also be careful not to place an illicit meaning on an innocuous item if the scene does not justify it.

4.4.2 Documentation

Documentation is essential to preserve the integrity of the evidence, since most of it will be disposed of fairly quickly because of its hazardous nature. To give a complete picture of the clandestine operation, a combination of techniques is necessary. One augments and complements another, thus providing a thorough record of the existence of the evidence even though it is no longer available. Methods of documenting a clandestine lab scene include photography, videotape, scene interviews, field notes, sketches and inventories. Each has its strengths and weaknesses.

The documentation of the scene should take place throughout the seizure process. It can be initiated during the evaluation/abatement phase but usually begins during the initial walk-through, where overall photographs and field notes are taken. A sketch of the scene is made to supplement the photographs. This is followed by an itemization and inventory of the specific items that are removed from the scene. A final documentation of how the seized items are disposed of and the remaining items are stored is done.

4.4.2.1 Photography

Photography is the traditional method of recording and documenting crime scenes. It captures the way the scene first looked and documents specific items and situations. The initial series of photographs should tell a story, walk the viewer through the scene and portray the scene just as the photographer observed it.

Photo documentation can commence once the scene has been secured, the hazards abated and the atmosphere deemed safe. General overall photographs should be taken to depict the lab area as it was originally encountered. These photographs should be taken from multiple angles before anything is moved. If possible, a series of photographs creating a panoramic from a fixed point should be taken. This creates an overall view of the lab as the photographer encountered it. This should be followed by close-ups of specific items in their original location. Reaction apparatus should be photographed before being dismantled.

A photograph should be taken of every item or group of items seized. This can be done in the warm zone under controlled conditions. These photographs serve as corroboration to the written field notes and official inventory. They should be as clear and accurate as possible because they are the official record. Photographs should be taken of the samples with their original container to demonstrate where the sample item was removed.

Individual item photographs should contain identifying information. This information should minimally include the case number, the exhibit or item number, and the date. A ruler scale should be included if volume calculations will be made using the geometry of the container at a later time.

A photo log should be maintained. This documents who took the photographs, when they were taken, what the photograph depicts and why the photo was taken. The investigator must remember that the photographs are more than pictures used to enhance his memory at some point in the future. They are evidence and as such, have the same requirements for maintaining the chain of custody as does any other physical evidence seized from the scene. Photographs are discoverable evidence and must be treated as such. To complete a photography packet, the back of each photo should be labeled with the case number, the photographer's ID, the photo number and the total number of photos taken at the scene. This information should correspond to the information on the photo log.

The introduction of digital photography into clandestine lab processing has provided the investigator with tighter control over the images he generates during his scene investigation. It is suggested that investigators using digital photography copy the images from their camera directly to a "read only" file format on a removable storage format (compact or floppy disc) to serve as the permanent record, in much the same way as negatives serve as the permanent record for traditional film photography.

4.4.2.2 Videotape

Videotaping the scene has become a popular method of visually documenting clandestine lab scenes. It is a real-time way to demonstrate how the site was originally found and gives the viewer a sense of actually being at the location as the photographer pans through the scene.

The audio component of videotape can be a help or a hindrance. If the person providing the commentary is knowledgeable about what is being viewed, it can augment the visual presentation. However, as exhibited in the initial crime scene example in the Practical Application chapter, the audio support did not accurately describe what was being viewed. This conflict caused problems during the expert's trial testimony. The videographer should therefore refrain from commenting if he is not sure what he is looking at.

As with a traditional photograph, videotape records of the scene are discoverable evidence. Every effort needs to be made to preserve the unaltered tape and maintain its chain of custody. Copies can be made as needed. However, the original tape must be fixed so that it cannot be altered or erased. It should also be labeled with the pertinent case information.

4.4.2.3 Field Notes

Field notes are taken on or about the time the clandestine lab is processed. They are another form of documentation that is used to augment the others. They are used to supplement the visual images of the photographs or videotape. Field notes can be written at the scene, dictated into a recorder and transcribed at a later time, or they can be written at the office immediately after the fact.

Field notes can take a variety of forms and are used to address who, what, when, where, why and how questions concerning the crime scene investigation. Who assisted in processing the scene and had access to the evidence? What items were seized? When was the lab entered or seized? Where were the seized items located? Finally, why were certain items seized and not others?

The three basic forms field notes take are worksheets, narrative descriptions and sketches. Each has its strengths and weaknesses. Most clandestine lab investigators utilize some combination of the three.

All forms of field notes must be treated as part of the written record and subject to discovery. Each page should have the same basic information as a photograph and identify the author (name, initials or ID number), case number, date, page number and total number of pages.

Worksheets are unique to a specific task. (See Appendix G.) They provide a convenient method of documentation and a format to walk a person (and later a jury) through the various phases of a clandestine lab seizure. There are worksheets or checklists for pre-raid planning and the hazard evaluation/abatement phases. Some investigators have designed worksheets to assist in documenting specific items seized from the lab. Worksheets are good for routine responses that require repetitive tasks or seize the same evidentiary items. However, worksheets can be ineffective if their parameters do not address the scenario of the scene. Worksheets with a narrative section can be used to provide additional information in situations that may be outside the scope of the worksheet.

A photo log form is an example of a worksheet used to supplement the photographs taken at the scene that contains both fill-in-the-blank and narrative sections. There can be sections for time, date, photographer ID, photo and roll number. The section that describes the photograph and its significance is a short narrative.

There is no right or wrong way to take narrative notes. Notes can be as short or as long as the author desires. They can be short phrases used to assist the investigator's recall at a later time, or they can be multi-sentence explanations of how the author perceived the operation functioned. As long as the author can place his thoughts on paper in such a way that he can decipher them at a later time, the goal has been achieved. The only rules regarding narrative notes are that they are preserved and that each page contains the appropriate case information.

Photographs and videotape document the actual presence of items at the scene. However, because they are two-dimensional, they do a poor job of demonstrating where things were located in relationship to other items. Wide-angle lenses or panoramic film might provide the answer, but their costs are rarely within the budgets of most law enforcement departments. Sketches provide a means of demonstrating a spatial relationship that photographs do not. By combining these two, two-dimensional formats can be used to develop a three-dimensional perception that accurately depicts the scene.

Sketches can be as detailed as the drafter deems necessary. They can be used to provide a basic spatial relationship between items at the scene using visual approximations to determine distances. Sketches can also be extremely detailed, using exact measurements from fixed points to create a scaled drawing of the scene. As with all forms of field notes, information concerning the case and the drafter's ID should accompany the sketch.

4.4.3 Site Control

Although a clandestine lab is a crime scene, it is also a hazardous materials incident. Minimizing the exposure of personnel to toxic chemicals as well as containing the hazardous materials until their disposal can be properly addressed are priorities. Therefore, establishing and implementing a site control plan is essential in attaining this goal.

Figure 4.1 is a generic site control diagram that applies to all haz-mat scenes and can be applied to clandestine lab scenes as well. It is comprised of three work zones: hot, warm and cold. The temperature designation of each zone is based upon the amount and type of hazardous materials contained within each.

The cold zone is an uncontaminated area that contains the command post and staging area. The command post is a location where administrative duties are performed. The staging area is an area where support personnel congregate prior to performing their function. Fire, EMS, crime laboratory and waste disposal personnel are those who commonly wait at the staging area.

Site Control Plan

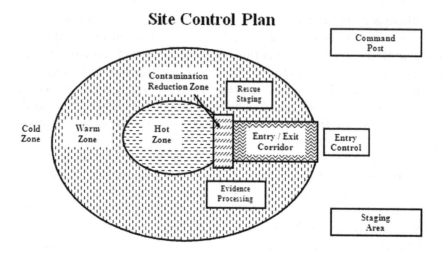

Figure 4.1 Generic site control diagram.

The warm zone is an area immediately outside the hot zone that is subject to contamination and requires controlled access. It is an area in which hazardous materials will be encountered, but in a more controlled environment than that experienced in the hot zone. This reduced exposure risk and the controlled nature of the operations within the warm zone allow a downgrading of the PPE required for tasks performed within.

The warm zone contains the contamination reduction zone, rescue staging and evidence processing (Figure 4.2). The contamination reduction zone is a transition area between the hot and the warm zones where personnel are decontaminated prior to leaving the hot zone. The rescue staging area is a location within the warm zone where personnel donning PPE are staged in preparation for a rescue operation if one is required. The evidence processing area is a segregated area within the warm zone, which is used to process evidence and store items removed from the hot zone prior to disposal.

The hot zone is the area of greatest contamination, generally the location of the manufacturing operation. However, the hot zone can include chemical storage areas or waste disposal sites (Figure 4.3).

Access to the hot zone must be controlled because of the hazardous nature of the site, coupled with the need to maintain the integrity of the evidentiary items identified and collected within. As such, an access corridor with entry control should be established. Entry control establishes a single entry point to the crime scene. The place and time personnel enter and leave the crime scene area are documented. The entry/exit corridor is within the warm zone and is the only authorized path between the hot zone and the cold zone.

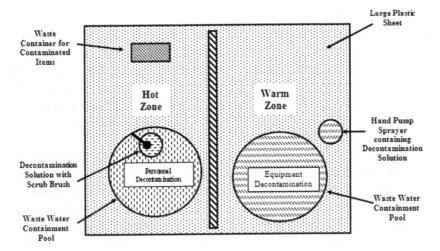

Figure 4.2 Contamination reduction zone.

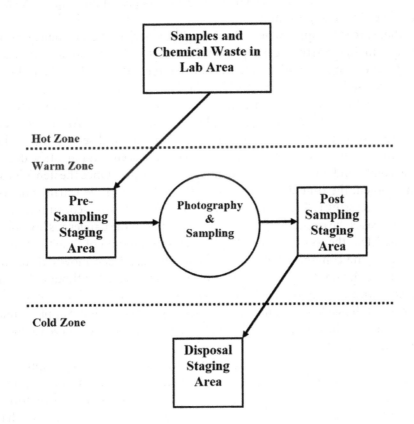

Figure 4.3 Sample waste flow.

4.4.3.1 The Search

Hot zone searches are handled differently from the balance of the scene because of the presence of hazardous materials and possible contamination. People who have not received specific clandestine lab training can process searches of warm and cold zones adjacent to the lab area using standard crime scene search techniques. However, a clandestine lab expert should be available to provide technical advice when clandestine lab–related items are found in these areas.

Once the original condition of the scene is documented, the search and inventory can commence. The search is necessary to locate and identify the items of physical evidence necessary to establish the clandestine manufacture of controlled substances. Experts in a variety of clandestine manufacturing techniques should conduct the search. These experts are able to recognize the evidentiary significance of ordinary items at the location. The searchers should also have been provided with the safety training that is required by 29 CFR 1910.120. Again, the hot zone is a hazardous materials scene that may also be a confined space, so the number of people conducting the search should be kept to a minimum.

There are three approaches concerning the search of the lab scene. First is simply utilizing a methodical approach. Second is a "clear the area" approach. Third is the "sample in place" approach. Personnel safety and legal considerations are used in selecting a search method.

First and foremost, a clandestine lab is a crime scene. Every effort should be made to preserve the integrity of the physical evidence. To this end, an organized search of the lab area is needed. A methodical search allows the processing team to sequentially move through the scene and identify, document, collect and preserve each evidentiary item located. Once the item's location is noted, the time it is moved from the hot zone to the warm zone for further processing is documented. This approach can be time-consuming and thus increases the length of time the search team is in the dangerous hot zone.

The "clear the room" approach minimizes the amount of time the search team is in the contaminated environment. This approach moves all of the items from the hot zone to the warm zone for processing without detailing the exact location and time each item is removed from the scene. The overall photos and the sketch document the scene's original condition. Supplemental photos during the inventory and sampling in the warm zone serve to document individual items.

The third method is a compromise of sorts. The "sample and go" approach reduces the search team's contact with the hazardous materials by not moving them from the hot zone to the warm. The scene is searched, inventoried and sampled without significantly moving any items. Once the team has completed its search and documentation of the hot zone, it is turned over to the waste disposal company. Using this method can increase the time in the hot zone but minimize the situations in which accidents while handling hazardous material can occur.

The search of uncontaminated areas adjacent to the lab area can be conducted using traditional crime scene search techniques. However, a forensic chemist and EOD personnel should be readily available. It is not uncommon for searchers to find chemical or explosive devices in an area surrounding the location. Hazardous materials have been located in children's bedrooms, bathrooms, closets, attics, outbuildings and vehicles parked on the street. The specialists are needed to evaluate the evidentiary significance of the items, take the action necessary to address any hazards, and also preserve the evidence.

4.4.3.2 Inventory

The need for an accurate inventory of the chemicals and equipment seized from a clandestine lab goes beyond establishing the elements of the crime. The inventory is used to establish facts and render opinions concerning the lab that are not apparent at the scene. Evaluating the inventories of the seized items can identify manufacturing methods used and create estimates of production. (See Chapter 6: Opinions.)

All law enforcement agencies have some type of form (worksheet) for documenting the items seized during the execution of a criminal search warrant. Depending on the agency's policies and procedures, these forms may or may not adequately depict the type and amount of chemicals and equipment seized. Therefore, it is suggested that a separate detailed inventory be taken to accurately document the types and amounts of chemicals and equipment seized so that experts can objectively evaluate the information away from the hectic environment of the seizure and render opinions concerning the overall manufacturing operation.

4.4.3.3 Sampling

Even small operations can create a significant number of evidentiary items. The hazardous nature of the chemicals and contaminated equipment requires special storage conditions. Most law enforcement agencies' property rooms are not equipped to handle the volume of evidence that can be generated by a clandestine lab. They also do not have the ventilation and safety equipment necessary to safely store and dispose of such hazardous waste once it is seized. Therefore, taking evidentiary samples from the scene of a clandestine lab is a necessity. The balance of the items can be properly disposed of once properly documented.

It is highly recommended that a forensic chemist who specializes in clandestine labs perform the sampling. His knowledge of the various clandestine manufacturing methods allows him to select the significant items while at the same time keeping the samples to a manageable number for the forensic laboratory. He can recognize subtle differences in unknowns and select samples that will accurately depict the seized operation. He also has the training to allow him to safely handle the hazardous chemicals that require sampling and recognize which materials should be handled with kid gloves or not at all.

Local statutes and case law will determine what to sample and how much to take. The following suggests guidelines to indicate which items should be sampled. Suspected controlled substances should be seized in total and submitted for examination, while a sugar cube–sized sample of suspected explosives is more than sufficient. All suspected precursor chemicals should be sampled and the amount measured or estimated. Their identity assists in determining the synthesis route used. The estimated or measured amount of precursor is used to calculate the amount of final product that could be produced. Taking samples of reagents and solvents should be left to the discretion of the on-scene chemist. The identity of reagents assists in determining the synthesis route, and the solvents' identity can assist in determining the extraction method used.

The fate of labeled containers is not as well defined as one would think. It is a good assumption that the contents of a factory-sealed commercial container are what the label reports them to be. Opened containers with commercial labels may or may not contain what is on the label, and the on-scene chemist should treat them accordingly.

The only difference between an unlabeled container and a container that the operator has labeled is that the label reflects only what the operator originally placed into the container; the contents of an operator-labeled container may have been changed any number of times since the label was originally applied. Therefore, the container should be treated as an unknown.

The following is a general guide for sampling chemical containers. It is unnecessary to sample factory-sealed commercially labeled containers if they have been properly documented. The sampling of open containers with commercial labels is up to the discretion of the on-scene chemist. Commercially labeled containers whose contents are not consistent with the label should be sampled. Samples should be taken of all containers with altered commercial labels or handwritten labels or without labels, and identification of such labels should always be indicated.

Reaction mixtures, extraction mixtures, waste liquids and sludge contain a wealth of information in a single location. Lack of a label of any type indicates nothing. These mixtures may contain many, if not all, of the precursors, reagents, final products and by-products for a given synthesis route. This wealth of information from a single source necessitates the sampling of these mixtures. Nothing should be ignored.

Glassware and equipment found at a clandestine lab scene may or may not contain residues that will provide information concerning what was being produced or the synthesis method being used. For safety reasons, the entire piece or a number of pieces of glassware or equipment must be submitted to the forensic laboratory if the identity of the residue is deemed critical. This will provide a controlled environment for the sampling and examination.

The search and inventory method used will determine where the samples will be taken. The "sampling in place" method is just that: samples are taken at the items' original location. Other methods move the items to an evidence-processing location where the conditions of the sampling can be more controlled. Trained personnel wearing the appropriate PPE should do any sampling in a well-ventilated area.

The hazardous nature of the samples requires special packaging. Packaging should take into account the chemical properties of the samples and be designed to minimize the hazardous exposures and cross-contamination if a sample container is broken. Finally, the packaging should provide information concerning the sample's identity. This information should include the case number, item number, date, and any relevant information concerning the sample, such as its pH, field test information, original volume or weight. Even the item's proximity to another item may be important. Table 4.2 provides information concerning the appropriate packaging of clandestine lab samples.

The question of how many samples are required is invariably asked. As with every other aspect of clandestine lab investigation, there are a number of ways to approach the situation. All represent merely a difference in philosophy. The sampling camps can be divided into their extremes: one camp advocates taking only the minimum number of samples necessary to establish the elements of the crime, with the other advocating sampling everything in sight. What actually occurs is discretionary sampling by the on-scene chemist based on his experience and/or the law enforcement department's expectations based on similar venues.

Taking a minimum number of samples streamlines the scene-processing phase. Minimizing exposure to toxic materials, it potentially reduces the analysis time required by the forensic laboratory. The downside to this minimization is that if proper or sufficient samples are not taken at the scene, there is no backup documentation to the physical evidence. Additional samples required to supply information that would provide a different perspective to the operation are thereby forever unavailable, since once law enforcement leaves the scene, a hazardous waste disposal company usually removes bulk items.

Another consideration of minimization is the possible appearance of deception. The argument could be presented that by not "completely"

Table 4.2   Sampling Guide

General Considerations
- Consult local statutes and case law concerning the type and amount of samples required.
- Sampling should be conducted in a well-ventilated area, preferably away from the lab area.
- All samples should be photographed with the original container.
- Photographs should contain: Case number, Exhibit number, Date, Reference scale.

What to Sample
- Finished product
  - Contraband drugs: The entire amount
  - Explosives: Pea-size to sugar cube–size amount (< 10 g)
- All reaction mixtures should be sampled.
  - Suspected primary explosive reactions should not be sampled.
- Random samples of waste material should be taken.
- Unlabeled chemical containers should be sampled.
- Sampling of reagents and solvents is left to the discretion of the on-scene chemist.
- Each phase (top and bottom layer) of a multi-phase liquid should be sampled.
- The sampling of commercially labeled containers is left to the discretion of the on-scene chemist.
  - Factory-sealed containers can be assumed to contain what the label reports.
  - Open commercially labeled containers may or may not contain what the label reports.
- The sampling of glassware and equipment is left to the discretion of the on-scene chemist.

Sample Packaging
- Liquid and solid samples should be packaged in glass vials with acid-resistant screw cap and placed inside a sealed zip lock plastic bag.
  - Solid samples can be placed into a sealed zip lock plastic bag.
- All samples should be placed inside a second sealed zip lock plastic bag.
- The outer sealed zip lock plastic bag should be marked with the appropriate case information.
- Individual items should be placed into a single container filled with an absorbent material for transportation and storage.
- Explosives or explosive mixture samples should be placed into a portable magazine for transportation and storage.

sampling the scene, the investigators were attempting to hide evidence that would exonerate the suspect or might intentionally be misleading as to what was actually occurring at the location. Are these valid arguments? Probably not. Can they arise? Yes, they can and do.

The other extreme is to sample everything. If the total operation consists of three or four items, sampling everything is not unreasonable. However, sampling 30 unknown liquids with similar color and chemical characteristics may be excessive, expensive and time-consuming. Still, a guiding principle may be that it is better to have too many samples than not enough. The forensic laboratory may choose not to analyze a sample if the examiner feels that no additional information will be derived from the examination.

However, it should always be remembered that the laboratory cannot analyze what it does not have.

4.4.3.4 Field Testing

Field testing is used to address two basic questions: What is it? How much is there? However, their existence leads to follow-up questions of where and when is the appropriate place and time to answer the first two questions.

The "how much is there?" question needs to be addressed at the scene. Representative samples are all that remain for detailed laboratory examination. The hazardous nature of the chemicals and reaction mixtures located at a clandestine lab site preclude the transportation of the entire substance. As such, on-site estimates of the quantity of material must be performed. Table 4.3 includes items that should be included in a sampling kit.

In order to establish the amount of controlled substance that could be produced by the operation, the weights and volumes of the precursor and reagent chemicals need to be determined. The original volumes of reaction and extraction mixtures, combined with the results of laboratory analysis of the samples, can be used to calculate the amount of controlled substance that was in the container at the time of seizure.

The volume estimate of commercially labeled containers is relatively straightforward: it can be assumed to be what is reported on the label. Estimates can be made for commercially labeled containers whose contents

Table 4.3   Sample Kit

Sampling Items	Desirable Items
• Camera	• Chalk or dry erase board with markers
• Nitrile gloves	• Scales (1 kg and 100 kg capacity)
• 30 1–2 oz. wide mouth HDPE with screw caps	• Field testing
• 30 4 ml glass vials with HDPE screw caps	• pH paper
• 30 3–5 ml disposable transfer pipettes	• Refillable butane lighter with
• 30 25 ml disposable transfer pipettes	extended tip
• Pipette bulb	• Premixed reagents
• 60 4″ x 6″ zip lock plastic bags	• Colorimetric test instrument
• 30 disposable laboratory spatulas	• Raman spectroscopy
• 100 3″ x 5″ index cards	• Infrared spectroscopy
• Marking pens	
• Tape	
• Ruler or tape measure	
• Worksheets or note pads	
• 3–5 gal. plastic bucket with lid	
• Portable explosives magazine	
• Absorbent material:	
• Kitty litter	
• Vermiculite	
• Diatomaceous earth	

appear to be consistent with the labeling information (e.g., a 500-gram jar approximately half full or a 500-ml bottle approximately 25% full). Physically weighing the container and subtracting the weight of an empty container, or one of the same approximate size and type, can make more accurate estimates of solid weights. Measuring the dimensions of the liquid in the container and using basic geometry to calculate the volume of the liquid can obtain more accurate liquid volume estimates.

Determining the exact volumes of reaction and extraction mixtures and unknown liquids is more critical. These liquids potentially contain a controlled substance, the amount of which may be used in the sentence determination phase of a trial if the suspect is convicted. Knowing the original volume of a container will enable the forensic chemist to calculate the amount of controlled substance that was in it.

All equipment and chemical containers found at clandestine lab scenes can be divided into one of three basic geometric shapes: cylinders, cones or spheres. Cylinders (Figure 4.4) are the basic geometric shape of beakers, bottles and drums. Erlenmeyer flasks, vacuum flasks and separatory funnels are shaped like cones (Figure 4.5). Reaction flasks have a spherical shape (Figure 4.6). Knowing the basic shape of an object and the measurements of the liquids within it allows the actual volume to be calculated with preexisting mathematical formulas. The examples in Practical Application 4.5 demonstrate how these calculations can be used in the field.

The accuracy of the answer to the "what" question is not as important to know at the scene. The exact identity of any given item cannot usually be established under field conditions without specialized equipment. However, the ability to classify compounds and mixtures provides a means to efficiently group like substances and streamline the sampling process. For example, knowing the basic chemical and physical properties of a liquid can tell the sampler whether the sample is a reaction mixture or waste material. Knowing the relative density of an organic liquid will provide insight as to whether the final product will be on the top layer or the bottom. Certain

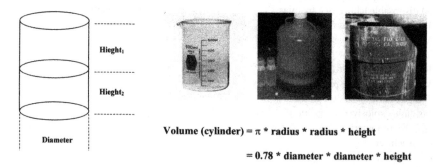

Volume (cylinder) = π * radius * radius * height

= 0.78 * diameter * diameter * height

Figure 4.4 Cylinder volume.

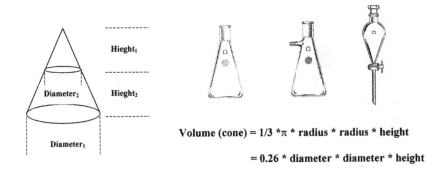

Volume (cone) = 1/3 *π * radius * radius * height

= 0.26 * diameter * diameter * height

Figure 4.5 Cone volume.

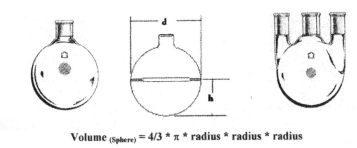

Volume (Sphere) = 4/3 * π * radius * radius * radius

= 0.52 * diameter*diameter*diameter

Flask Size (ml)	Diameter (inches)	Diameter (cm)
500	3.87	8.84
1000	4.88.	12.4
2000	6.14	15.64
3000	7.04	17.90
5000	8.35	21.22
10000	10.52	26.74
12000	11.18	28.40
22000	13.68	34.76
50000	18.00	45.70
55000	18.57	47.18
72000	20.32	51.62

Figure 4.6 Sphere volume.

chemical color tests can be used to provide presumptive information as to whether an item contains a controlled substance or can be used to indicate its reactive properties. For all of these determinations, a forensic chemist's experience on the scene is invaluable.

Appendix I and Table 4.4 describe the contents of a field test reagents kit and some of the generic results that can be obtained from reactions with chemical color tests in the kit. As with sampling, field testing should only be performed by trained personnel under controlled conditions. Trained

Table 4.4　Field Test Reactions

Reagent	Color	Indication
	Drugs	
Marquis	Orange	Phenethylamines, phenylacetic acid
	Purple	Opiates, MDA, MDMA
Cobalt thiocyanate	Blue	Cocaine HCl, PCP, meperidine, lidocaine
Copper sulfate	Blue	Ephedrine, pseudoephedrine, lidocaine
Dinitrobenzene	Purple	Phenylacetone (P2P)
pDMBA	Purple	LSD, indoles
	Explosives	
Diphenylamine	Blue	NO_3^-, ClO_3^-, ClO_4^-, nitro-compounds, oxidizers
Thymol	Green	NO_3^-
	Brown	ClO_3^-
	Red	RDX, HMX
	Blue-green	PETN
Sulfuric acid	Yellow-orange	ClO_4^-
Aniline sulfate	Blue interface	ClO_3^-
Nessler's	Orange ppt.	NH_4^+
Methanol/NaOH	Red to orange	TNT
	Blue to brown	DNT
Vapor pH	Basic	NH_4^+
Pyridine/NaOH	Blue to brown top layer	Sulfur
Silver nitrate	White ppt.	Cl^-, CO_3^{-2}, SO_3^{-2}
	Cream	Br^-
	Yellow	I^-, PO_4^{-3}
	Brown	OH^-
	Black	S^-
+ HNO_3	ppt. Remains	Cl^-, Br^-, I^-, S^-
	ppt. Dissolves	CO_3^{-2}, SO_3^{-2}, PO_4^{-3}, OH^-
Barium chloride	White ppt.	CO_3^{-2}, SO_4^{-2}, SO_3^{-2}, PO_4^{-3},
+HNO_3	ppt. Remains	SO_4^{-2}
	ppt. Dissolves	CO_3^{-2}, SO_3^{-2}, PO_4^{-3}

personnel know the sequence of field tests for a particular sample that is necessary to provide an accurate picture of the sample's physical and chemical properties as well as potential contents. They are also aware of the subtle differences between a positive and a negative test result. Many field tests involve a sequence of chemical reactions, and the use of trained personnel reduces the likelihood of an incompatible chemical reaction.

Advances in technology have taken the specificity field testing to a new level. Test results that previously required expensive instrumentation, conducted in a controlled environment by trained scientists, are now available for use in the field. The same colorimetric, Raman and Infrared technologies used by forensic laboratories have been downsized to handheld devices that can be effectively utilized by trained field investigators and field laboratories located near or behind battle lines.

The specificity of these handheld instruments exceeds that of wet chemical color tests. However, they are still considered presumptive tests that require confirmation. Their specificity is sufficient to serve as probable to obtain a search warrant, identify samples that require additional testing, or justify elevating the level of response in a military action.

The information generated by these instruments is accurate when used correctly. However, the samples must be examined by a forensic laboratory to establish the chemical composition beyond a reasonable doubt. A forensic laboratory is able to detect levels of trace elements and chemicals that the handheld instruments cannot report because of the limitations of the built-in libraries. Additionally, forensic explosives or drug chemists have technical knowledge concerning various manufacturing methods that will provide insight as to manufacturing methods and links to source of chemicals being used in the process.

4.4.3.5 Disposal

The final phase of a clandestine lab seizure must be considered long before the investigation begins. The following question must be asked and addressed: What is going to happen to the hazardous waste generated as a result of the seizure?

In the early days of clandestine lab investigations, waste disposal was not an issue. Chemicals were routinely poured down drains or on the ground, thrown into the trash, burned, blown up or simply submitted to the police property room for indefinite storage. These actions were taken ignorantly yet innocently without regard to chemical compatibility, the health and welfare of the people who would be subsequently exposed, or the environmental impact.

Since then, numerous safety and environmental regulations have been established to protect the health and welfare of workers, the general population and the environment. Law enforcement is not exempt from these regulations. In many cases, the lead agency legally becomes the "generator" of the hazardous waste produced as a result of a seizure. As the generator, they have a "cradle-to-grave" responsibility to ensure that the waste generated by the seizure is disposed of in such a way as not to adversely impact the environment or public health. In many cases, this means that the seizing agency takes on the financial burden of cleaning up the chemical contamination that resulted from the clandestine operation.

This potential liability has sometimes made small agencies reconsider clandestine lab enforcement operations. The cleanup costs for the smallest operations can easily run into the thousands of dollars. A medium -size operation could have devastating effects on the police department's budget. For this reason, task forces have been created and funds established to address the financial burden associated with clandestine lab seizures.

The generator is ultimately responsible for the waste that is generated from the lab. Therefore, it is imperative that the disposal company chosen be reputable in every way. When evaluating waste disposal companies, the low-bid approach taken by many government agencies may not be the optimum method of selection. The company selected should have the appropriate federal and local licenses to handle hazardous waste. They should be transporting it to approved facilities for proper disposal using approved methods. Finally, the personnel the waste disposal company sends to the site should have clean criminal histories, especially in the area of drug abuse.

4.5 Practical Application

4.5.1 Practical Example: Initial Crime Scene Evaluation

Everything done at a crime scene potentially has evidentiary value. Every action and word has the potential for finding its way into the court. Nowhere was this more apparent than in the O. J. Simpson case, in which the investigators spent hours on the witness stand explaining comments and personal opinions that were expressed during the crime scene investigation. This section will provide examples of how actions taken at the clandestine lab scene may have ramifications in later stages of the investigation.

A clandestine lab scene chemist and the lead investigator began processing the scene of a suspected clandestine drug lab after the scene had been secured and the hazards abated. The lead investigator, who had only minimal knowledge of clandestine manufacturing techniques, immediately began videotaping the scene and providing audio descriptions of how each chemical and piece of equipment would be used to manufacture methamphetamine.

The scene chemist, who had 10 years of experience in clandestine lab processing, was performing a preliminary walk-through at the same time. Within minutes, the scene chemist realized the operator was manufacturing explosives, not drugs. He stopped the processing and evacuated the lab area to revise the processing plan. Reading the physical evidence, the scene chemist realized the operator was manufacturing nitroglycerine, not methamphetamine. This piece of knowledge radically affected how the balance of the scene was processed.

4.5.2 Practical Example: Training and Experience

A forensic chemist who was not trained in clandestine lab manufacturing methods arrived at the scene of a suspected methamphetamine lab. Without evaluating the combination of chemicals and equipment that were present at the scene, he pronounced that the operation was manufacturing methamphetamine. He took a minimal amount of samples and left the scene. Using the scene chemist's opinion, without corroborating laboratory analysis, the lead investigator charged the operator with manufacturing methamphetamine.

A clandestine lab chemist subsequently evaluated the physical evidence from the scene. His laboratory examinations of the evidentiary samples, evaluation of the scene's photographs and review of the chemicals revealed that the operator was manufacturing diethyltriptamine, a hallucinogen, not methamphetamine. The hasty actions of the scene chemist and the lead investigator led to the dismissal of the charges against the operator.

4.5.3 Practical Example: Training and Experience

In contrast to the previous example, this practical example demonstrates the positive effects of having qualified personnel evaluating the physical evidence and adjusting the scene-processing protocols as more information is developed.

A clandestine lab response team properly secured and abated the hazards at the site of an educated commercial operator who had a history of manufacturing gamma hydroxybutyric acid (GHB). The operation under investigation was clean and non-operational. During their evaluation of the chemicals and the operator's notes, the scene chemists determined that the operator was experimenting with the manufacture of meperidine and fentanyl analogs. These compounds have been linked to Parkinson's disease. At this point, the hazard potential dramatically changed, as did the approach to the way the scene would be processed. The lab area was evacuated, and the scene-processing plan was revised.

4.5.4 Practical Example: Sampling

The following is an example of the ramifications of improperly sampling a clandestine lab scene. The scene chemist in this situation did not possess training in clandestine lab scene processing. He did an admirable job of photographing the scene, which allowed the clandestine lab chemist to render some opinions after the fact.

Three containers located at the scene of this clandestine drug lab were identified as containing an acid. One container was a plastic gasoline container containing a clear acidic liquid that produced white fumes when the container was opened. The two other containers were commercial 500-ml clear glass bottles with black caps. The labels on the bottles had been removed. The bottles contained a clear acid liquid with a yellow tint. Photographs were taken of each of the containers. Only the red plastic container was sampled. The subsequent laboratory analysis of the sample revealed that the contents were consistent with hydrochloric acid.

The following is an excerpt of the analytical chemist's testimony. The defense contended that the lack of the presence of hydriodic acid precluded the operator from manufacturing the controlled substance that the state contended. The cross-examination of the analytical chemist charged with analysis of the evidence proceeded as follows:

Defense Attorney:	You stated exhibit 12 contained hydrochloric acid?
Chemist:	Yes sir.
Defense Attorney:	So there was no hydriodic acid found at the scene?
Chemist:	No Sir, I cannot say that.
Defense Attorney:	But your report states that you found hydrochloric acid, not HI. How can you say that there was hydriodic acid present?
Chemist:	The items in exhibit 24 and exhibit 31 were not sampled. The packaging and the color of the liquid are consistent with hydriodic acid.
Defense Attorney:	But your report states that hydrochloric acid was the only acid identified.
Chemist:	That is correct. However, the items in exhibit 24 and 31 were not sampled so I could not analyze the contents. Without laboratory analysis I cannot comment on the contents.
Defense Attorney:	So you are saying you did not find any hydriodic acid?
Chemist:	What I am saying is that I cannot say that there was no hydriodic acid at the scene. The packaging and color of the liquid of items 24 and 31 is consistent with commercially packaged hydriodic acid.

This whole exchange could have been avoided if the items at the scene had been sampled properly. This would have provided the analytical chemist with the opportunity to identify the contents of each container.

4.5.5 Practical Application: Bottle Volume Estimates

In some jurisdictions, the penalty associated with crime is related to the amount of controlled substance seized. Many statutes use the phrase "at time of seizure" as the time benchmark. In simple possession or possession for sale cases, this value is easily determined by weighing the substance on a calibrated balance during the laboratory examination of the exhibit. This can be problematic in clandestine lab investigations, since the majority of the evidence is disposed of due to its hazardous nature. The 2-fluid ounce sample that is retained for laboratory examination may hardly be representative of the volume of substance that was seized. Without documentation to support a larger volume/weight, the only value the court can rely on to establish a sentence would be the weight/volume of the representative sample that was submitted for laboratory examination. If the dimensions of the containers and their contents are documented, simple geometry can be used to establish the original volume of the substance, thus giving the court the "at the time of seizure" value to use in establishing the sentence.

The volume of the contents of a clear glass bottle is needed (Figure 4.7). The cylindrical-shaped bottle has a diameter of 10 cm. The height of the liquid in the bottle is 50 cm. The volume of the liquid can be calculated using the simplified cylinder volume equation as follows:

$$\text{Volume}_{\text{cylinder}} = 0.78 * \text{diameter} * \text{diameter} * \text{height}$$
$$= 0.78 * 10\ \text{cm} * 10\ \text{cm} * 50\ \text{cm} = \textbf{3900 ml} = \textbf{4.1 qt}$$

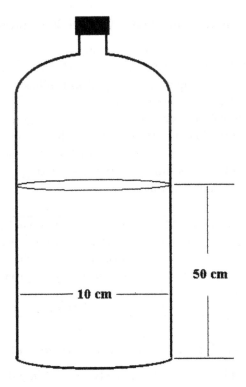

Figure 4.7 Bottle containing liquid.

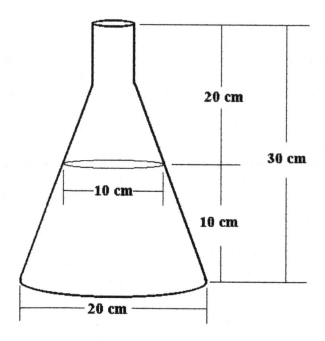

Figure 4.8 Flask containing liquid.

4.5.6 Practical Application: Flask Volume Estimates

The volume of a clear yellow liquid in an Erlenmeyer flask is needed (Figure 4.8). The Erlenmeyer flask has a conical shape 30 cm tall, with a base diameter of 20 cm. The height of the liquid in the flask is 10 cm, and the diameter of the flask at the top of the liquid is 10 cm.

Calculating the volume is a three-step process. The total volume of the flask is calculated first. Second, the volume of the air on top of the liquid is calculated. Finally, the air volume is subtracted from the total volume to determine the volume of the liquid in the flask. Using the equation from Figure 4.8, the calculation is as follows:

$$
\begin{aligned}
\text{Volume}_{total} \quad &= 0.26 * \text{diameter} * \text{diameter} * \text{height} \\
&= 0.26 * 20\ cm * 20\ cm * 30\ cm = \mathbf{3120\ ml} \\
\text{Volume}_{air} \quad &= 0.26 * \text{diameter} * \text{diameter} * \text{height} \\
&= 0.26 * 10\ cm * 10\ cm * 20\ cm = \mathbf{520\ ml} \\
\text{Volume}_{bottom} \quad &= \text{Volume}_{total} - \text{Volume}_{air} \\
&= 3120\ ml - 520\ ml = \mathbf{2600\ ml} = \mathbf{2.7\ qt}
\end{aligned}
$$

4.5.7 Practical Application: Separatory Funnel Volume Estimates

A separatory funnel is found containing a two-phase liquid (Figure 4.9). The bottom layer is 5 cm high, and the top layer is 10 cm high. The diameter of

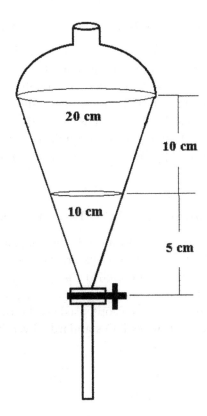

Figure 4.9 Funnel containing liquid.

the funnel at the point where the two liquids meet is 10 cm. The diameter of the funnel at the top of the top liquid is 20 cm. What is the volume of both liquids?

The basic shape of a separatory funnel is that of an upside-down cone. Therefore, the same method that was used to calculate the volumes in the flask is used for this calculation. The total volume in the funnel occupied by liquid is the first thing to be calculated. The next step is to calculate the volume in the liquid phase closest to the apex of the cone. Finally, the lower-layer volume is subtracted from the total volume to establish the volume of the upper layer.

Volume$_{total}$	= 0.26 * diameter * diameter * height
	= 0.26 * 20 cm * 20 cm * (10 cm + 5 cm) = **1560 ml = 1.64 qt**
Volume$_{bottom}$	= 0.26 * diameter * diameter * height
	= 0.26 * 10 cm * 10 cm * 5 cm = **130 ml = 0.5 cup**
Volume$_{top}$	= Volume$_{total}$ – Volume$_{bottom}$
	= 1560 ml – 130 ml = **1430 ml = 1.51 qt**

4.5.8 Practical Application: Reaction Flask Volume Estimates

The calculations to establish the volume of a sphere that is filled with liquid are more complicated than the simple mathematics used in the previous examples (Figure 4.10). In July 1991, the DEA published a table of partial sphere volumes based on reaction flask size and liquid height. This table has been reproduced in Appendix H. The diameter of the reaction flask is used to establish its volume. The height of the liquid in the flask is then cross-referenced on the table in Appendix H to obtain the amount of liquid in the reaction flask.

4.5.9 Practical Application: Volume Estimate Accuracy

Regardless of the method used to establish the original volume of vessels containing liquid, all volume calculations are estimates. The primary reason is that there is no means to establish the uncertainty of the measurements used in the calculation. It is highly unlikely that the measuring device used to establish the dimensions used in the calculation has been calibrated in a manner that meets ISO standards. Even if the calibration of

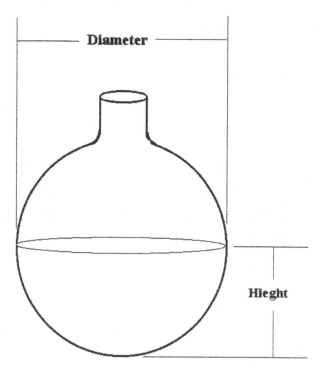

Figure 4.10 Sphere containing liquid.

the measuring device's calibration is current, the conditions under which the measurement(s) were obtained bring into question the repeatability and reproducibility of the measurements in question. As such, all volume calculations should be reported as "approximately" to account for the uncertainty of the measuring device.

4.6 Summary

The seizure of a clandestine lab often goes beyond the scope of a traditional crime scene. The dynamics of hazardous materials involved must be taken into account, and the safety of the personnel processing the scene must be paramount. However, the process cannot lose sight of the goal of preserving the physical evidence that indicates the existence of a clandestine laboratory. This balancing act can be accomplished through the use of specialized teams with specific functions. The number of people involved can give the process a circus-like atmosphere; however, with a documented set of policies and procedures delineating the responsibilities of each of the teams in place, order can be derived from what appears to be total chaos. The array of arbitrary unknown items can be sequentially identified, documented, preserved and properly disposed of in such a way that the health and safety of all parties is protected while at the same time conducting a forensic investigation that will prove or disprove the existence of criminal activity.

Laboratory Analysis

5

Abstract

The analysis of samples from clandestine labs involves the use of a variety of scientific techniques to chemically identify substances. These techniques range from simple chemical color tests to the use of X-ray and infrared energy to elicit the compound's chemical fingerprint. The type of test used depends upon the information desired from the sample and the burden of proof required to establish its identity. The origin of the samples, be it from a crime scene or a site exploitation, is irrelevant.

This chapter specifically addresses the techniques used to analyze evidentiary samples from clandestine labs. A number of technical issues will be presented in a basic format to provide an understanding of the analytical process to a broad range of readers. This chapter's purpose is not to provide a detailed discussion concerning the theory of a particular examination technique. It is simply to present the options available to the analytical chemist. This chapter will provide the investigator with an understanding of what examinations to request when submitting his evidence for examination. It will also assist the attorneys involved in the case by providing them with information concerning why certain tests were used as opposed to others.

This chapter addresses the role of the chemist performing the laboratory examinations and how they should specialize in clandestine lab analysis. The clandestine lab chemist has training focused on clandestine manufacturing techniques as well as in inorganic analysis, expanding their analytical scheme to identify the manufacturing process, not just the drugs or explosives involved.

Additionally, there are two schools of thought concerning the role of the clandestine lab chemist in analyzing the samples that enter the laboratory. One school has the chemist who processes the crime scene analyzing the samples, essentially a cradle-to-grave approach. The other school has an independent chemist analyze the samples once they reach the laboratory. Each philosophy has its advantages and disadvantages. The introduction of bias into the analytical process is also discussed.

The analytical process will be discussed in a sequential manner, starting with nonspecific tests and gradually increasing their specificity. Each technique's applicability to the realm of drug and explosive analysis will be

DOI: 10.4324/9781003111771-5

presented. The chapter will conclude with a series of practical examples of the analysis of clandestine lab samples, based upon the analysis of actual samples submitted to a forensic laboratory for analysis.

5.1 Introduction

The analysis of samples from clandestine labs involves the use of a variety of scientific techniques to chemically identify substances. These techniques range from simple chemical color tests to the use of X-ray and infrared energy to elicit the compound's chemical fingerprint. The type of test used depends upon the information desired from the sample and the burden of proof required to establish its identity. The origin of the samples, be it from a crime scene or a site exploitation, is irrelevant.

This chapter specifically addresses the techniques used to analyze evidentiary samples from clandestine labs. A number of technical issues will be presented in a basic format to provide an understanding of the analytical process for a broad range of readers. This chapter's purpose is not to provide a detailed discussion concerning the theory of a particular examination technique. It is simply to present the options available to the analytical chemist. This chapter will provide the investigator with an understanding of what examinations to request when submitting his evidence for examination. It will also assist the attorneys involved in the case by providing them with information concerning why certain tests were used as opposed to others.

The results of the laboratory analysis of samples taken from the clandestine lab scene is the link between the investigation and the opinions. It provides the scientific proof that corroborates the investigator's theories and is used as the basis of the opinions rendered in reports, deposition and testimony. Without complete and thorough laboratory analysis, the case may become unresolved.

The laboratory analysis of evidence is more involved than the simple chemical identification of a substance. The identification of the components of the sample matrix may be just as important. The complete analysis is important in establishing the manufacturing method or providing information that can be used to establish the source of the precursor used in the process. It is not absolutely necessary. However, if the chemist's analysis is not complete, it may be implied that he is not qualified to perform the analysis or he has something to hide. The lack of a complete analysis may also affect other aspects of the investigation or prosecution that the chemist is not aware of.

It is not sufficient to say that the clandestine lab operator was using a particular method simply because some or all of the ingredients were believed to be found at the site. The presence or absence of precursor or reagent chemicals must be established beyond a reasonable doubt via laboratory examination. The relabeling or lack of labels on the containers at the scene makes the actual identity of the chemicals at the location questionable.

The specificity of the results of handheld instrumentation used in field testing has increased exponentially. Their results, generated in the chaotic environment of a crime scene or site exploitation, are subject to misinterpretation. They must be confirmed in the controlled environment of a forensic laboratory, using dedicated instrumentation operated by forensic chemists or criminalists with expertise in the clandestine manufacture of contraband drugs and explosives.

The laboratory examination of reaction mixtures is even more important. The chemist should identify the ingredients within the reaction mixture. The fact that chemicals or chemical containers were located at the scene does not establish their presence in a reaction mixture. It only provides the chemist with information he can utilize in developing an analytical scheme that will isolate and identify the individual components of the mixture within the limitations of the available instrumentation.

5.2 The Chemist

The chemist performing the laboratory examinations should specialize in clandestine lab analysis. In bookkeeping, all certified public accountants (CPAs) are accountants, but not all accountants are CPAs. The same is true with forensic chemists. All clandestine lab chemists are forensic chemists, but not all forensic chemists are clandestine lab chemists. The clandestine lab chemist has additional training in clandestine manufacturing techniques as well as in inorganic analysis. This allows him to expand his analytical scheme to identify all the chemicals used in the manufacturing process. His analytical scheme is geared to identifying the manufacturing process, not just the controlled substance involved.

The clandestine lab chemist's role in a clandestine lab investigation requires a different thought process when approaching his analysis. He approaches each sample as if he has to tell a jury what components are in the sample and how they fit into the manufacturing process. From an investigative standpoint, his analytical approach is geared toward profiling the sample to provide the investigators with information concerning the sample's composition so that the investigators know what components to look for. From an intelligence standpoint, the chemist is looking for trace components that may provide information that would lead to the source of the precursor chemicals used.

There are two schools of thought concerning which forensic clandestine lab chemist analyzes the samples once they enter the laboratory. One school has the chemist who processes the crime scene analyzing the samples, essentially a cradle-to-grave approach. The other school has an independent chemist analyze the samples once they reach the laboratory. This school theorizes that it should not matter who does the analytical work as long as the person is trained in clandestine lab analysis. Section 5.9 contain examples of actual situations demonstrating the ramifications that need to be considered when addressing how many chemists are assigned to process a clandestine lab case.

5.2.1 Single Chemist

Having a single chemist process the scene and subsequently analyze the samples can streamline the analytical process. The scene chemist understands the relationship between samples and the importance of each in the investigation. This broad understanding produces an intuitive prioritization of the samples based upon the direct knowledge of the sample's origin. If a sample's analytical results are consistent with the chemist's on-scene theories, analysis of similar subsequent samples may not be necessary. If there are no theories, analytical schemes and opinions may need to be modified to follow the direction in which the evidence leads.

The scene information is extremely useful to the analytical chemist, who uses this information to devise his analytical scheme. The scene chemist makes mental notes concerning what he believes was the process the operator was using. His sampling scheme is affected by the observations made. Each sample should be geared to address a specific question(s) that are used to establish that a manufacturing operation was in fact taking place at the location. Unless the scene chemist prepares a detailed written report, the information concerning his intuitive impressions of the operation will not be effectively relayed to the analytical chemist.

When the analytical results differ from the on-scene theories, the chemist gains a different perspective on what actually was taking place at the scene. The differing results may address questions the scene chemist had at the scene but could not rectify without a laboratory analysis of the item. The additional knowledge allows the analytic chemist to adapt his analytical scheme and mold his opinions to conform to the new information.

Courtroom presentations should also be considered when addressing how many chemists should be involved. The use of a single chemist provides continuity during courtroom presentations. He can explain the sampling scheme, transition into the laboratory analysis and finally, tie the two together and provide an opinion concerning the operation. All the forensic information can be provided from a single source. The jury receives a less fragmented presentation that walks them through the process. A single chemist addresses what was found at the scene, why samples were taken and the subsequent laboratory results. Finally, as an expert in clandestine manufacturing techniques, he ties all the information together and renders an opinion concerning the totality of the circumstances in the case.

From a case management standpoint, using a single chemist can reduce the overall time necessary to process the samples once they reach the laboratory. As the scene chemist processes the scene, he has subliminally prioritized the samples. Once the samples reach the laboratory, he can analyze only the samples he believes would be necessary to establish the facts of the case. Without a detailed report or specific directions from the scene chemist, the analytical chemist is compelled to analyze each sample. This may lead to unnecessary analysis and longer turn-around times for the investigator.

The use of a single chemist is not practical in the realm of battlefield forensics and site exploitation. The location and volatile nature of these scenes generally exclude the participation of laboratory-based chemists in the processing of the scene and the collection of relevant samples. Additionally, chemists in front-line laboratories are not trained or authorized to participate in military actions that occur in the areas of active conflict. However, a chemist may be called to process a site after it has been secured or sample exhibits that have been moved to a safe location at a time subsequent to the seizure of the clandestine lab.

5.2.2 Independent Analytic Chemist

The independent analytic chemist does not have specific knowledge concerning the history of the samples from a clandestine lab operation. Philosophically, it is believed that he will provide objective analytical results. Theoretically, he would not be inclined to skew the analysis to meet the opinions formed at the scene.

The independent analytic chemist does not have independent knowledge of the sample history of the case; he may be obligated to analyze every sample. Unanswered questions may lead to other problems by not doing so. Assumptions concerning the facts of the case can be avoided by providing the analytic chemist with a detailed report and a complete set of the scene photographs. This information should provide an understanding of the thought process used by the scene chemist at the time the samples were taken. Proper scene documentation should convey this information adequately to avoid as many problems as possible.

The case management philosophy of the forensic laboratory will dictate the use of the scene chemist or an independent analytic chemist to analyze clandestine lab evidence. The proper processing of a clandestine lab scene is a time-consuming process. It can remove a chemist from the bench effectively for one day or more per scene. The skills required to process a clandestine lab scene are different from those required to analyze the samples. Having chemists trained in specific areas of forensic clandestine lab investigation may provide for a more efficient flow of the case through the forensic system.

Documentation and the flow of information are essential to the effective forensic investigation of clandestine lab cases. No matter whether a single chemist or multiple chemists are used, communication is critical. A single chemist must document his activities completely to justify his conclusions at any point during the investigation. A qualified chemist should be able to review the scene chemist's scene documentation and arrive at the same conclusion. Therefore, providing a copy of this documentation to the analytical chemist should provide the information necessary to perform a complete evaluation of the evidence.

5.2.3 Analytical Bias

On a subconscious level, cognitive bias occurs when people see what they want to see. This distorts a person's perception or thought process, impacting their decision-making and influencing their findings. Training and experience can lead to what may be considered deviations from a standard of rationality or good judgment. A person's preferences and experiences unconsciously impact the way they acquire and process information. If there is a lapse in judgment, an illogical conclusion can be drawn.

The extent to which cognitive biases may influence decision-making in forensic science is an important question with implications for the validity of the results generated. The injection of cognitive bias into the forensic investigation of the clandestine manufacture of contraband drugs and explosives permeates the process. It begins with the processing of the crime scene or exploitation site. The on-site investigator, chemist or not, uses their training and experience to select the samples required to demonstrate the presence of a clandestine manufacturing operation and potentially not sampling items of exculpatory value. This bias may be compounded by the forensic laboratory's policies, procedures and accreditation scope, which focuses the chemist's analysis and report on the presence of specific classes of compounds.

The literature indicates that issues associated with misidentification due to cognitive bias are limited to the forensic disciplines associated with pattern matching, i.e., fingerprints, firearms and tool marks, footwear impressions and hand writing comparisons. The reports generated in these disciplines are opinion based, with the conclusions living in a shade of grey based upon the number of unique individual characteristics that the questioned and known samples share.

By contrast, chemistry reports identifying the presence of specific chemical compounds are more black or white. Either the chemical meets identification criteria or it does not. Additionally, the literature does not provide examples of cognitive bias related to the identification of specific chemical compounds utilizing techniques generally accepted by the scientific community. The science and the data associated with the identification are beyond reproach. The perceived bias is introduced when the chemist, because of integrity issues, manipulates the data to achieve a predetermined result. These are deliberate actions, not ones performed on an unconscious level.

It could be argued that there is bias in the interpretation of the relevance of the combination of chemicals identified in a suspected clandestine manufacturing report. Interpretations are expert opinions, based on the known facts, generated by qualified individuals. Different interpretations of the same facts do not mean that there was cognitive bias involved.

Consider the following two statements. X, Y and Z were identified. The combination of X, Y and Z will produce an explosive/drug. The first statement is based upon scientific fact and is uncontested. The second statement, although true, could be perceived as being biased because there is no overt recognition that the combination of X, Y and Z could have a legitimate purpose. Conceding that the second statement is biased for the reason given, the statement was consciously worded in such a manner as to answer a specific question posed by the investigator. Therefore, cognitive bias does not exist.

Cognitive bias associated with the chemical analysis of samples submitted to the forensic laboratory can occur when the full suite of analytical tests is not performed on a given sample. Not performing all relevant tests on a given sample or samples can lead to mischaracterization of the combination of the identified chemicals. Lack of all available information leads to the mischaracterization of a combination of chemicals, not an unconscious desire for a specific outcome.

Consider the following. X and Y were identified. The combination of X and Y will produce an explosive. Testing for Z was not performed. The presence of Z would negate the explosive potential of X and Y. Was the decision not to perform the testing that would confirm or deny the presence of Z cognitive or not cognitive? Was the decision associated with the identification of Z made in an effort to bias the opinion generated by the interpretation of the results, or as a result of an effort to expedite the report by not performing all the required examinations?

Bias in some form exists in every step of the forensic clandestine lab investigation. Some decisions are made consciously. Some are made unconsciously. The forensic clandestine lab investigator must be vigilant to recognize their innate biases and utilize their experience and training to do the right thing for the right reason.

5.3 Types of Analysis

The analysis of samples from clandestine labs involves a broader range of analytic techniques than are traditionally used by the forensic controlled substance chemist. Many of the same instrumental and wet chemical techniques are used. The difference is the way the techniques are applied and the information is interpreted. Both organic and inorganic examinations can be performed on any individual evidentiary sample. Each type of analysis provides insight into the manufacturing process used by the operator. An individual examination type is necessary to establish the identity of a specific chemical used in the manufacturing process. That is, organic analysis is used to establish the identity of specific precursor chemicals or inorganic analysis to identify reagent chemicals. A combination of the two types of analysis may be required to establish the manufacturing method used, i.e., using a combination of organic and inorganic analysis to establish the presence of the components of a reaction mixture.

The burden of proof required to identify a particular chemical varies with its role in the manufacturing process. Controlled substances have the highest burden of proof, "beyond a reasonable doubt," because their possession is regulated in some manner. The burden of proof for the presence of precursor chemicals varies with the circumstances. The "beyond a reasonable doubt" standard may apply if possession of the precursor chemical is illegal under a given set of circumstances (e.g., possession with the intent to manufacture a controlled substance). A preponderance of evidence may be all that is required if the chemical identification is associative evidence; thus, the burden of proof may be less.

There is no stated burden of proof level associated with the examination of samples acquired through site exploitation. One school of thought is that a preponderance of evidence may be all that is required by the intelligence community, since they are simply gathering information to build profiles of the groups that are the center of the terrorist activities. Additionally, the information is used to identify potential sources of the precursors used for the manufacture of explosives so that the supply can be disrupted, leading to inability to produce the explosive due to lack of raw materials.

The other school of thought believes that site exploitation samples require identification beyond a reasonable doubt. The rationale behind this belief is that the ultimate use of the information derived from the analysis of site exploitation samples is to identify specific individuals or sites responsible for the manufacture of explosives or other weapons of destruction. More importantly, once identified, these people or places are treated with extreme prejudice. As such, it is extremely important that the forensic information generated from the site exploitation samples is as accurate as possible due to the nature of the consequences of decisions made using that information.

The burden of proof determines the level of testing required. "Beyond a reasonable doubt" requires specific confirmatory tests that will provide a chemical fingerprint of the substance under examination. These fingerprints can be obtained through the use of mass spectroscopy (MS), X-ray diffraction (XRD) or infrared spectroscopy (IR). Techniques such as nuclear magnetic resonance (NMR) and Raman spectroscopy are considered confirmatory tests but are not as widely used by the forensic chemist analyzing samples from clandestine labs.

A series of nonspecific tests indicating the presence of the chemical in question may be sufficient to meet the burden of proof required for establishing a preponderance of evidence. These examinations can include one or more chemical color tests, microcrystalline tests or instrumental examinations that produce nonspecific results. The following is an example of a series of nonspecific tests that can be used to establish the identity of a chemical without a specific test.

Under low-power microscopic examination, a white powder is found to have granules containing a spherical appearance (test one). The granules are water-soluble (test two). A chemical color test indicates the presence of nitrate ions (test three). A microcrystalline test indicates the presence of ammonium ions (test four). When combining the information from these four nonspecific tests, a chemist could reasonably conclude that the substance is consistent with ammonium nitrate (NH_4NO_3), a common fertilizer.

The presence of ammonium nitrate has little probative value in a contraband drug manufacturing case. Therefore, the combination of these nonspecific tests may be adequate to establish the identity of the compound within the requisite burden of proof.

By contrast, ammonium nitrate is the primary component of numerous fuel/oxidizer explosive mixtures. Therefore, determining its identity beyond a reasonable doubt is an essential component of establishing the elements of the composition of the explosive mixture. Techniques such as XRD and the use of X-ray detectors could provide specific information concerning the identity of the compound.

5.4 Identification Requirements

Two organizations have established benchmark requirements for the identification of chemicals associated with the clandestine manufacture of drugs and explosives. The Scientific Working Group for the Identification of Contraband Drugs (SWGDRUG) has a three-tiered system (see Table 5.1) used to determine

Table 5.1 SWGDRUG Test Hierarchy

Category A	Category B	Category C
• Infrared spectroscopy • Mass spectrometry • Nuclear magnetic resonance spectroscopy • Raman spectroscopy • X-ray diffractometry	• Capillary electrophoresis • Gas chromatography • Ion mobility spectrometry • Liquid chromatography • Microcrystalline tests • Thin layer chromatography • Ultraviolet spectroscopy	• Color tests • Fluorescence spectroscopy • Immunoassay • Melting point • Pharmaceutical identifiers

Table 5.2 TWGFEX Test Hierarchy

Categories 1 and 2	Category 3	Category 4
• Infrared spectroscopy (IR) • Gas chromatography/mass spectrometry (GC/MS) • Energy dispersive X-ray analyzer (EDX) • Raman spectroscopy • X-ray diffraction (XRD) • Liquid chromatography/mass spectrometry (LC/MS)	• Gas chromatography (GC) • Gas chromatography thermal energy analyzer (GC-TEA) • Liquid chromatography (LC) • Liquid chromatography thermal energy analyzer (LC-TEA) • Ion chromatography (IC) • Capillary electrophoresis (CE) • Thin layer chromatography (TLC) • Ion mobility spectrometry (IMS) • Polarizing light microscopy (PLM) • Stereo light microscopy (SLM)	• Burn test • Flame test • Spot test • Melting point

the specificity of tests used to positively identify contraband drugs and by extension, the chemicals associated with their manufacture. This system is codified in the American Society of Testing Materials (ASTM) method E2329-17, Standard Practice for Identification of Seized Drugs.

The Technical Working Group for Fire and Explosives (TWGFEX) has a four-tiered system similar to SWGDRUG's. (See Table 5.2.) Tiers one and two are combined, giving the TGWFEX matrix a similar profile to that of SWGDRUG. ASTM E3253-21, Standard Practice for Establishing an Examination Scheme for Intact Explosives codifies the explosives' examination process in the same way E2329 has codified the examination process for contraband drugs.

Each identification tier has a defined level of specificity. The tier one tests have the most selectivity, which is obtained through structural information. These tests are defined by SWGDRUG as Category A or by TWGFEX as Category 1 and 2. Tier two tests obtain their selectivity through the compound's chemical and physical characteristics and are delineated by SWGDRUG as Category B or by TWGFEX as Category 3. Selectivity through general or class information defines the third tier of tests, which fall into SWGDRUG Category C and TWGFEX Category 4.

Although the SWGDRUG and TWGFEX hierarchies of tests are essentially the same, the testing required for positive identification differs slightly. SWGDRUG recommends the following criteria for identification:

- A validated Category A technique in conjunction with least one other Category A, B or C technique, or
- At least three different validated techniques, two of which shall be based on uncorrelated Category B techniques.

By contrast, TWGFEX recommends:

- A single Category 1 technique is sufficient for identification.
- A Category 2 technique requires one more supporting techniques for identification.
- A Category 3 technique requires two more supporting techniques for identification.
- A Category 4 technique requires three more supporting techniques for identification.

5.5 Inorganic Analysis

Many reagent chemicals and components of explosive mixtures are considered inorganic chemicals (i.e., the molecule does not contain carbon). Their ability to dissolve in water, the resulting pH, and their physical and chemical properties provide the first insight into their identity. Identifying the inorganic chemicals involved in a clandestine lab or the inorganic components of an explosive mixture or reaction/waste mixture enables the clandestine lab chemist to definitively establish the reaction methods that the operator employed or the step in the manufacturing process the operation was in at the time of seizure.

Inorganic analysis is not something that is routinely performed by the forensic drug chemist but is a routine part of explosive analysis. Drug and explosives analysis share many of the same techniques and instruments. The types of tests that can be performed on inorganic compounds include chemical color tests, microscopic examinations, ion chromatography, IR spectroscopy and the use of X-ray energy.

5.5.1 Chemical Color Tests

Chemical color testing is one of the oldest methods of chemical identification. It is a method that can rapidly establish or exclude the presence of certain categories of compounds or ions. The specificity of the results varies with the test and the ions under examination.

In a chemical color test, a chemical reagent is added to the unknown. The color of the resulting mixture indicates the presence or absence of a group of compounds. For example, a white precipitate resulting from the addition of a 1% solution of silver nitrate to an aqueous solution containing the unknown indicates the presence of chloride ions. Additional testing would be necessary to exclude borate and carbonate ions, which also form a white precipitate with the silver nitrate reagent.

Chemical color tests provide a method to identify inorganic acids. The use of a series of three chemical color tests can reveal the identity of a clear acidic liquid. These tests have both laboratory and field application. However, caution should be taken when conducting these tests in the field due to the potentially violent nature of the reactions. Nitric acid is a reagent chemical that reacts violently with certain organic acids and has been known to cause ignition when it comes in contact with methamphetamine reaction mixtures containing phosphorus.

Simple chemical color tests can be used to quickly provide presumptive information concerning the identity of acidic liquids. The same color test reagents described in Table 4.4 can be used for the examination of acidic solutions under controlled conditions in a laboratory setting. Table 5.3 correlates the various color reactions with the inorganic acids commonly encountered in clandestine lab operations.

In some laboratories, a simple chemical color test is the only test available to establish the presence of some inorganic compounds (Table 5.4). For example, the reaction between hydrolyzed starch solution and iodide ion produces the characteristic blue color seen in the elementary school science experiment. This may not be a specific identification of iodine. However,

Table 5.3 Acid Test Color Reactions

Acid	Silver Nitrate Reagent[a]	Silver Nitrate + Nitric Acid	Barium Chloride Reagent[a]	Barium Chloride + Nitric Acid	Diphenylamine Reagent[a]
Hydrochloric Acid (HCl)	White Precipitate	Precipitate Remains	No Reaction	No Reaction	No Reaction
Hydriodic Acid (HI)	Yellow Precipitate	Precipitate Remains	No Reaction	No Reaction	No Reaction
Sulfuric Acid (H_2SO_4)	White Precipitate	Precipitate Dissolves	White Precipitate	Precipitate Dissolves	No Reaction
Nitric Acid (HNO_3)	No Reaction	No Reaction	No Reaction	No Reaction	Blue

[a] See Appendix I for reagent composition.

Table 5.4   Color Test Reactions for Inorganic Compounds

Reagent*	Color	Indication
Silver Nitrate	1. White	1. $Cl^-, CO_3^{-2}, SO_3^{-2}$
	2. Cream	2. Br^-
	3. Yellow	3. I^-, PO_4^{-3}
	4. Brown	4. OH^-
	5. Black	5. S^-
Barium Chloride	• White	• $CO_3^{-2}, SO_3^{-2}, SO_4^{-2}, PO_4^{-3}$
Diphenylamine	• Blue	• NO_3^-, ClO_3^-, ClO_4^-, nitro compounds, oxidizers
Thymol	1. Green	1. NO_3^-
	2. Brown	2. ClO_3^-
Nessler's	• Orange	• NH_4^+
Starch	• Blue	• I^-
Sulfuric Acid	• Yellow Orange	• ClO_4^-

*See Appendix H for reagent composition.

to a trained forensic chemist, the color reaction is characteristic enough to establish the preponderance of evidence of its existence. The chemist cannot make a statement concerning the existence of iodine in a sample without any testing to support his conclusion. This simple color test provides that support for those situations where the laboratory does not have access to the instrumentation that can establish the presence of iodine beyond a reasonable doubt.

5.5.2 Microscopic Techniques

Microscopic examinations of inorganic compounds are the second type of testing that can be performed on inorganic compounds. As in chemical color testing, the specificity of the results depends upon the compounds being examined and the tests being performed.

The three types of microscopic examination involve observation of the compound's basic optical properties, recrystallization and microcrystal examinations. Each type of microscopic examination requires different levels of microscopic expertise. Each method takes practice for the examiner to be able to recognize crystal structures as a specific to a given ion (Table 5.5).

Observing the optical properties of pure compounds under the microscope can be used as an identification tool. Combining information concerning the compound's color, crystal form and index of refraction can be used to make a specific identification. Appendix K contains a table of the optical properties of inorganic compounds found in clandestine laboratories. The physical structure and optical properties of a compound or mixture of

Table 5.5 Microcrystal Development Techniques

Crystal Structure/Optical Properties
- Add a few crystals of the unknown to a liquid the substance IS NOT soluble in and observe the crystalline structure and optical properties.

Recrystallization
- Dissolve a few crystals of the unknown in a liquid and observe the crystals that develop as the liquid evaporates.

Micro Chemical Reactions
1. A drop of reagent solution is caused to flow into the test drop. Observe the crystals that develop at the interface of the two drops, OR
2. A drop of the reagent is added directly to the test drop (or vice versa). Observe the crystals that develop, OR
3. Reactions take place in a capillary tube. Observe the crystals that develop at the interface of the solutions, OR
4. A fragment of solid reagent is added to the test drop containing the analyte. Observe the crystals that develop.
   - Scratching or mixing the reagent and test drop mixture may induce crystal formation.

Hanging Drop
- A drop of the reagent is suspended over a test drop (or vice versa).
  - A drop of acid, base or solvent may be added to the test drop to assist in the volatilization.
- Observe the crystals that develop in the reagent drop.

compounds can be observed by placing the unknown in a drop of nonvolatile organic liquid such as mineral oil or the Cargile liquids that are used to establish the refractive index. The use of polarized light and optical filters can assist the chemist in visualizing the various crystals as well as providing information concerning their birefringence and other optical properties.

Recrystallization is a method in which the inorganic components of pyrotechnics have been identified. The component is dissolved in a minimal amount of deionized water and placed under a microscope. As the water evaporates, the solution reaches the saturation point, and the compound begins to crystallize around the edge of the drop. The shape of the resulting crystals is characteristic of the compounds in the sample. As with other microscopic techniques, the use of a polarized light microscope is beneficial but not absolutely necessary.

Microcrystal tests are conducted in a manner similar to chemical color tests. A chemical reagent is added to the substance under examination. Instead of observing the resulting color, the examiner looks for the formation of characteristic crystals under the microscope. The use of a polarized light microscope is not necessary but can be beneficial. Caution should be exercised because a number of anions may produce similar crystals using the same chemical reagent. However, a series of microcrystal tests can be considered a specific identification for an anion. Appendix J lists the various reagents used for inorganic microcrystal examinations.

Microcrystal tests can be used on pure compounds as well as mixtures. There is a limitation. A single test will only identify one half of the inorganic compound. Only the cation (Y^+) or the anion (X^-) component is identified using a given reagent. Additional tests need to be undertaken to identify the other half of the compound. In mixtures, this can be problematic if there is more than one set of cations and anions present. The burden is on the examiner to establish which cation is paired with which anion. Assumptions can be made using information from the scene. However, if the only information the analytic chemist has is the sample in front of him, he may be hard pressed to "prove beyond a reasonable doubt" the original cation/anion pairing.

The use of microscopic techniques as a means of identification can be problematic if the testing is not documented. All testing used to make an identification should be documented in such a manner as to allow an independent expert to evaluate the results. Instrumental techniques such as MS and IR spectroscopy provide a record of the results of the examination. This demonstrates that the analytic chemist actually performed the test, documents the results and provides a means of independent evaluation if that becomes necessary.

When the analytic chemist utilizes microscopic techniques as a means of identification, he should follow the same protocol of documenting his test results as he is obligated to when utilizing instrumental techniques. It is recommended that microscopic examinations used to identify compounds be documented through the use of photomicrographs. This provides a record of the examination used to identify the compound. It also provides a means for an independent evaluation of the results. If the chemist cannot use photomicrographs to document his examination, he should sketch the crystal forms he observed and used to make his identification. Each photomicrograph or sketch should have documentation correlating the resulting crystal form to the sample preparation technique used.

5.5.3 Infrared Spectroscopy

IR spectroscopy has long been used as a method of positively identifying organic compounds. A compound's IR spectrum has been called its chemical fingerprint. It has not been extensively used for the identification of inorganic compounds in the forensic arena. This may be due to the broad bands and lack of detail in the spectra. Even with these handicaps, IR spectroscopy can be used to identify inorganic compounds.

The two keys to the use of IR for inorganic compound identification are sample preparation and peak identification. As with any analytical technique, sample preparation is the key to obtaining a usable spectrum. The exact location of peaks in the IR spectrum is critical in identifying the salt form of an inorganic compound.

The broad absorbance bands of inorganic IR spectra make it difficult to identify the maximum absorbance of the peak. The sample concentration should be diluted until a definite peak is observed in the primary absorbance band. This will allow the examiner to identify the maximum absorbance of each peak of the spectrum.

The determination of the maximum absorbance of each peak in the spectrum is critical. A shift in the maximum absorbance of 10 wavenumbers can be the difference between the sodium and potassium salts of a compound. These minor shifts may seem insignificant, but if they are reproducible in properly prepared samples, they provide the specificity required for identification purposes.

Sample preparation is critical when using IR to identify and differentiate inorganic compounds. Many inorganic compounds are efficient absorbers of IR radiation and easily overload the test sample. It is essential that the primary absorbance band of the resulting IR spectra has a well-defined peak with a resolvable maximum absorbance value, as opposed to a broad nondescript band with a rounded or flat maximum absorbance area. Broad rounded absorbance bands do not show the subtle absorbance shifts needed to differentiate between salt forms of an inorganic compound.

Many anions have characteristic absorbance bands in the IR spectrum. These can be used as a screening tool to classify the type of inorganic compound the examiner is dealing with. The presence of particular anion absorbance bands in a mixture can provide information that can be used to categorize the types of compounds that may be present. Table 5.6 contains a list of anions and their corresponding IR absorption wavelengths (Table 5.6).

In some instances, the results from IR analysis cannot distinguish between salt forms of a given compound. In these situations, the analytic chemist may be able to use the results of other techniques to make a specific identification. For example, the IR spectra of sodium and potassium cyanide are almost indistinguishable. However, sodium and potassium are easily distinguishable using microcrystal techniques. Combining the results of these examinations provides the analytic chemist with the information to render an informed opinion.

5.5.4 Ion Chromatography

Ion chromatography (IC) is an instrumental method that allows the chemist to identify both the anion and the cation of an inorganic substance or mixture. The analysis is a three-step process. First, the cation is determined. Then, the anion is determined. Finally, the results are combined, and the compound is determined (e.g., $Na^+ + Cl^- \longrightarrow NaCl$). Additional testing may

Table 5.6   Anion IR Absorbance Table

Anion	Absorbance (cm^{-1})
BO_2^-	1300–1360
$B_4O_7^-$	990–1000, 1340–1375
CO_3^{-2}	1425–1455
HCO_3^-	690–710, 830–840, 990–1010
SCN^-	2020–2100
SiO_3^{-2}	950–1010
NO_2^-	1225–1250
NO_3^-	1340–1390
NH_4^+	1390–1440
PO_4^{-3}	1000–1040
HPO_4^{-2}	1010–1060, 840–900
$H_2PO_4^-$	1025–1090, 900–950
SO_3^{-2}	920–980
SO_4^{-2}	1075–1130
HSO_4^-	845–855, 1030–1070, 1160–1175
$S_2O_3^{-2}$	950–1000, 1100–1120
$S_2O_5^{-2}$	975–990, 1175–1190
$S_2O_8^{-2}$	700–710, 1050–1075, 1275–1300
ClO_3^-	940–970
ClO_4^-	1050–1120
BrO_3^-	795–805
IO_3^-	725–750
$Cr_2O_7^{-2}$	795–860, 860–875
$Cr_2O_7^{-2}$	725–775, 845–890
WO_4^{-2}	790–825
MnO_4^-	890–915
$CN-$	2077–2100
OCN^-	2100–2180

be necessary to establish the hydration state of a compound that contains multiple hydration states.

IC is very effective in separating and identifying cations and anions. However, when there are multiple components in a solution, IC cannot distinguish which cation is associated with which anion, or the forms of the compounds originally added to the mixture. Here is where the chemist uses his knowledge of clandestine manufacturing methods along with the chemical inventory from the clandestine lab scene to establish the most probable combination of chemicals that would produce the results obtained from an IC run on a complex mixture.

An example of how IC can be used to propose a reaction mechanism would be the analysis of a basic aqueous solution from a clandestine lab that contained a trace of methamphetamine. Anion analysis reveals the presence of iodide with small amounts of chloride and carbonate present. Cation analysis reveals the presence of sodium. From this information, the chemist proposes that the iodide originated from HI, and the chloride originated from the HCl salt of the ephedrine precursor. The sodium came from sodium hydroxide, which was used to neutralize the HI. The odd trace of carbonate was a result of the sodium hydroxide reacting with the carbon dioxide in the air to produce sodium bicarbonate.

5.5.5 X-Ray Analysis

The use of the X-ray detector on various instruments provides the analytic chemist with a method of identifying the elemental composition of a compound or mixture. This is accomplished by recording the energy that is emitted from the substance that has been exposed to a beam of electrons. Each element releases a characteristic wavelength(s) of X-ray energy as it releases the energy it absorbed from the electron beam. The instrument also calculates the percentage of each element in a sample. This information can be used to determine the molecular formula of most inorganic compounds.

The drawback to this method is that many X-ray detectors cannot detect hydrogen, carbon, nitrogen and oxygen. This limits their use in organic analysis and in the determination of hydrogen, oxygen, nitrogen and certain low–molecular weight metals. For example, lithium, which can be used in the Birch reduction method, is outside the detectable range of most X-ray detectors.

Mineral acids are an example of compounds that cannot be directly identified using X-ray technology. Besides the acid being corrosive and detrimental to internal instrument parts, the detector cannot detect hydrogen. However, derivatization techniques can be used to compensate for this problem. The chemist replaces the undetectable H with a detectable element like potassium or sodium by reacting the mineral acid with a strong base. The water is evaporated, and the remaining solid is analyzed (e.g., $HI + NaOH \longrightarrow NaI + HOH$).

X-ray techniques can be used to analyze solid inorganic waste to determine the elemental makeup of the mixture. The dried waste solids are essentially the same inert inorganic salts that are created when a mineral acid is neutralized with a strong base. However, these waste solids can contain other inorganic by-products that may complicate the data interpretation. Practical Application in Section 5.9 demonstrates the utilization of this technique.

X-Ray fluorescence (XRF) and XRD are techniques employed by explosives chemists as part of their routine analytical scheme. These techniques are generally used in tandem. The subsequent results can not only positively identify a specific inorganic compound but in some instances, determine the hydration state (the number of attached water molecules) of the compound (Figures 5.1 and 5.2).

XRF is used to identify the elements that are within the sample. It cannot provide information concerning how the different atoms are attached.

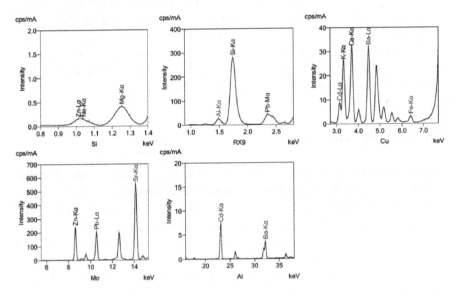

Figure 5.1 X-ray fluorescence data.

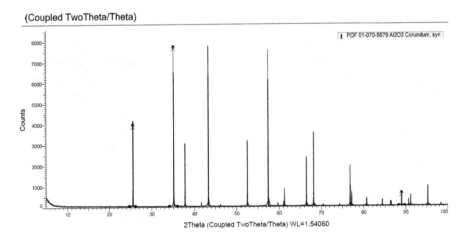

Figure 5.2 X-ray diffraction data.

However, the ability to confirm or eliminate the presence of a specific element is a powerful tool in itself.

XRD can positively identify inorganic compounds. It can distinguish between compounds that may be too similar in structure to identify using IR (e.g., potassium vs. sodium nitrate). More importantly, XRD has the ability to positively identify mixtures of inorganic compounds that appear to be a single chemical using other techniques. The following example demonstrates the power of XRD and the need for laboratory testing.

A sample submitted for testing was identified as ammonium nitrate using a hand instrument. Subsequent laboratory testing using XRD identified small amounts of calcium carbonate within the ammonium nitrate sample that had been undetected. The presence of the calcium carbonate provided the investigators with information concerning the source of the ammonium nitrate that would have been lost without the XRD analysis.

5.6 Organic Analysis

The examination of organic compounds is the analysis that is most familiar to the forensic chemist who analyzes controlled substances. It is also a routine part of the explosive chemist's analytical scheme. The methodology and instrumentation used to analyze drugs of abuse and explosives are routinely used in the analysis of organic compounds. Gas chromatography (GC), IR spectroscopy, MS, XRD, chemical color and microscopic techniques all can be used to analyze organic compounds. As in inorganic analysis, each technique has advantages and limitations.

5.6.1 Test Specificity

The analysis of individual organic chemicals is relatively straightforward. The process begins with the visual examination, followed by presumptive testing and concluding with a confirmatory examination. However, before this process can commence, the analytic chemist must decide what degree of certainty is required for the identification of this exhibit. The degree of specificity needed to identify a solvent may be different from that needed to identify a precursor chemical or a controlled substance. The degree of specificity needed will determine the type of examination sequence used to identify the compound.

Solvents are an example of chemicals that have a low degree of specificity needed in their identification. The positive identification of the type of solvent used is generally unnecessary to establish the elements of a manufacturing charge. The solvent's identity only assists the chemist in formulating his opinion on the manufacturing method being used. Generally, the scene chemist can establish the probable identity of a solvent by comparing a container's contents with the label. If the contents look, feel, smell, taste and act like the solvent on the label, generally, that will be sufficient to establish the probable identity of the solvent. If it does not react as expected, the scene chemist should perform presumptive tests on the substance to confirm or refute the label and possibly sample the substance for laboratory analysis.

Clear liquid solvents present a unique problem. Laboratory analysis is the only way to determine whether a controlled substance is dissolved in them. What appears to be an unused liquid may in fact contain the final product or other components that would shed light on the manufacturing process. Therefore, the analytic chemist should minimally perform a screening examination on unknown solvents and not disregard them just because the liquid sample appears to be unaltered.

The introduction of handheld instruments that utilize Raman spectroscopy to the tools available to the clandestine lab investigator has removed some of the on-site mystery concerning the identity or contents of clear liquids. However, as with all field tests, the results using this technology should be confirmed in a laboratory under controlled conditions.

Reagents are an example of chemicals that require a moderate degree of certainty in establishing their identity. Reagents themselves are not controlled, but they are used to create a controlled substance. Their identification may help identify the manufacturing route. An example of using a reagent chemical's identity to establish a manufacturing method would be the use of acetates to manufacture phenylacetone. Both sodium and lead acetate can be used to manufacture phenylacetone. Each is used in a specific synthesis route, and they are not interchangeable. A simple microcrystal test for the presence of sodium or lead along with a chemical color test for the presence of an acetate can be used to identify the type of acetate, which helps to identify the manufacturing route the operator is probably using.

Nonspecific presumptive tests include chemical color and microscopic techniques, ultraviolet (UV) absorbance, gas chromatographic retention time, index of refraction and density. Each test has its own degree of specificity. A combination of these techniques is necessary to rule out other compounds. If positive identification is required, XRD analysis would easily differentiate the two.

Precursor chemicals and the controlled substances they form as a rule need to be specifically identified. This usually involves the use of instrumental analysis to confirm any presumptive testing that may have been done. The possession of precursor chemicals may not be controlled. However, the necessity for their positive identification increases when they are possessed in conjunction with the appropriate reagent chemicals or equipment, which creates the potential ability to produce a controlled substance. In that situation, the identity of the precursor chemical(s) should be specifically established as well as the identity of any controlled substance.

The specific tests used in most forensic laboratories include IR spectroscopy, MS and XRD. NMR and Raman spectroscopy are also considered specific tests; however, the book will not address those techniques. Each technique is considered specific for the identification of a compound, but each has its own limitations. For example, MS cannot determine the salt form of a compound. MS also may have problems differentiating between stereo- and geometric isomers. The resulting ion patterns can be almost indistinguishable. Without retention time data or derivatization before analysis, complete identification is not possible with MS alone. IR spectroscopy cannot distinguish between optical isomers. XRD requires the compound to have crystalline structure. Therefore, it cannot be used on liquids or amorphous substances such as polymers and plastics.

A significant portion of the samples submitted for laboratory analysis require the identification of organic compounds. The composition of the samples may vary, but the procedure remains the same. Each sample requires a screening step, an extraction or sample preparation step, and a confirmatory step. These steps can be subdivided into wet chemical and instrumental procedures. Wet chemical procedures are used as a screening method or for sample preparation. Instrumental procedures are used for screening or as a confirmation tool.

5.6.2 Wet Chemical Procedures

Wet chemical procedures are used in the initial stages of the organic chemical identification process. These nonspecific tests provide a method to quickly indicate whether a controlled substance may or may not be present within a sample. These procedures can also be used to isolate controlled substances for confirmatory testing using instrumental techniques. Wet chemical procedures consist of chemical color tests, microscopic techniques, thin layer chromatography and various extraction techniques. A series of these tests can be used to deductively identify a compound or mixture.

5.6.2.1 Chemical Color Tests

Chemical color tests are chemical reactions that provide information regarding the structure of the substance being tested. Certain compounds or classes of compounds produce distinct colors when brought into contact with various chemical reagents. (See Appendix I for a list of color test reagents and their compositions.) These simple reactions can indicate the presence of generic classes of compounds.

Chemical color tests are generally conducted by transferring a small amount of the substance being tested to the well of a spot plate or into a test tube. The test reagent is added to the substance. Some tests may be conducted in a sequential fashion utilizing multiple reagents. The results of each step in the sequence are observed and noted. Positive and negative controls should be run on a regular basis to ensure the reliability of the testing reagents.

There is a certain amount of subjectivity when a color is reported. It is not uncommon for two people to describe the same color differently. The colors produced can also be influenced by the concentration of the sample, the presence of diluents and adulterants, or the age of the reagent. The length of time for which the reaction is observed may also influence the color reported. Color transitions and instabilities are not unusual. Allowances should be made for these differences.

5.6.2.2 Microscopic Techniques

Microscopic techniques are used as a screening tool to confirm a diagnosis made using other testing methods. Many of the same microscopic techniques used for inorganic analysis have organic applications as well. They are fast, simple to administer and can be highly specific. There is a debate as to whether they are specific enough to be used as a confirmatory test.

The microscopic crystal structures of a compound can be used to tentatively identify components within a mixture. The examiner can obtain a profile of the various components within the mixture by placing a sample into a liquid test drop in which most, if not all, of the components are insoluble. (Mineral oil works well for this type of analysis.) The component's physical and optical characteristics are then observed under plain or polarized light. Commonly encountered components can be tentatively identified and quantitated through this technique.

Microcrystal tests can also be used to determine the optical isomer of a compound. Single isomer compounds (d or l) produce a different crystal form than a racemic mixture (d and l) of the same compound. Single isomer crystals will form if a substance with the same isomer is added to the test solution prior to the test reagent. Racemic crystals will form if the opposite isomer configuration is added to the test solution prior to analysis.

Mixed crystal examinations can give insight into a compound's optical orientation. They are performed by seeding a sample with a known isomer of the substance under examination. The crystals that result from the addition of reference material with the same optical orientation will result in single isomer crystals. Racemic (mixed) crystals are formed if the optical orientation of the two compounds is different.

Microcrystal tests involve observing the crystals formed when the questioned sample is reacted with a test reagent. The test reagent and the sample can be combined using any of the methods described (refer to Table 5.3). A reaction between the component of interest and the test reagent forms a solid compound that is not soluble in the test drop. This solid forms uniquely shaped crystals that can be observed with a microscope.

Microcrystal identification relies on the comparison of the crystals formed by the unknown with those formed by a reference standard using the same reagent. Difficulties in obtaining an exact match between the crystals of the unknown and those of the reference sample may arise. Impurities in the unknown sample may lead to the formation of deformed, irregular or unusual crystals. These problems can be overcome by utilizing a cleanup procedure such as TLC, extractions or particle picking prior to microcrystal analysis.

Polymorphism can occasionally be a source of trouble. Sample concentration and reagent age can lead to the creation of different microcrystalline forms. This reemphasizes the comparative nature of microcrystal identification. The comparison should be done using the same sample concentration with the same crystal reagent.

Differences in crystal appearance can arise from the concentration of the solution. The crystals in highly concentrated test drops develop rapidly, resulting in a distortion of the classic crystal shapes. Concentrated test drops should be diluted to a concentration that produces classic crystal forms that are conducive to comparison and identification.

The reagent's age will also affect crystal development. Therefore, unknown and reference samples should be run using the same reagents, under the same conditions and at approximately the same concentration. Reagents should also be checked on a regular basis to ensure not only that they will produce crystals with reference standards, but that the crystals that are produced are consistent with the accepted crystal form for the reaction between the reagent and the substance in question.

5.6.2.3 Thin Layer Chromatography (TLC)

TLC is a wet chemical test used to screen for the presence of drugs and explosives. It is a separation technique that utilizes molecular mobility and solvent compatibility to separate and distinguish compounds within a mixture. That is, the way a component dissolves in the TLC solvent and how it reacts with the coating on the thin layer plate as the solvent travels over it affect the separation. Compounds are separated by their size, shape and reactivity with the solvent, similarly to rocks flowing down a river. Small compact molecules will travel across the TLC plate at a different rate than large rambling molecules.

The typical TLC procedure places a sample of the unknown toward the bottom of a glass plate containing a thin layer of silica gel. A sample(s) of a reference compound(s) is placed at the same distance from the bottom of the plate. The TLC plate is placed into a tank containing a solvent (or mixture of solvents). As the solvent travels up the TLC plate, the various components within the sample are separated. When the solvent migration is stopped, the TLC plate is removed from the tank, and the solvent is allowed to evaporate. The compound's movement is then visualized through observation under UV light or through development with a chemical color reagent(s) designed to react with various compounds.

The Rf value is used to establish the identity of the spots on the TLC plate. The use of Rf values for a known solvent system only provides a generic insight as to the identity of the unknown spot. They should not be relied upon for confirmation of unknowns. A known reference sample, run on the same TLC plate, should be used for comparison.

Rf values can be affected by many factors. The adsorbent uniformity on the thin layer plate, sample concentration (spotting is too weak or strong), room temperature during the mobile phase, and development distance of the solvent during the mobile phase will all affect the results. Care should be taken to eliminate variances in the method caused by any of these factors. Placing a reference sample containing the suspected compound on the TLC plate with the questioned sample reduces the variables involved in TLC comparisons.

5.6.2.4 Extractions

Extractions are not a screening test per se. However, the fact that the compound was isolated as a result of the extraction indicates that the compound had certain chemical characteristics. These are class characteristics that can be used to deductively support the confirmatory test.

Extractions are then used to separate the compound of interest from the rest of the sample. The type of extraction used will depend upon the compound of interest and the matrix in which the compound is located. In some cases, multiple extraction techniques are necessary to separate the substance of interest from the remainder of the sample. In other instances, instrumental analysis is the only way to separate compounds with similar chemical properties for confirmation.

The screening techniques used should be designed to identify as many of the components of the sample matrix as possible. This allows the examiner to select the extraction technique that efficiently and effectively isolates the component of interest from the rest of the compounds. Misidentified or unidentified components within a sample mixture may lead to the selection of an inappropriate extraction technique, which in turn may affect the results of the confirmatory test.

5.6.2.5 Wet Chemical Documentation

Wet chemical tests are traditionally considered a non-documentable technique. There is no independent record of the performance of the test. The test documentation solely rests on the examiner's handwritten notes. Therefore, the chemist should describe his observations as completely as possible. A (+) or (–) notation next to a test name does not provide a peer reviewer with insight as to the examiner's observations during the performance of the test.

The colors or transition of colors that were observed during the course of a chemical color test should be described. Photographing a chemical color test may or may not be a solution to the documentation issue. Photography does demonstrate the color that was observed during the examination. However, it may only preserve a portion of the test. Many chemical color tests have a transition of colors from the beginning of the test to the end. Photographs may not adequately reflect the totality of the examiner's observations.

No supporting documentation is usually generated with microcrystal examinations. Therefore, the examiner's description of his observations should be as complete and accurate as possible. When definite crystals are formed, their form and habit should be noted (described, sketched or photographed). Table 5.7 is a list of descriptive terms with diagrams that can be used to describe the observed crystals.

Lack of supporting documentation may be less significant if microcrystal tests are used as a screening tool. However, if they are to be used as a tool to specifically identify a compound's identity or isomer configuration, steps should be taken to provide reliable documentation verifying the examiner's observation. Photomicrographs should be taken of the microcrystals that were used to make the identification as well as positive controls performed at the same time as the questioned sample was examined. The photomicrographs of both known and unknown samples should be included in the examiner's notes for peer review when necessary.

As with chemical color and microcrystal examinations, no supporting documentation is usually generated using TLC. Accurate notes regarding the type of solvent system(s) used should be included in case notes, along with the Rf calculations for the spots used for compound identification. Any deviations from the referenced method or unusual occurrences noted should also be documented. The examiner should thoroughly describe the observations used to make his conclusions, including the colors and patterns observed on the TLC plates as well as any observations made under UV light.

Photography provides a solution to the conundrum associated with documenting observations. Although it may not completely capture the totality of the chemist's observations, it is a means to replace subjective written descriptions concerning shape, size, color and texture with digital images that objectively depict these conditions at the time of the observation. Images

Table 5.7 Microcrystal Descriptions

Crystal	Shape	Description
Blade		Broad needle
Bunch/Bundle		Cluster with the majority of the crystals lying in one direction
Burr/Hedgehog		Rosette, which is so dense that only the tops of the needles show
Cluster		Loose complex of crystals
Cross		Single cruciform crystal
Dendrites		Multibrachiate branching crystals
Grains		Small lenticular crystals
Needles		Long thin crystals with pointed ends
Plates		Crystals with the length and width that are of the same magnitude
Prisms		Thick tablet
Rod		Long thin crystals with square cut ends
Rosette		Collection of crystals radiating from a single point
Sheaf		Double tuff
Splinters		Small irregular rods and needles
Star		Rosette with four or six components
Tablet		Plates with appreciable thickness
Tuff/Fan		Sector of a rosette

that contain positive and negative controls not only provide a means of documenting test results but also document the performance of the requisite quality assurance procedures. Although not required, this method of documenting observations that are used to formulate the conclusion or opinion has become the expected norm.

The extraction phase of the analysis is not used for preliminary or confirmatory identification purposes. However, it is a means to those ends. As such, it should be documented. Peer reviewers should be able to evaluate the extraction technique used to prepare the sample for any subsequent testing.

5.6.3 Instrumental Examinations

Instrumental examinations are documentable testing methods. This point is key to the confirmation process. It is not enough for the examiner to be able to say that the compound had the same chemical fingerprint as the substance in question. He has to be able to demonstrate it beyond a reasonable doubt. This includes subjecting the examination to peer review. Instrumental examinations, and their associated documentation, provide the vehicle for this review.

There are four basic instruments that are routinely utilized by forensic chemists analyzing clandestine lab samples. The UV spectrophotometer and the gas chromatograph are used as screening and quantitative tools. Liquid chromatography is utilized as a screening tool, but not as widely. The IR spectrometer and the mass spectrometer are instruments used to confirm the identity unknowns. As stated previously, NMR and Raman spectroscopy are used as confirmatory tools but do not have as broad a base of use. XRD can be a useful tool for identifying organic compounds with a crystalline structure. It is commonly used for the identification of inorganic explosives components but is rarely encountered in the traditional forensic drug laboratory.

5.6.3.1 Ultraviolet Spectroscopy

UV spectroscopy is an instrumental technique that provides compound classification. It is a screening tool and not a confirmatory test. Although some compounds exhibit unique UV spectra, the spectra are considered class characteristics and do not contain sufficient detail (individual characteristics) to be considered a compound's chemical fingerprint.

The two general uses for UV spectroscopy in the controlled substances unit are general screening and quantitation. The shape of the spectrum provides insight into the identity of the compound. The amount of UV light absorbed can correlate to the amount of substance in the sample.

UV spectroscopy is a useful tool for single component analysis of samples with a known or suspected composition, such as pharmaceuticals. The UV spectrum can confirm or reject the composition of the preparation under examination. However, if compound identification is required, it should be done using a specific test such as IR or MS.

Mixtures of compounds capable of absorbing UV energy can present an analytical problem. Compounds have differing capacities to absorb UV light. A strong UV absorbing substance mixed in with a controlled substance that is a weak UV absorber may result in a UV spectrum that does not reflect the presence of the controlled substance.

Quantitation is another venue in which UV spectroscopy is useful. To be effective, the sample should contain a single UV absorbing component. If there are multiple UV absorbers in the sample, the component of interest should have distinct, resolvable absorption bands. The quantitation procedure can be as simple as comparing the concentration of the suspected tampered sample with that of a known unaltered sample. The UV absorbances should be the same if the concentrations and compositions of the samples are identical. An in-depth analysis can determine the concentration of the substance in question. The absorbance value of the test sample is compared with the absorbances of a series of known solutions. The concentration of the test sample can be derived from the graph of concentration versus absorbance values of the reference samples.

5.6.3.2 Gas Chromatography

GC is a documentable chromatography form that can be used in lieu of TLC. It is not a specific confirmatory test for controlled substances. However, dual column techniques and the evaluation of alkaloid peak patterns can be used for identification purposes. The gas chromatograph also is used as a separation device for confirmatory examinations such as MS (Figure 5.3) and Fourier transform infrared spectroscopy (FTIR).

The gas chromatograph separates compounds by their size, shape and reactivity with the chemical coating of the GC column in a similar manner to rocks flowing down a river. The carrier gas acts as the water and the column coating as the riverbed. The small molecules travel through the chromatographic column more rapidly than larger molecules. Their shape and their reactivity with the column's coating separate molecules of the same size.

GC chromatograms are used to identify unknowns based on the retention time or relative retention time of a peak under certain operating conditions. The retention time (Rt) is the time it takes a compound to travel from the injection port of the gas chromatograph to the detector. The relative retention time (RRt) is the ratio of the retention time of the substance to the retention time of an internal standard that is added to the sample.

The RRt is considered a more reliable value. The use of an internal standard provides a reference point to calculate RRt values. It also demonstrates

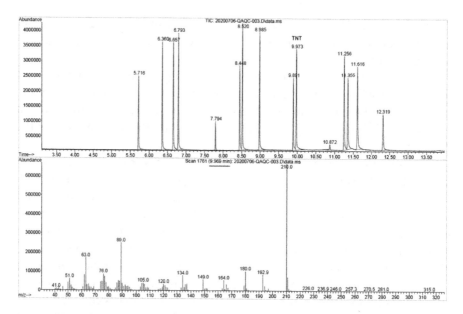

Figure 5.3 Explosives test mix GC/MS data.

the precision and accuracy of the instrument at the time of the test. The internal standard eluting at the proper time indicates that the gas flow and oven conditions are operating properly. The size of its peak indicates proper operation of the detector if the concentration of internal standard is known.

GC can be used to differentiate geometric isomers. An example of the use of GC retention times to differentiate between isomers is the identification of the cis- and trans-phenylaziradines that are a by-product of the HI reduction of ephedrine to methamphetamine. Even though these compounds have essentially the same mass spectrum, the GC retention times are significantly different. On a non-polar GC column, the cis isomer has a noticeably lower retention time than the trans isomer. Baseline resolution of the three isomers of dinitrotoluene can be achieved through the use of GC.

Analysis by GC alone is not generally considered confirmation of a controlled substance. More than one compound could possibly have a given Rt or RRt. Therefore, with conventional detectors (i.e., flame ionization, electron capture, nitrogen/phosphorus, etc.), the chemist cannot definitively tell what compound elutes at a given Rt or RRt. The specificity increases with the specificity of the detector. For example, a nitrogen/phosphorus detector will only see compounds containing nitrogen or phosphorus.

Dual column GC has been used as a confirmatory test. A single sample is injected into a gas chromatograph that divides the sample into two chromatographic columns. Each column contains a different liquid phase (the interior coating that causes compound separation). A compound is considered identified if the compound has the proper Rt or RRt on both columns.

Commonly, GC is used as the separation tool for the confirmatory tests of MS and FTIR. GC separates the compounds, and MS or FTIR provides information concerning the chemical properties of each of the compounds as they elute from the chromatographic column.

Quantitation is another use for GC. This can be accomplished by analyzing a series of diluted samples using a method similar to that used in UV analysis. The other method uses the relative response of the item in question to that of an internal standard.

As a quantitation tool, GC has an advantage over UV. The effects of multiple components within the sample are reduced or eliminated because it separates the components of the sample during the analysis. The UV absorptivity does not affect the results either. Each component has a similar delectability range with a given detector.

5.6.3.3 Mass Spectroscopy

MS is the workhorse instrument used by the forensic chemist. It uses the pattern of molecular pieces (ions) that is produced when a molecule is exposed to a beam of electrons as a means of identification. This characteristic pattern is called the mass spectrum. It is considered one of a compound's chemical fingerprints (Figure 5.4).

The mass spectrometer exposes the compound under analysis to a beam of high-energy electrons that shatters the molecules. The mass spectrometer

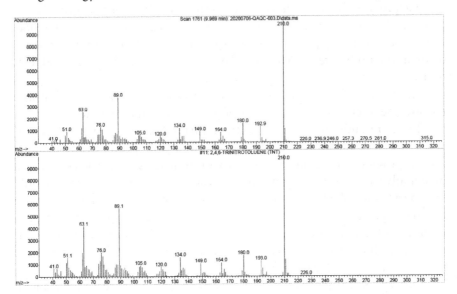

Figure 5.4 Mass spectrometry data.

then sorts and counts the resulting pieces (ions) and produces a pattern, the mass spectrum. When the energy of the electron beam remains constant, the molecule will produce the same mass spectrum, which is considered one of the compound's chemical fingerprints.

MS alone has its limitations. It cannot differentiate among certain types of isomers. Stereoisomers and geometric isomers may produce mass spectra that are essentially identical. Stereoisomers (molecules that are mirror images of each other) have identical mass spectra. Without additional information, i.e., GC retention time data, the chemist may not be able to say that the compound was one isomer or the other. Ephedrine and pseudoephedrine are an example of two compounds that have essentially the same GC retention times and mass spectra.

Geometric or positional isomers will also produce similar, if not the same, mass spectra. Many times, they can be differentiated by the chromatographic retention time of the compound. Other times, there are one or two clusters of ions that have ion ratios that are specific to a particular isomer. Methamphetamine and phentermine are two geometric isomers that can be differentiated through the use of MS.

The mass spectrometer generally cannot distinguish between the salt and free base forms of a drug. The salt portion of the compound is generally outside the detection range of the MS. The detector sees only the free base portion of the compound.

The information obtained from the mass spectrometer can be used to establish the synthesis route used to manufacture the controlled substance. Each reaction produces by-products as part of the reaction. In some instances, the by-products that are produced are specific to a particular manufacturing method. Even if the detected by-products are not specific to a reaction, their presence can be used to corroborate other information, i.e., notes, chemicals on hand, etc., regarding the method of manufacture. Appendix M is a five-peak table of drug precursor chemicals, the controlled substances and by-products. The table also includes the reaction that the presence of these compounds indicates.

There are a number of mass spectra libraries that are commercially available to assist in the identification of unknowns. The spectra in these libraries can provide insight into the identity of numerous components that can potentially be within these mixtures. However, final confirmation should only be accomplished by comparing the mass spectra of the unknown with the mass spectrum of a traceable reference standard. The reference spectra should be obtained on the same instrument, under the same operating conditions. The burden of proof required in a given situation will dictate how the information from these libraries should be used.

Library spectra should not be used for proof beyond a reasonable doubt. The variations in operating parameters between the instrument used to obtain the sample's spectrum and the one used to obtain the library spectra will differ. These deviations may be subtle, but they can be significant enough to eliminate the compound as the source.

Library spectra do provide a preponderance of evidence concerning the identity of a compound. Many of the by-products of clandestinely produced controlled substances do not have traceable primary standards that can be used for positive identification purposes. However, their probable identity, established through a mass spectral library search, can be used as associative evidence to render opinions concerning the manufacturing methods used in the operation.

5.6.3.4 Infrared Spectroscopy

IR spectroscopy has been the traditional method of confirming the identity of a controlled substance. Traditionally, the sample went through a series of screening tests to establish the compound's suspected identity, and the identity of any adulterants or diluents was determined. The controlled substance was then extracted and purified. Finally, an IR spectrum was obtained. Modern technology has introduced instrumentation that can obtain an IR spectrum from a single particle or from a peak in a GC chromatogram. This has eliminated the need for the nonspecific tests used as screening tools and the extractions necessary to isolate the compound of interest.

IR spectroscopy uses a compound's ability to absorb IR light as a means of identification. The bond of each of the molecule's functional groups will absorb specific wavelengths of IR radiation. The exact wavelength will depend on the arrangement of the functional groups on the molecule. The pattern that results from charting the absorbance or transmittance of IR light that is passed through (or reflected from) a sample is considered a chemical fingerprint (Figure 5.5).

The ability to differentiate between isomers is a benefit of using IR as a confirmation tool. Compounds with isomers that are indistinguishable by MS may be differentiated though the use of IR. The position of the functional groups on the molecule dictates how they will vibrate, which affects the wavelength of IR radiation that is absorbed. The stereoisomer of the compound may allow or hinder the vibration of a particular functional group. This allows the chemist to differentiate between stereoisomers such as ephedrine and pseudoephedrine (Figure 5.5). However, optical isomers, i.e., d-ephedrine and l-ephedrine, do not exhibit any significant differences in their IR spectra.

In some jurisdictions, a compound's salt form may be important in determining the sentence after a conviction is obtained. IR spectroscopy can be used to identify a compound's salt form. Figure 5.6 differentiates between free base cocaine (crack) and cocaine hydrochloride. The specific salt form

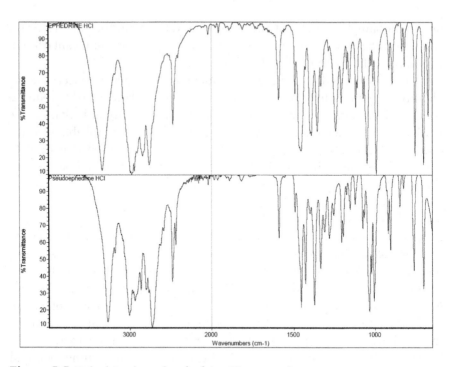

Figure 5.5 Ephedrine/pseudoephedrine IR comparison.

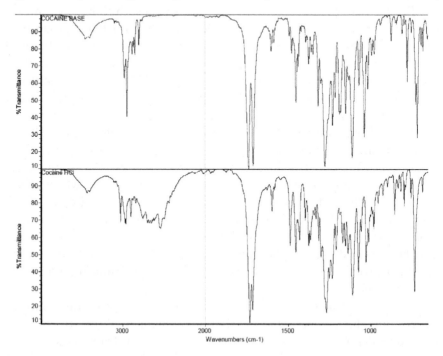

Figure 5.6 Cocaine base/HCl IR comparison.

can be used to establish a manufacturing method. Figure 5.7 contains the IR spectra of the HI and HCl forms of methamphetamine. In both examples, the most obvious difference is demonstrated in the spectra's front portion (4000 to 2000 cm^{-1}).

IR spectroscopy is also useful in differentiating structural and geometric isomers that MS cannot without derivatization, retention time data or both. Changing the position of a functional group on an aromatic ring will change the IR spectrum enough to allow easy identification. Figure 5.8 is an example of the IR differences of the three structural isomers of dinitrobenzene. The only difference is the position of one nitro (-NO$_2$) group on the aromatic ring.

Traditionally, IR confirmation has been limited to compounds that have gone through some type of extraction to produce a pure compound prior to analysis. Analysis time could be lengthy, depending on the resolution the analyst desires. Advances in technology have reduced the time required for sample preparation and analysis. With the advent of FTIR analysis, the time has fallen from minutes to seconds. The ability to obtain an IR spectrum instantaneously has allowed FTIR detectors to be used in conjunction with gas chromatographs. This allows the chemist to identify the components of a mixture by IR without first separating each of the components from the

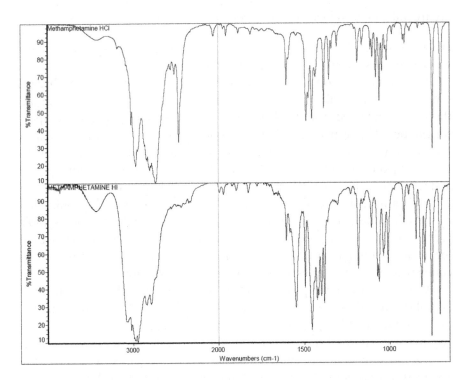

Figure 5.7 Methamphetamine HCl/HI IR comparison.

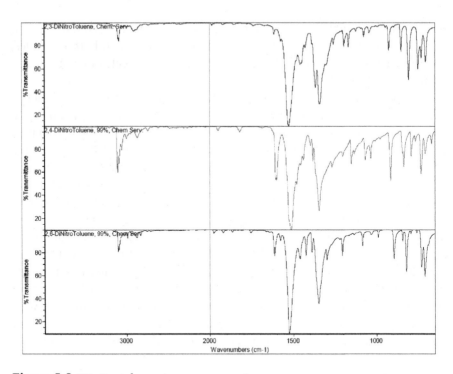

Figure 5.8 Dinitrotoluene isomer comparison.

others. Additionally, micro FTIR can isolate and obtain IR spectra of individual particles within a mixture.

Sample preparation is a key element in IR examinations. The physical state of the sample will significantly affect the resulting spectra. For example, the spectra obtained from GC/FTIR will be in the vapor phase. These spectra will be different from the liquid or solid phase IR spectra that a chemist traditionally uses for identification purposes. Pellet spectra of solid samples will vary from thin film spectra of the same sample. Transmission spectra and reflectance spectra of the same compound will have variations. There can even be significant variation between thin film spectra of the same compound as a result of polymorphism when the compound crystallizes. Each sample preparation technique produces a unique reproducible result that can be used for identification purposes. However, the analytical chemist must be sure to compare apples with apples when making an identification.

Computerization has produced the ability to compare the IR spectrum of an unknown with those in various libraries. As with the MS library searches, they should not be used for identification purposes for compounds that require proof beyond a reasonable doubt. The compound's physical state, type of detector used and sample preparation techniques will all affect the spectrum that is obtained and the results of a computerized library search.

Identification should only be made by comparing the spectra from the questioned sample with the spectra of a sample from a traceable reference that was prepared under the same conditions using the same instrument.

5.6.3.5 Documentation

Instrumental techniques are documentable in that they generate analytical data in a form that demonstrates that the analysis was performed. The data itself is objective and can be subjected to peer review as part of a quality assurance program or independent evaluation at a later date. The interpretation of this data is less subjective than in other areas of the forensic laboratory. However, it is still subject to interpretation.

For peer review purposes, case notes or instrument printouts should include the operating conditions of the instrument during the analysis. This allows the reviewer to evaluate whether instrumental results are consistent with the analytical conditions. If necessary, an independent examiner should be able to achieve the same results under the same test conditions. All data should contain at a minimum the examiner's initials, case number, solvent information and date of the analysis. The examiner should have the instrument print this information on the spectra at the time of analysis if the instrument has the capacity to do so. For GC analysis, the calculated RRt value should be on the chromatogram or on the printout of the peak retention times. The divisions of the mass value axis on MS data should be such that the examiner can easily determine the mass value of each of the ions of the spectra. The wave number of the significant peaks of an IR spectrum should be labeled or should be easily determined by a peer reviewer. The examiner should have the instrument print this information at the time of analysis if the instrument has the capacity to do so.

5.7 Analytical Schemes

The analytical schemes used to examine clandestine lab samples can generically be divided into solid and liquid schemes. Solids are usually precursors, reagents or controlled substances and should be treated as unknown controlled substances. Liquid samples can be either organic, aqueous or a mixture of the two. They can be pure chemicals, reaction mixtures or waste products and should be treated as if they contain a controlled substance.

It is common practice for operators to remove the labels from chemical containers, repackage chemicals into different containers, and place waste or finished product into empty chemical containers. Therefore, a chemist may find that a container's label gives little more than insight into what chemicals the operator possessed at one time. A container labeled ethyl ether may just as well contain a brown liquid with a chlorinated solvent odor as the clear volatile liquid it is supposed to. The chemist will need to modify his analytical scheme to identify the compound(s) in this situation as opposed to the confirmation of a chemical from a labeled container.

All samples should be screened for the presence of controlled substances. The screening method used is up to the chemist and the capability of his laboratory. The screening method should not only detect the presence of controlled substances but should include techniques that would tentatively or positively identify the presence of precursor chemicals, reagent chemicals or reaction by-products commonly used or encountered in the manufacture of the controlled substances. If a controlled substance is detected, its identity should be confirmed. If the substance appears to be a precursor, reagent or solvent, its identity should be established to the degree of certainty the circumstances dictate.

5.7.1 Solid Samples

The analysis of solid samples follows the same analytical scheme as for a controlled substance unknown. The same systematic analytical approach is used as if the sample were an unknown, even if the identity of the substance is suspected. Sample analytical schemes should include screening for the presence of controlled substances, confirming the identity of controlled substances that have a high burden of proof, and confirming the identity of other substances to the extent that their burden of proof dictates (Figure 5.9).

The sampling scheme for solid samples from labeled containers and loose solid material is basically the same. The samples are initially screened using a combination of nonspecific tests. At this point, a decision is made. In the case of known samples, the following questions are asked: Is the sample consistent with the label? Does the sample contain a compound that needs a confirmatory examination? In the case of unknown samples: What does the sample contain? Does it need a confirmatory examination? The answers to these questions and the level of specificity needed for the particular identification will guide the flow for the balance of the testing.

5.7.2 Liquid Mixtures

The analysis of liquid samples from clandestine labs is a deviation from the sample form normally encountered by the forensic chemist who deals with controlled substances. These samples, however, can give the chemist the most complete picture of the type of synthesis that the operator was using. An in-depth analysis of these liquids can produce information about the product and by-products of the reaction as well as the precursor and reagent chemicals

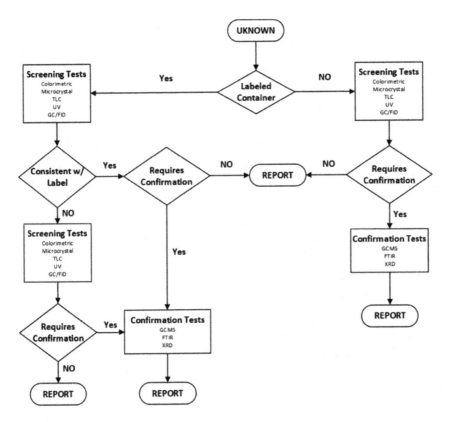

Figure 5.9 Unknown solid flowchart.

that were used in the synthesis. By evaluating all the information that can be obtained from these liquid samples, the chemist can also determine what step in the process the operation was in at the time of seizure (Figure 5.10).

Liquid samples come in organic and aqueous forms. Both forms can be analyzed using the same analytic tools that were utilized for solid sample analysis. Organic samples are usually extraction solvents that may or may not contain a controlled substance. They are treated simply as a controlled substance exhibit in a liquid substrate. The analysis of unknown aqueous liquids requires a combination of organic and inorganic analytical techniques.

5.7.2.1 Organic Liquids
Organic liquids from clandestine labs can be reaction mixtures; extraction solvents (ether, Freon, chloroform or petroleum products) that contain the final product; extraction solvents from which the final product has been removed, leaving only trace quantities of product; wash solvents that contain

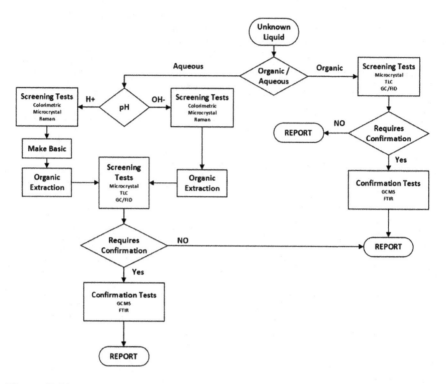

Figure 5.10 Unknown liquid flowchart.

reaction by-products with traces of the final product; or clean unused solvents, reagents, precursors or the final product itself.

The only way for the chemist to definitively say that a clear liquid contains more than a single component is to analyze its contents. Just because a liquid look, smells, feels and tastes like Freon, this does not mean there is not something dissolved in it. In almost every case, if the final product came into contact with the organic liquid, there will be a detectable amount of that substance in the liquid. It is the chemist's challenge to find it if it is there.

The analysis scheme of organic liquids mirrors that of organic solid samples. Using chromatographic techniques, the sample is screened to establish whether or not it contains a controlled substance. A profile of the sample's contents is used to establish the liquid's place in the manufacturing process. Confirmatory tests are performed on compounds whose burden of proof requires it. Finally, a report is generated.

5.7.2.2 Aqueous Liquids

Aqueous liquids can be reaction mixtures or waste products. Each type of liquid contains a wealth of information concerning the manufacturing process and will require both organic and inorganic examinations. The inorganic profile of the liquid will guide the analysis.

Determining the pH of an aqueous liquid is the initial analytical step that provides the analytical chemist with information concerning the liquid and the compounds that it may contain. As a general rule, acidic liquids are reaction mixtures, and basic liquids are waste material. Each type of liquid will have its own characteristic organic and inorganic composition.

Once the pH of the liquid has been established, the chemist must determine what type of information he desires from the sample. He should ask himself: Does he simply want to isolate and identify any controlled substances from the sample? Does he want to extract all of the organic constituents from the solution and try to establish a synthesis route? Does he want to identify the inorganic components of the aqueous solution? Or, does he not want to analyze the sample because it is similar to 12 other samples from the same location? The answers to these questions will determine the analytical sequence as well as any extraction techniques that may be used.

Establishing the inorganic profile of an aqueous sample is the next segment of the analysis. It can be accomplished by using the same methods described in the inorganic analysis section. The chemist is looking for the type of acid or base that was used in the reaction as well as any inorganic reagents that may indicate a particular reaction route. A series of chemical color tests and microcrystal examinations can provide the chemist with a sense of what the sample contains and where it fits into the manufacturing process. Three drops of sample may be all that is necessary to give the chemist a complete inorganic profile of the aqueous liquid.

As a general rule, the chemist should screen all aqueous liquids for their organic component content. Acidic liquids are generally reaction mixtures and will potentially contain large quantities of the controlled substance. Basic liquids are generally waste products that will contain reaction by-products characteristic of the manufacturing process used as well as a detectable amount of the controlled substance that was manufactured. The fact that there were no organic components detected in the sample is significant information.

Most contraband drugs and precursors are soluble in acidic aqueous liquids and may be visually undetectable. Thus, the analytic chemist should expect high concentrations of these chemicals in acidic solutions. To remove the controlled substance from the aqueous solution, he should change the pH of the solution and extract it with an organic solvent. This will remove the basic and neutral organic compounds from the aqueous solution. The organic extract can then be analyzed as if it were an unknown organic liquid.

There may be instances in which the acidic and neutral compounds that the aqueous liquid may contain will be of significance. In that situation, the acidic liquid should be extracted with an organic solvent prior to the aqueous solution being made basic to extract any controlled substances that may be there. Again, the organic extract is analyzed as if it were an unknown organic liquid.

The analysis of the organic extracts of aqueous samples mirrors that of organic liquid unknowns. Chromatographic techniques are used to screen the sample to establish whether the sample contains a controlled substance. The chromatographic profile of the sample's contents is used to establish the liquid's place in the manufacturing process. Confirmatory tests are performed on samples that contain compounds whose burden of proof requires it. Finally, a report is generated.

5.7.2.2.1 Unknown Bulk Explosives (UBE)   The analysis of UBE presents a unique challenge to the forensic chemist, who traditionally has focused on the analysis of contraband drugs using organic techniques. The nature of explosives and explosive mixtures requires the forensic chemist to expand his focus and range of possibilities associated with the contents of the sample under examination. Explosives are not simply organic compounds such as trinitrotoluene or nitroglycerine. Fuel/oxidizer explosives can be purely inorganic, e.g., chlorate/phosphorus mixtures used as primary explosives in detonators. They can also be an organic/inorganic combination of materials as found in ammonium nitrate fuel oil (ANFO), commonly used in the mining industry.

The information generated from the analysis of UBE samples recovered during site exploitation events reaches beyond that of a criminal prosecution. The identification of the specific explosives or combination of chemicals within an explosive mixture is still relevant. However, the identification of all the components within the mixture has as much importance to the intelligence analyst who is attempting to source the precursors used in the manufacturing process or to identify the cell of individuals who are associated with the manufacturing technique used. Therefore, a more comprehensive approach to the examination of UBE is required.

Figure 5.11 is an example of the sequence of activities used to examine an UBE. Although this work flow emphasizes the use of instrumentation, nonspecific tests can be inserted at the appropriate intervals. The work flow simply depicts steps on the process and the types of data to be collected. It does not place any value on specific pieces of information or how the data is combined to reach a conclusion or generate an opinion.

The process begins with a visual inspection of the sample. Gross visual observations coupled with scrutiny using a stereomicroscope are used to characterize the sample. Intact particles are identified, mechanically extracted and appropriately examined using the solid sample work flow.

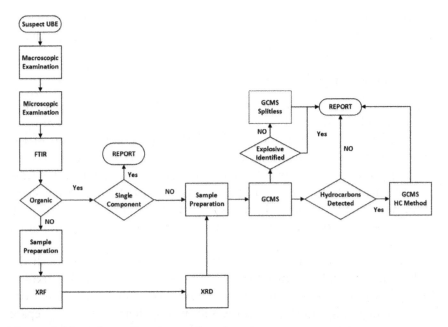

Figure 5.11 Unknown explosive flowchart.

A sample of the bulk material is analyzed via FTIR. This allows the chemist to characterize the sample as organic, inorganic or a combination of the two. If the sample is organic, a determination is made to establish whether the spectrum is of a single compound with sufficient detail required for identification so that a report can be generated. If the sample is a mixture of organic compounds, the sample is prepared for GC/MS analysis.

If FTIR analysis indicated that the sample is inorganic in nature, it is prepared for analysis via XRF and XRD to identify the component or mixture of components. If the sample is a mixture of organic and inorganic components, an extraction is performed to isolate the organic components and prepare them for analysis via GC/MS. The washed inorganic material is subsequently prepared for examination via XRF and XRD.

GC/MS methods used to analyze explosives differ from the methods used to analyze drugs. Many organic explosives thermally degrade at the operating temperatures commonly used to analyze drugs. As such, the presence of nitroglycerine will not be detected if the sample was examined using a method designed to detect methamphetamine. Therefore, explosives-specific methods that compensate for the thermal degradation of explosives must be used.

Numerous explosive mixtures contain hydrocarbon components of various boiling ranges as fuels or binders. These components range from flammable liquids such as gasoline to combustible solids such as wax or petroleum jelly. These components also require specialized methods to define the chromatographic patterns consistent with specific classes of petroleum products.

5.7.3 Chromatographic Screening

Chromatographic techniques include gas and liquid chromatography. They are useful tools for the analysis of organic liquid samples. These techniques allow the chemist to obtain a profile of the organic composition of the unknown with a single test. They can determine whether a liquid contains one component, a mixture of a dozen or nothing at all. Using retention time data, the chemicals and reaction by-products associated with various clandestine manufacturing methods can be identified. The chromatogram's peak areas can be used to establish concentration ratios leading to quantitation estimates.

Chromatographic analysis allows the chemist to quickly establish the number and probable identity of the components in an organic mixture. Under general screening parameters, the unknown's solvent will elute from the chromatography column with the solvent the chemist uses to prepare his sample for analysis. If the analyte is a clean solvent, the resulting chromatogram will appear to be a blank. The identity of the analyte solvent can be determined by modifying the chromatograph's acquisition parameters to enable the separation of the low–boiling point solvents.

The forensic chemist determines the significance of the peaks in the chromatogram. The sample preparation technique used will impact the relative amounts of the compounds detected. All samples should be examined using an established sample preparation scheme so that analytical results from all the samples of a case can be compared. For example, a sample of an extraction solvent will contain a large amount of final product, possibly overloading the chromatographic column. If the sample is a waste solvent, there will be only trace amounts of product present. If the chromatographic peaks of samples that were prepared identically produce different peak areas for a peak that elutes at the same time as methamphetamine (e.g., sample 1 area is 500 counts, sample 2 area is 500,000 counts), the chemist could infer that sample 1 was a waste solvent and sample 2 was an extraction solvent containing product.

The symmetry of a chromatographic peak can provide information concerning the sample. In some instances, it can be used as a presumptive test to establish whether the compound in a solution is the free base or in a salt form. Generally, the peaks of free base and neutral compounds produce sharp, symmetrical GC peaks when using non-polar columns, commonly used in drug and ignitable liquid analysis. Sulfate salts and HCl salts of low-molecular weight compounds chromatograph poorly, producing asymmetric peaks that can tail badly. High–molecular weight salt compounds do not demonstrate this tendency.

If there is an indication of a controlled substance, its identity must be confirmed by a documentable technique. The type of information desired will determine the confirmatory route taken. If only the identity of the compound is desired, purification extraction prior to confirmation may be desirable. If the identity of all of the components of the mixture were desired, a simple dilution and analysis would be appropriate.

The chromatograms from similarly prepared samples can be used to establish a common origin or manufacturing technique. The pattern and ratio of product, precursor and by-product peaks can be used to determine common origin of samples or position in the reaction sequence. Samples from the same case can be compared in a similar manner to the way a chemist compares ignitable liquids and the residues extracted from fire debris.

5.7.4 Extractions

Extractions are used to separate the compound of interest from the rest of the sample. The type of extraction used will depend upon the compound of interest, the sample matrix and the information desired from the resulting analysis. In some cases, multiple extraction techniques are necessary to separate the substance of interest from the remainder of the sample. In other instances, instrumental analysis is the only way to separate compounds with similar chemical properties for confirmation.

In devising an extraction scheme, the chemist must decide what he wants to isolate and the form it will be in after it is isolated. The answers to these questions are in the statutes under which the chemist is working. Some statutes only require the presence of the controlled substance without regard to its salt form, isomer status or purity. Other statutes are very specific when it comes to identifying controlled substances by their salt form, structure, isomer form or purity.

The basic types of extraction include physical extractions, dry washing, dry extractions and liquid/liquid extractions. This section will describe the generic applications of the different types of extraction. Appendix L describes specific extraction procedures.

5.7.4.1 Physical Extraction

Physical extractions are the simplest. They involve physically removing the particles of interest from the balance of the sample. The isolated particle is then analyzed by the technique(s) the examiner deems appropriate.

This technique is appropriate when the examiner observes particles of different size, shade and consistency within the sample. The particles are physically or manually separated from the bulk sample by using stereomicroscopes, tweezers, sieves or other devices designed to physically isolate particles of different sizes.

5.7.4.2 Dry Wash/Extraction

Dry washes and dry extractions are different versions of the same process. The only difference is the substance that is removed from the sample matrix. A dry wash uses a solvent to dissolve and remove adulterants and diluents from the sample matrix, leaving the compound of interest. A dry extraction uses a solvent to dissolve and remove the compound of interest from the sample matrix.

5.7.4.3 Liquid/Liquid Extractions

The ability of a substance to dissolve in a liquid can change with the liquid environment. Liquid extractions utilize these solubility characteristics to separate a substance from a mixture. Appendix L lists the general solubility rules and procedures used for liquid/liquid extractions.

During a liquid/liquid extraction, the sample is initially mixed into an aqueous solution. The aqueous liquid is washed with an organic solvent in which the compound of interest is not soluble but the diluents and adulterants are. The organic liquid is separated, and the pH of the water is changed in such a way as to make the compound of interest insoluble in the water solution. An organic solvent is used to separate the substance from the water.

Care must be taken when selecting the acidic environment and the organic solvent used in liquid/liquid extractions. Some drugs are subject to ion pairing. This means that the hydrochloride salt of the drug is soluble in chlorinated solvents (e.g., chloroform) and will choose the chlorinated solvent over an acidic environment with a high chloride concentration (i.e., HCl).

Ion pairing can be used to the examiner's advantage when there are multiple basic drugs within a matrix that need to be isolated. If one of those drugs is subject to ion pairing, it can be isolated from the other drugs that under normal circumstances could not be separated.

The pH of the aqueous solution can also influence the recovery of the compound that is the target of the extraction. Amphoteric drugs and compounds like morphine exhibit characteristics of acid and basic drugs. The efficient extraction of these compounds requires the pH of the aqueous solution to be weakly basic (pH between 8 and 9 on a scale of 0–14, with 7 being neutral). Amphoteric compounds will preferentially remain in strongly basic aqueous layers, which are generally used for the extraction of basic drugs.

In some instances, the compound of interest cannot be isolated because the sample matrix contains multiple drugs of the same salt type. In these instances, a combination of techniques may be necessary to isolate the component of interest. An example of a combination extraction would be performing a TLC separation of the final extract of a liquid/liquid extraction. The silica gel around the spot corresponding to the compound of interest is physically removed from the TLC plate. A dry extraction or another liquid/liquid extraction is performed to isolate the substance from the silica gel.

5.7.5 Isomer Determination

Once the identity of a substance has been confirmed, the analysis is usually complete. Most statutes are written to include isomers, salts, and salts of isomers when defining a controlled substance. However, there are instances when the statute is very specific in defining the controlled substance. It specifically defines the structural configuration or the optical isomer of the compound that is controlled. In these instances, additional work may be necessary to satisfy the statutory definitions.

"Isomer" is a generic term that can encompass a number of different meanings. Isomers are compounds that have the same molecular formula but a different structural formula. The differences can be obvious, as in the case of structural isomers, or very subtle, as with stereoisomers.

Structural isomers are compounds that have the same molecular formula but a different structural formula. An example of structural isomers is ethyl ether and ethanol. Each compound has the molecular formula C_2H_6O. However, the structural formulas are CH_3OCH_3 and CH_3CH_2OH, respectively. Their different structures give them different chemical and physical properties that allow them to be differentiated through various instrumental techniques.

Geometric isomers are isomers that result from the positioning of two different functional groups attached to different ends of a double bond. The double bond prevents rotation, creating a cis (functional groups on the same side of the double bond) and a trans (functional groups on opposite sides of the double bond) configuration. Geometric isomers have similar chemical properties but different physical properties. GC and IR can be used to differentiate them through analysis.

Optical isomers are compounds that have the same structural formula. The only difference is the arrangement of functional groups around a chiral (asymmetric) carbon. This difference affects the rotation of plane polarized light. One configuration will rotate light to the right (d, dextrorotatory) and the other to the left (l, levorotatory). Otherwise, the chemical and physical properties of these isomers are identical. Microcrystalline tests and instrumental analysis of the derivatized compound are two methods available to forensic labs to differentiate optical isomers.

5.7.5.1 Microcrystal Examination

Microcrystal examinations to determine the orientation of optical isomers are a quick analytical method requiring only a microscope and the necessary reagent chemicals. The compound in question reacts with an inorganic reagent chemical to form a complex that is insoluble in the test solution. The resulting complex has a characteristic crystal shape that can be observed under the microscope. Racemic mixtures (mixtures that contain both optical isomers) produce different microcrystals than the single isomer compounds. The microcrystals of a single optical isomer generally cannot be distinguished from the microcrystals of the other optical isomer.

If the microcrystals of a single optical isomer are observed, and the chemist needs to know the exact optical orientation of the compound, he can perform a mixed crystal test. This involves placing an equal amount of the compound with a known optical orientation with the unknown sample (e.g., a small amount of known d-methamphetamine is combined with the same amount of single isomer methamphetamine unknown) The microcrystal test is performed on the known/unknown mixture. If the resulting crystals are single isomer crystals, the unknown has the same optical orientation as the known. If the resulting crystals are the crystals obtained from a racemic mixture, the unknown is of the opposite optical orientation.

5.7.5.2 Derivatization

The other method of determining the optical orientation of a compound is through derivatization. In this technique, the derivatizing reagent reacts with the compound at a reactive site on the molecule, usually at a nitrogen site or at a hydroxyl group. The addition of the derivatization agent to the molecular structure of the compound alters the chromatographic properties of the compound to such an extent that optical isomers and stereoisomers can now be chromatographically separated.

Not only does derivatization alter the chromatographic properties of the derivatized compound, but the resulting ion patterns of the mass spectra are differentiable between the derivatized isomers Not only are the mass spectra differentiable, but also, each is distinguishable chromatographically.

5.8 Quantitation

Once the identity of the controlled substance has been established, it may become necessary to determine the exact amount of that substance that is in the sample. This may be necessary for a number of reasons. The governing statutes may require the exact amount of controlled substance to be determined. The percentage of the sample that is a controlled substance may influence the chemist's opinion as to whether the substance is finished product, waste material or something in between. Or, the chemist just wants to know.

As a general rule, there is no statutory requirement to perform a quantitative examination on controlled substance samples. Quantitation is used as an investigative tool or is done as part of a laboratory's internal security policy. With a few exceptions, criminal statutes only regulate the possession of a given substance. The concentration does not affect guilt or innocence.

The concentration of a sample may become an issue during the sentencing phase of a trial. Some statutes provide enhanced penalties for possession of a substance over a given quantity. The words "possession of X grams of compound Y" are distinctly different from "X grams of substance containing compound Y." This wording may affect whether a quantitative exam is required to establish a sentence of 1 year or 10.

There are numerous quantitative techniques that the chemist can use to determine the concentration of a substance in a sample. Before the chemist can begin his quantitative analysis, he must first determine the type of information he is trying to obtain. Does he want to accurately know how much controlled substance is in a given sample? Or, does he want his analysis to reflect the amount of substance the operator could obtain from the sample? The answer to these questions will determine the type of quantitation method used.

The four basic methods of quantitating the amount of controlled substance in a sample are microscopic examination, gravimetric comparison, UV analysis and GC analysis. The following are generic descriptions of the various quantitation methods.

5.8.1 Microscopic Examination

The quickest and most subjective solid sample quantitative method is accomplished through microscopic examination. In this technique, a sample is placed on a microscope slide and diluted with a solvent in which the components are insoluble. The examiner estimates the percentage of crystals of the various substances in the sample under observation based upon the differences in shapes and optical properties. This is the most subjective, least precise and least accurate method. It is subject to the examiner's ability to recognize the microscopic crystalline form of the controlled substance under consideration. The uniformity of the bulk sample also affects the accuracy and reproducibility of the results.

5.8.2 Gravimetric Techniques

Gravimetric analysis provides a rapid means to determine the approximate amount of controlled substance in a sample. This technique can be used on both organic and aqueous samples. This technique also mimics the method operators use to extract the final product from reaction mixtures or extraction solvent. Therefore, it provides a practical approximation of how much of the final product the operator could expect to recover from the sample.

Gravimetric techniques can be performed in conjunction with the extraction phase of an analysis. The examiner weighs or measures the volume of the sample to be extracted prior to the extraction process. He obtains a weight of the extracted substance prior to any confirmatory tests being performed. The ratio of the post-extraction weight to the pre-extraction weight provides the percentage of the item that is the controlled substance.

A limiting factor to the precision and accuracy of this technique is the efficiency of the extraction solvents. If they do not effectively remove the diluents and adulterants, the calculated controlled substance percentage will be high. If the solvents do not efficiently and completely isolate the controlled substance, the percentage will be low.

An advantage of gravimetric techniques is that the identity and composition of the final extract can be confirmed. If all the diluents and adulterants have been removed from the matrix, the resulting residue can be analyzed for purity and identity.

The examiner must be aware of the salt form the controlled substance is in before and after the extraction process. This will affect the percentage calculated, because the molecular weights of the salt forms differ from the molecular weight of the free base. For example, a 100% pure sample of cocaine hydrochloride contains 89.38% by weight free base cocaine. The examiner must take into account the mass of the salt when calculating the percentage of controlled substance in the sample, or qualify the conclusion by stating the salt form of the substance identified.

5.8.3 UV Techniques

The use of UV light provides an effective method to quantitate a sample if it has a single UV absorber. If the sample has components with overlapping UV absorbances, the instrument cannot determine which compound is contributing to the absorbance. Compounds also absorb UV radiation at different rates. Therefore, UV is not conducive to quantitating mixtures.

Simple yes/no concentration comparisons can be accomplished with the use of UV. These comparisons are conducted in association with a product tampering case in which the product in question may have been diluted or altered. A comparison of the UV spectra of the item in question to a known reference sample can indicate whether the unknown has been diluted or altered. The composition of both samples should also be confirmed through a separate examination.

A detailed examination can determine the concentration of the substance in question. To accomplish this, the examiner obtains the UV spectra for a series of solutions with a known concentration of the substance in question. The absorbance values are placed on a concentration versus absorbance graph. A solution of the unknown is prepared and analyzed. The absorbance value is plotted on the graph to determine the concentration of the substance in the solution. This value is then used to calculate the percentage of substance in the unknown.

UV techniques, done properly, are precise. However, the accuracy of the results of multi-component mixtures may be in question because of the interference of the UV absorbance of other compounds in the sample.

5.8.4 GC Techniques

The use of GC for quantitation provides the greatest accuracy and precision of the analytical techniques discussed. This technique provides the examiner with the ability to isolate and quantitate a specific compound in a single method. The identity of the chromatographic peak can be confirmed at the time of the analysis or by analyzing the test solution with GC/MS. The same chromatographic conditions should be used during the confirmation test so that a direct correlation between the two techniques can be made.

Traditionally, GC quantitation uses a concentration versus peak area plot to establish the concentration of an unknown solution. If the peak areas of the serial dilutions of a substance are charted, the concentration of an unknown solution can be determined through from its instrumental responses. This method uses the relationship between the concentration of a sample and the instrumental data to calculate the concentration (i.e., doubling the sample concentration will double the GC peak area). A series of diluted samples are prepared and analyzed by GC. The resulting peak areas are plotted on an X/Y graph with their corresponding solution concentration. The unknown solution is analyzed in the same manner. Its concentration is obtained using the graph generated by the known solutions.

The increase in the precision and accuracy of modern instrumentation has allowed the analytical chemist to reduce the number of reference samples necessary for GC quantitation. The relative response GC method of determining sample concentration uses the ratio of the compound's and internal standard's peak areas, known sample concentrations and algebra. Two GC injections using this method can provide the same results as multiple injections using the serial dilution method. This procedure is based on the predictable relationship between sample concentration and the peak area of a chromatogram; that is, doubling the sample concentration will double the resulting peak area of the chromatogram.

The use of area concentration ratio to determine the concentration of a solution is dependent on the precision of the volumes injected into the gas chromatograph for analysis. Small deviations in injection volume will affect the accuracy and precision of the analysis.

To compensate for any deviations in injection volumes that may occur, a known concentration of internal standard is added to both the standard and unknown solutions prior to analysis. The concept of a given concentration producing a given peak area is just as true for the internal standard as for the samples it is added to. This being the case, the ratio of peak area of sample to peak area of internal standard will not change for a solution, no matter what volume is injected into the GC.

When quantitatively analyzing organic liquids, the chemist must dilute the sample until the unknown sample produces approximately the same compound to internal standard ratio that exists in the standard solution. With the known dilution factor, the chemist can then calculate the original concentration. If the sample in the previous example had a 20:1 dilution factor, the concentration of the original sample would have been 25.2 mg/ml.

By converting the concentration term into its basic units of weight (W) and volume (V), the concentration equation can be manipulated into an equation that describes the percentage of the unknown that contains the target compound.

5.9 Practical Application

5.9.1 Data Interpretation

5.9.1.1 Practical Example 1

The laboratory analysis of samples taken at the scene of a clandestine laboratory can generate a significant amount of information. The analytical methods traditionally utilized by the forensic chemist analyzing the samples can provide the answers to the questions the expert needs to provide an informed opinion concerning the operation under investigation. However, the analytical chemist needs to know what questions should be answered so that he can apply the appropriate analytical technique to the samples being analyzed. This section provides a few examples of how the information provided by the analytical chemist is used by the clandestine lab chemist to reach his conclusions.

The inorganic analysis of an off-white crystalline substance from a suspected methamphetamine lab reveals the presence of sodium, phosphorus, iodine and chlorine. The chemist must account for the presence of each of the elements. From the notes seized at the scene and the chemical inventory, it is the chemist's opinion that the operator was probably using the HI/red phosphorus method of reducing ephedrine to methamphetamine. Using this as a basis, the chemist reasons that:

- The iodine originated from the HI.
- The phosphorus was from the red phosphorus.
- The sodium hydroxide used to neutralize the HI contributed the sodium.
- Finally, the chlorine could be assumed to have originated from the hydrogen chloride salt of the ephedrine that was used as a precursor chemical. However, this assumption cannot be validated, as the methamphetamine analyzed was in the hydrogen chloride salt form.

5.9.1.2 *Practical Example 2*

A red sludge material that was recovered from a trash search was submitted for laboratory analysis to help develop the probable cause necessary for a search warrant for the location of a suspected clandestine lab. The chemist is asked what method is being used to manufacture the methamphetamine without being supplied with a list of chemicals or recipe from the operator. To provide the requisite opinions, the clandestine lab expert combines the information from the organic and inorganic chemical profiles to determine the controlled substance being manufactured and the method of manufacturing.

The organic analysis of the sample revealed the presence of methamphetamine, ephedrine and a trace of phenylacetone. The ratio of these components indicated that the operator was reducing ephedrine to methamphetamine using hydriodic acid. This was also consistent with the chemist's expectations from his visual inspection of the sample prior to analysis.

The inorganic analysis revealed the presence of potassium, iodine, phosphorus and chlorine in relatively equal amounts with a small amount of sulfur present. This inorganic profile presented a quandary. It was not consistent with the information provided by the organic analysis. The unusually high concentrations of chlorine and potassium, in combination with the presence of sulfur, were of concern to the clandestine lab chemist.

To rectify the discrepancy between the organic and inorganic results, the clandestine lab chemist had to think outside the box, utilize his knowledge of organic synthesis, and incorporate alternative methods that have been used by clandestine lab operators in the past. He referred to a clandestine method that substitutes a strong acid and a source of iodide, usually from an iodine salt, to reduce the ephedrine. Using this information, he proposed that the potassium and iodine were from a potassium iodide salt, the chlorine was from hydrochloric acid and the phosphorus was from the red phosphorus. The small amount of sulfur was attributed to the sulfate salt form of the ephedrine that was used as a precursor. When the operation was seized, the chemical inventory and the operator's notes confirmed the chemist's proposal.

5.9.1.3 Practical Example 3

All samples should be given the same value and treated the same analytically. It is easy for the analytical chemist to become complacent and dismiss a sample out of hand because it "looks" a certain way. There is some validity to the initial observations of a seasoned analytic chemist. However, his "gut feeling" should not be substituted for a documentable scientific analysis. The following example demonstrates how misleading gut feelings can be.

An analytic chemist received a sample of a clear liquid seized from a freezer located in a clandestine lab that was operating in a government laboratory. The sample was from a labeled container whose odor and general appearance was consistent with the label. The chemist was prepared to dismiss the sample as "contents consistent with the label" without the analytical data supporting his conclusion. However, much to his surprise, he detected a significant amount of methylenedioxyamphetamine (MDA) when he performed a routine gas chromatographic screen on the sample.

5.9.2 Extractions

5.9.2.1 Dry Extractions

There are a number of different extraction techniques available to the analytic chemist. Each has its place in the chemist's toolbox. The challenge is for the analytic chemist to choose the one that is most appropriate for the sample under examination. The following practical examples are applications of various extraction techniques used in the analysis of clandestine lab samples.

Cocaine HCl is generally found in a powder matrix mixed with one or more adulterants or diluents. Many of these compounds can be removed through one or more dry extractions. The following are two techniques that can be used to isolate cocaine HCl without altering its salt form.

5.9.2.1.1 Dry Extraction 1

- Physically isolate particles that appear to be pure cocaine HCl with tweezers.
- Analyze the particles using IR, which identifies the cocaine as well as establishing the salt form. The sample can be analyzed via GC/MS if the salt form of the cocaine is not an issue.

This particle licking technique is useful in the analysis of post blast debris. Particles of unconsumed explosives are commonly observed during the initial inspection of the samples using stereo microscope. Particles of suspected explosives are mechanically removed from the bulk material and analyzed using one of more instrumental techniques.

5.9.2.1.2 Dry Extraction 2

- Dissolve powder sample in chloroform (Cocaine HCl is soluble in chloroform, most common diluents and adulterants are not.)
- Separate the chloroform from the powder.
- Evaporate the chloroform.
- Analyze the residue via IR to confirm the presence of cocaine as well as establish the salt form. The sample can be analyzed using GC/MS if the salt form of the cocaine is not an issue.

5.9.2.1.3 Dry Extraction 3

- Dry wash powder sample with ethyl ether. (This removes assorted impurities, i.e., niacinamide and any free base cocaine that might be present.)
- Dry wash the sample with acetone. (This will remove common diluents, i.e., lidocaine HCl.)
- Analyze the residue via IR to confirm the presence of cocaine as well as to establish the salt form. The sample can be analyzed using GC/MS if the salt form of the cocaine is not an issue.

These extractions may or may not remove all of the auxiliary components from the sample, allowing it to be confirmed as cocaine. However, enough of them should be eliminated to establish whether a salt form exists and if so, which one it is.

5.9.2.2 Methamphetamine Extraction

Just as the salt form of cocaine may be significant, the salt form of methamphetamine may affect how the defendant is initially charged or ultimately sentenced after conviction. The salt form may additionally provide information as to the manufacturing method the operator was using. Again, the chemist must refer to the statutes he is working under to determine whether or not a salt determination is necessary.

The differentiation of pure methamphetamine base and a methamphetamine salt, usually HCl, is straightforward. The free base form of methamphetamine is an oily liquid. The HCl salt form is a crystalline solid. The free base is soluble in most organic solvents. The HCl salt is a solid with varying solubility in common solvents like methanol and chloroform but insoluble in ether, acetone, Freon and hexane.

The simplest way to determine the salt form of methamphetamine in a volatile organic solvent is to evaporate the solvent and examine the residue. If the residue is an oily liquid, chances are the methamphetamine is in the free base form. If a solid residue remains, some sort of salt form is indicated. In either case, IR analysis would confirm the methamphetamine and identify the salt form if the sample is not contaminated with reaction by-products.

5.9.2.3 Methamphetamine Extraction
Methamphetamine salts are soluble in acidic aqueous solutions, i.e., reaction mixtures. There are times when the salt form of the methamphetamine can be used to establish a manufacturing route. The following extraction can be used to remove methamphetamine HI from a reaction mixture without altering the salt form.

- Wash an acidic aqueous liquid, or red reaction sludge, with ether to remove many of the neutral organic by-products.
- Wash the sample with chloroform to remove the mineral acid salts of methamphetamine that may be soluble in chloroform.
- Isolate and evaporate the chloroform.
- Analyze the residue for methamphetamine and its associated salt form via IR.

5.9.2.4 Ephedrine/Pseudoephedrine Separation
Ephedrine and pseudoephedrine can both be reduced to methamphetamine. Their GC retention times and their mass spectra are essentially the same under the conditions generally used in forensic drug identification. Without derivatization, they are generally considered indistinguishable. Most other screening tests also cannot differentiate the two. Their solubility differences can be used to separate them (pseudoephedrine HCl is soluble in chloroform; ephedrine HCl is not). The following extraction can be used to separate pseudoephedrine HCl from ephedrine HCl.

- Dry wash of powdered mixtures of ephedrine HCl and pseudoephedrine HCl with chloroform.
- Analyze the dry washed solid via IR for ephedrine HCl.
- Isolate and evaporate the chloroform.
- Analyze the residue via IR for pseudoephedrine HCl.

5.9.2.5 Methamphetamine By-Product Profile Extraction
Powdered methamphetamine samples can be a wealth of information concerning not only the salt form of the methamphetamine but the presence of

reaction by-products and any adulterants or diluents that may be present. The following is a series of extractions that can be used on powdered methamphetamine samples.

- Dry wash the sample with acetone to remove niacinamide and reaction by-products.
- Isolate the acetone.
- Add hexane to the isolated acetone to precipitate out any existing niacinamide.
- Isolate and analyze the precipitate by IR.
- Analyze the acetone/hexane mixture for the reaction by-products by GC/MS or GC/FTIR.
- Dry wash the solid sample with chloroform to remove the methamphetamine salts.
- Isolate and evaporate chloroform.
- Dry wash residue with acetone.
- Analyze residue by IR to confirm methamphetamine and determine the salt form.
- Dry wash the sample with methanol.
- Isolate and evaporate the methanol.
- Analyze the residue for ephedrine.

5.9.2.6 Black Powder Substitute Components

Black powder and its associated substitutes are commonly used as the low-explosive filler in pipe bombs and other improvised explosive devices. Identifying the components of the mixture of a particle with the morphology of black powder, or similar product, allows the investigator to place a signature on the device as well as providing information that can be used in identifying the source of the low-explosive filler.

The components used in the manufacture of black powder and its related alternatives can be isolated and identified by using their differences in solubility. The following is a multi-step dry extraction process that can be used for this purpose.

- Grind test sample particles into a powder.
- Dry extraction with dichloromethane.
  - Analyze via GSMS for sulfur.
- Dry extraction with methanol.
- Transfer methanol to a clean watch glass (A) and evaporate to dryness.
- Dry extract residue with acetone.

- Transfer acetone to a clean watch glass (B) and evaporate to dryness.
- Analyze residue in watch glass A (acetone INSOLUBLE) via FTIR for sodium benzoate.
- Analyze residue in watch glass B (Acetone SOLUBLE) via FTIR for dicyandiamide or sodium-3-nitrobenzoate (sodium-3-nitrobenzoic acid).

5.9.3 Practical Applications: Quantitation

The determination of the actual amount of controlled substance in a sample is not generally required to establish the facts of the case. However, this knowledge does provide investigators and prosecutors with information that can be used to develop investigative leads or demonstrate what portion of the process the operation was in at the time of seizure. Quantitation information can also be used as part of a quality control mechanism within the laboratory itself. The quantitation method used depends on the information desired and the level of accuracy required. This section will briefly describe how different quantitation methods can be employed.

One application concerns determining the amount of pure controlled substance at the time of seizure. As a general rule, the analytic chemist only sees a representative sample of an exhibit. To determine the amount of controlled substance that was in the original container, he must determine the concentration of the sample presented to him. He then uses this value and the original volume to calculate the amount of controlled substance in the original container.

5.9.3.1 Gravimetric Quantitation

To obtain the concentration of the sample, the chemist begins with a 10 ml sample. He performs a series of extractions to isolate the previously identified methamphetamine as the hydrochloride salt. He then divides the weight of the resulting residue (0.10 g) by the volume of the sample to obtain the concentration. The concentration value of the sample can then be multiplied by the original volume of liquid of the item seized (1000 ml) to obtain the amount of methamphetamine in the original container. The following is the resulting calculation sequence:

$$
\begin{aligned}
\text{Concentration}_{\text{Sample}} &= \text{Weight}_{\text{Extracted Methamphetamine}} / \text{Volume}_{\text{Sample}} \\
&= 0.10 \text{ g} / 10 \text{ ml} \\
&= \mathbf{0.010 \text{ g/ml}} \\
\text{Weight}_{\text{Original Container}} &= \text{Original Volume} * \text{Concentration}_{\text{Sample}} \\
&= 1000 \text{ ml} * 0.010 \text{g/ml} \\
&= \mathbf{10.0 \text{ g}} \text{ in Original Container}
\end{aligned}
$$

5.9.3.2 Beer's Law Extrapolation

Plotting a graph of the sample concentration versus the instrumental response of a gas chromatograph produces a line, which can be used to calculate the concentration of an unknown sample. This calculation technique is demonstrated by using the following scenario (refer to Figure 5.12).

A series of three solutions with a known concentration of compound A are prepared. Their concentrations are 3 mg/ml, 2 mg/ml and 1 mg/ml, respectively. A 3 mg/ml solution of the questioned sample, which contained an unknown amount of compound A, was also prepared. The peak areas from the GC analysis were 3000, 2000, 1000 and 1275, respectively. The concentration of the unknown mixture can be extrapolated directly from the graph. In this case, the concentration is approximately 1.2 mg/ml. The accuracy and precision of this method are subject to the size of the graph paper and the eye of the chemist.

The concentration value of the sample is required to establish a calculated amount of compound in the bulk substance at the time of seizure. However, there are times when use of percentage is a more effective means of

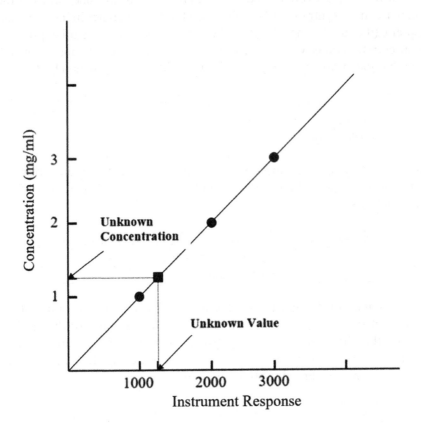

Figure 5.12 Beer's Law plot.

relaying a concentration value to a lay audience. This simple calculation can be performed as follows:

$$
\begin{aligned}
\text{Concentration}_{\text{Percentage}} \quad &= (\text{Concentration}_{\text{Substance}}/\text{Concentration}_{\text{Original sample}}) * 100 \\
&= (1.2 \text{ mg/ml}/3.000 \text{ mg/ml}) * 100 \\
&= \mathbf{40.0\%}
\end{aligned}
$$

5.9.3.3 Mathematic Calculation

The modern gas chromatographs produce precise analytical data that generates a linear response on a graph of concentration versus instrument response. The line tends to travel through the origin, making the use of mathematical calculations an option. Therefore, the mathematical equation of a line ($y = mx + b$) can be used to calculate the value of the concentration of the unknown. This removes the subjectivity introduced by the size of the graph paper and the thickness of the pencil that are used to chart the concentration in the previous example.

The first step is to establish the slope of the line (the m value), which is the difference in concentration (the rise) divided by the difference in instrumental response (the run). Using the extreme values of the standard solution provides a concentration range within which the unknown sample will most likely fall. Using the data from the previous application, the slope calculation becomes:

$$
\begin{aligned}
\text{Slope} \quad &= (\text{Concentration}_{\text{Max}} - \text{Concentration}_{\text{Min}})/(\text{GC Response}_{\text{Max}} - \text{GC Response}_{\text{Min}}) \\
&= (3 \text{ mg/ml} - 1 \text{ mg/ml})/(3000 - 1000) \\
&= \mathbf{0.001 \text{ mg/ml}}
\end{aligned}
$$

The concentration of the unknown can now be calculated by inserting the instrumental data of the unknown solution into the basic equation of the line as follows:

$$
\begin{aligned}
\text{Concentration (y)} \quad &= \text{line slope (m)} * \text{peak area (x)} + \text{Y intercept (b)} \\
&= 0.001 \text{ mg/ml} * 1275 + 0 \\
&= \mathbf{1.275 \text{ mg/ml}}
\end{aligned}
$$

The percentage of the mixture that contains the compound is a ratio of the amount of substance in the sample divided by the amount of the original sample. The ratio can be of the weights, as in gravimetric quantitations, or concentrations, as in this instance. In either case, using the data from the previous examples, the calculations are as follows:

$$
\begin{aligned}
\text{Concentration}_{\text{Percentage}} \quad &= (\text{Concentration}_{\text{Substance}}/\text{Concentration}_{\text{Original sample}}) * 100 \\
&= (1.275 \text{ mg/ml}/3.000 \text{ mg/ml}) * 100 \\
&= \mathbf{42.5\%}
\end{aligned}
$$

5.9.3.4 *Single Standard Calculation*

The precision and linear response of modern instrumentation can be used to further simplify the quantitation process. The quantitation process can be simplified into a single calculation using data from the unknown sample and a single reference standard by utilizing the premise that the ratio of the sample's concentration and its instrumental response is a constant and is relative to the concentration of the substance. With this in mind, the quantitation calculation can be written as:

$$\text{Concentration}_{unknown} = \left(\text{Area}_{unknown} * \text{Concentration}_{Standard} \right) / \text{Area}_{Standard}$$

When the calculation is performed utilizing this relationship and the data from the previous application, the same result is obtained.

$$\text{Concentration}_{unknown} = 1275 * 3\text{mg/ml}/3000 = 1.275\text{mg/ml}$$

Using an internal standard in the test solutions increases the accuracy of the results as well as introducing a quality control step into the analytical method. If the same internal standard solution is used to prepare the unknown sample and the reference standard, the concentration calculation can be written as:

$$\text{Concentration}_{unknown} = \left(\text{Area}_{unknown} * \text{Area}_{\text{Internal Standard of Standard}} * \text{Concentration}_{Standard} \right) /$$
$$\left(\text{Area}_{Standard} * \text{Area}_{\text{Internal Standard of Unknown}} \right)$$

The following concentration calculation demonstrates the effect of introducing an internal standard. The only additional information required is the peak areas for the internal standard in the reference standard and the unknown solutions, which were 1510 and 1525, respectively.

$\text{Concentration}_{unknown}$	$= (\text{Area}_{unknown} * \text{Area}_{\text{Internal Standard of Standard}} * \text{Concentration}_{Standard}) / (\text{Area}_{Standard} * \text{Area}_{\text{Internal Standard of Unknown}})$
	$= (1275 * 1510 * 3.000 \text{ mg/ml})/(3000 * 1525)$
	= 1.262 mg/ml
$\text{Concentration}_{Percentage}$	$= (1.262 \text{ mg/ml}/3.000 \text{ mg/ml}) * 100$
	= 42.0%

The less than 1% difference between the two methods is due to the use of the internal standard. The values obtained in the initial analysis were subject to variations due to sample volumes. The use of an internal standard compensates for variations in injection volumes and will provide a more accurate representation of the actual concentration of the solution.

As demonstrated in the following, each analytical technique produced approximately the same result. The level of accuracy required will drive the analytical method used to establish the concentration of a test sample that will be used to extrapolate the amount of the compound of interest that was in the container from which the sample was acquired. Modern instrumentation will produce satisfactory results regardless of the method chosen, as long as good laboratory practices are utilized.

- Single Standard Calculation = **42.0%**
- Mathematical Calculation = **42.5%**
- Beers Law Extrapolation = **40%**

5.10 Summary

The comprehensive analysis of clandestine lab samples is an essential portion of the forensic investigation. The information derived from the testing can only be obtained through the scientific examination of the evidence. It is used to meet the burden of proof necessary to establish the presence of a controlled substance beyond a reasonable doubt. The cumulative effect of the examinations that can only establish a preponderance of evidence can be used to formulate expert opinions concerning the operation.

The tools the clandestine lab chemist uses to analyze these samples are the same tools that the forensic chemist uses to analyze samples during their normal course of business; the difference being that the clandestine lab chemist may apply certain techniques in a method that will produce a more detailed profile of the sample under examination. The scope of the analysis goes beyond the forensic chemist's desire to determine and identify the presence or absence of specific compounds. The clandestine lab chemist needs to know the sample's composition so that he can develop opinions concerning the operation.

Opinions

6

Abstract

This chapter will deal with clandestine lab opinions that can be generated by the expert studying the forensic evidence that is collected. This chapter will further address the evaluation process and hopefully provide some insight into the opinion-making process. Sources of information that are needed to formulate an opinion will be listed. Questions the forensic expert needs to derive from that information will be addressed. These questions most significantly include: *What* is the operator making? *How* is the operator making it? And, *How* much (quantity) could the operation produce?

This chapter will discuss the variety of sources of information used to form a strong objective opinion. The calculations used to make production estimates, and the limitations associated with those calculations, will also be discussed. The topic of contextual bias and the need for qualitative uncertainty statements in an effort to present opinions that are neutral and based upon the known facts will also be presented.

6.1 Introduction

A clandestine lab is a Pandora's Box of illegal and anti-social activities. Drugs and explosives are produced using household, industrial and agricultural chemicals mixed using ordinary utensils in what some have called a "kitchen of death." What appears at first glance to be simply atrocious housekeeping or even just a hobby gone awry may actually be the final step in the production of many of the drugs sold on the street or the explosives used in various forms of terrorism.

Everyone has their own image of what a clandestine lab would look like. The man on the street, from which a typical pool of jurors is drawn, will more than likely report images of smoking and boiling chemical reactions using scientific equipment in a hidden laboratory such as Frankenstein's birthplace. As has been demonstrated in the previous chapters, this is far from the case.

Experts are required to bring order out of the chaos. Initially, they are called to bring a basic understanding of what is initially observed at the scene. Experts must be able to draw a mental picture for others. The picture they draw must be as vibrant to the responders processing the scene as it is

DOI: 10.4324/9781003111771-6

striking to the jury in a criminal action or the intelligence analyst fighting the war on terror.

All experts are, first, investigators, but not all investigators are able to become experts. Investigators collect the pieces of the puzzle and organize them to some degree. Experts have the knowledge, skills and abilities to arrange the pieces of the puzzle into a picture that provides clarity to the impartial observer.

A forensic expert must first assemble enough pieces of physical evidence just to demonstrate that a clandestine lab even exists. He is able to combine the known facts to present a scenario of the "What? Where? Why? and How?" of the operation. His knowledge base is broad enough to acknowledge that other explanations could exist for the combination of chemicals and equipment found at a location. He considers the totality of circumstances of the case and concludes with a schematic of the most likely arena in which the manufacture of a controlled substance took place.

Forensic evidence and the opinions it generates are used to supplement or answer the basic *Who? What? Where? When? Why?* and *How?* questions involved in any criminal investigation. They are able to address objective questions. However, there are limits to the ability of evidence and opinions to answer subjective questions with the degree of specificity required by the criminal justice system or the intelligence community. The expert's job is to use the forensic evidence to compile, evaluate and render an opinion concerning the facts of each case as presented. The expert may be involved in the case from the beginning, or he may be brought in at the end to evaluate other expert opinions. Either way, the process is basically the same.

Chapters 4 and 5 present the theme of the forensic chemist who is trained in clandestine lab issues as the ultimate expert. He does have a significant knowledge base to draw from in presenting expert testimony and generating opinions. However, he is not the only source of expert opinions. Criminal investigators, bomb technicians and hazardous materials specialists who regularly deal with clandestine lab issues can all generate qualified opinions. While such individuals may not know the specific theories concerning the chemistry involved, their experience and training provide the knowledge base that a given set of chemicals and equipment can be used to produce a controlled substance.

This chapter will deal with clandestine lab opinions that can be generated by the expert studying the forensic evidence that is collected. This chapter will further address the evaluation process and hopefully provide some insight into the opinion-making process. Sources of information that are needed to formulate an opinion will be listed. Questions the forensic expert needs to derive from that information will be addressed. These questions most significantly include: *What is the operator making? How is the operator making it?* And, *How much (quantity) could the operation produce?*

6.2 The Questions

Clandestine lab experts are forensic investigators who come from a variety of functions within the criminal justice, military and intelligence communities. They may be peace officers, soldiers, sailors, civilians or forensic chemists. They may or may not carry a gun and have arrest authority. Whatever their function, their mission is basically the same: to collect and objectively evaluate the information concerning a clandestine lab operation.

The expert must deal with the same *Who? What? Where? When? Why?* and *How?* questions that the investigators do. Experts simply approach the question from a different perspective. Combining the information concerning the physical evidence and scientific principles to objectively evaluate the operation is their forte. They use this approach to the case to provide their opinions. Some questions can be answered at the scene. Other questions can only be addressed and determinations made after the laboratory examination is complete. It is the combination of answers to all of the questions that the expert uses to form his ultimate opinion concerning the operation as a whole.

In a criminal prosecution, the opinion process begins with the affidavit for a search warrant. Military operations have different names for the same process of compiling information to justify site exploitation. In both situations, the lead investigator provides the expert with information concerning the site in question. The expert renders an opinion as to whether there is sufficient information to establish the existence of a clandestine lab. The expert's opinion is a key factor in this part of the legal process and is used to support the investigator's conclusion of the facts. If flawed, in a criminal investigation, these opinions could later be raised in court by defense counsel as "motions to suppress evidence" due to lack of probable cause. Flawed opinions in a military action can lead to much graver consequences. It is therefore necessary to exhibit extreme care when determining that certain conditions do exist and that they point irrevocably in the direction of a clandestine lab operation.

Experts should document conversations with investigators concerning clandestine lab opinions. What on the surface may appear to be a casual conversation may be a fishing expedition for an opinion concerning a situation where not all the facts may be presented. As a result, the expert's name may be placed into an affidavit without his knowledge. The statements attributed to him may be totally wrong, misinterpreted or even never uttered in the first place. His own supporting documentation of what was said during the preliminary conversations may become necessary if the expert is asked about statements he made in the search warrant affidavit. This obviously could crucially affect his future credibility.

The search of the suspected clandestine lab site is the pinnacle of the investigation. Emotions run high, people may be stressed and tired; answers are demanded. Before an expert gets out of his car, investigators want to know: *What are they making in there? How are they making it? How much product is there? How much could they make in a ...?* The unfortunate thing is that many novice clandestine lab investigators really expect answers to these questions immediately.

In these instances, the expert does not have sufficient direct knowledge of the situation to render an opinion. At best, he should only make qualified generic statements concerning what could be made and possibly how it was being made using the limited information he has. The chaos of a clandestine lab scene does not provide an atmosphere to render any type of objective opinion. Opinions concerning the specific details of the operation should be rendered only after the physical evidence can be evaluated in an objective manner, which is after the fact only.

Many types of opinions can only be generated from the laboratory analysis of evidentiary samples. Some are a result of generalities that do not require the support of analytical data. For example, just because a red powder is found at the scene of a suspected ephedrine reduction lab, this does not make the powder the critical red phosphorus. A fine silver gray powder in a suspected explosives manufacturing operation does not mean it is aluminum, the critical fuel of a fuel/oxidizer explosive. These may be valid assumptions that can be used to establish the requisite safety protocols for scene processing as well as guiding the collection and preservation of physical evidence. However, the ultimate opinions associated with the operations under investigation must be generated after all the facts have been evaluated and must have the analytical data to support them.

The forensic expert must remember the laboratory analysis, and his opinions should be able to withstand peer review. A component chemist or other forensic expert should be able to review the facts of the case or the laboratory data and draw the same conclusion that the original expert did. Alternative opinions can and do exist, as is evidenced by prosecution and defense differences. But the information must support the opinion, or the opinion is worthless.

6.2.1 Who?

One of the initial questions in any investigation is: *Who was involved?* This simple investigative question can be answered to one degree or another by looking at the people detained or living at the scene. Placing an individual at the scene is one thing, but connecting them to the lab operation may take a forensic expert. Latent prints or DNA can be used to establish who had access to the lab equipment and chemicals. In some instances, laboratory analysis of a suspect's clothing can be done to detect drug, explosive and chemical residues, which can be used to connect him to a lab operation. The analysis of the handwriting on paperwork associated with the operation can also be used to establish a link between the operation and people who were nowhere near it at the time of seizure.

6.2.2 What?

The most common "what" question is *What are they making?* Asking the operator is the easiest way to obtain this information. However, operators have been known to be less than truthful or uncooperative, or they may simply not know what the final product is. That is where the forensic expert comes into play. He combines the information from the paperwork, available chemicals and equipment as well as the laboratory analysis of items associated with the operation to provide an objective opinion concerning what was being produced.

Some lab operations only perform a portion of the process. This is done to avoid compromising the entire operation. In these instances, the *What step in the process are they at?* question may play a significant role in the investigation. The expert utilizes the same information to determine what they are making and then rolls that information into an opinion concerning what stage of the process the operation is in.

6.2.3 When?

When were they cooking? This is not a question that is conducive to traditional forensic techniques. Unless they are caught in the act, the forensic expert cannot provide much insight. Traditional investigative techniques are the best method of answering this question.

6.2.4 Where?

Where was the lab? The location of all or any given segment of the operation can be determined through forensic investigation. The position on the scene of the lab, the chemicals or where the equipment was located can easily be documented. Forensic investigation and analysis can demonstrate where each portion of the process was conducted as well as where the suspect was putting his waste products. Even if the lab has been dismantled and removed from a location, analysis of the residues left on floors, or even stains left on the walls and counters, can be used to determine where the lab was located and what was being produced.

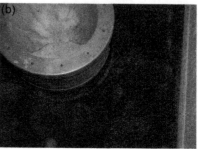

Figure 6.1 Heating mantle location.

Figure 6.1 is an example of how the stains observed in a closet were used to identify the location of a methamphetamine lab. A heating mantle was located in an outbuilding on the property. The heating mantle's circumference was a match to the stain on the carpet. Laboratory analysis of the stained carpet revealed high concentrations of iodine as well as detectable amounts of methamphetamine and pseudoephedrine. The combination of the physical evidence and the subsequent laboratory analysis led to the opinion that a portion of the methamphetamine manufacturing operation was located in the closet as some point in time.

Where was he getting his chemicals and equipment? or *Where is he storing excess chemicals and equipment?* can be answered by reviewing documents seized from the scene or other locations associated with the operation. Operators are generally packrats and save everything. Where you will find evidence concerning the operation is only limited to the imagination of the operator and the investigator searching the scene.

Establishing where the clandestine lab operator is acquiring the chemicals used in the manufacturing process can be an important piece of information. Stemming the flow of precursor chemicals at the source is a primary means of disrupting the manufacturing operation, whether it be contraband drugs or explosives. Simply, label information of chemical containers located at the scene can provide insight as to the source of the chemicals used. Detailed forensic chemical analysis may be able to identify unique characteristics within the samples that allow the investigators to identify the origin of the precursor chemicals used, in some instances down to the country or region of origin.

6.2.5 Why?

On the surface, *Why?* is a simple question to answer. Money, drugs, political ideology or some combination of the three is the answer to the basic *Why are they manufacturing?* question. However, the forensic expert needs to dig a little deeper and ask a few more questions in this same vein: *Why did the operator use this method of manufacture? Why does he use a particular chemical supplier? Why did he use the equipment he used?* The answer to most of these *Why?* questions is subject to conjecture. However, experts by their very nature have the latitude to speculate. They are expected to use their training and experience to evaluate the facts of the case and present an educated theory concerning the "why's" of the operation.

As with the "What is he making?" question, the operator can provide answers to many of the "Why?" questions. Detailed interviews of the operators can provide a wealth of information or can be an exercise in futility. The operator may be cooperative and willing to provide answers to all questions, or he may be obstinate and unwilling to cooperate in any manner. In many instances, the operator does not know "why" because he is operating from written directions obtained from a known or unknown source. "The Internet said to ..." may be a truthful, but frustrating, response to inquiries as to "why ...."

6.2.6 How?

The answer to many of the "How?" questions requires input from a technical expert. All of these questions have the potential to be asked in court and should have scientific corroboration. The lead investigator may want on-the-spot answers to many of the same questions. Unfortunately, many of them can only be completely answered after laboratory analysis and data interpretation. The "how" questions include: *How were they making the product? How much product was there? How much could they make per batch or over a given period of time? How much could they make with the seized chemicals and equipment?*

6.3 Information

Information is the key to answering any question. Answering questions without information is like putting a puzzle together without enough pieces. With a puzzle, the more pieces that exist, the closer to the complete picture is created. The more information an expert knows about a clandestine operation, the closer he can come to painting a complete picture of the operation.

The answer to investigative questions and their associated opinions can change over time. The original answer or opinion was not wrong or intentionally misleading. They are the best answers/opinions based upon the information that was available at the time. Answers and opinions can change as more objective information is developed through investigation and laboratory analysis.

The following is an example of how a premature opinion can lead to a problem in court. An inexperienced clandestine lab chemist processing the scene of a clandestine drug lab rendered an opinion that led the investigators to charge the suspect with the manufacture of methamphetamine, a dangerous drug under the local statutes. Laboratory analysis of samples from the scene coupled with scrutiny of the list of chemicals that were seized led the senior clandestine lab chemist to the opinion that the operation was manufacturing diethyltryptamine, not methamphetamine. This dichotomy of opinions lead to a problem during the arraignment, as the prosecutor had charged the suspect specifically with the manufacture of methamphetamine based upon the on-scene chemist's statements at the scene and input from the principal investigator. The charge was dropped, and the suspect was immediately arrested and charged with the manufacture of a dangerous drug, that is, diethyltryptamine, a hallucinogen.

Based upon the totality of circumstance, both opinions were accurate within a relative degree of certainty. In the first instance, inexperience and situational bias contributed to the opinion of the manufacture of a specific drug as opposed to the manufacture of a class of drugs that encompassed both methamphetamine and diethyltryptamine. The chemist in the second instance had laboratory results and more time to objectively evaluate the data to render an opinion concerning the specific drug that was being produced and the method of synthesis that was being used. Two different valid opinions of the same situation, based upon two different fact patterns.

The clandestine lab expert has a needs triangle similar to the clandestine lab operator to be able to render an opinion (Figure 6.2). The clandestine lab operator needs chemicals, equipment and knowledge to make the operation work. The clandestine lab expert needs information from the scene, from the laboratory analysis and from the knowledge gained from his training and experience. The scene information provides knowledge concerning the operation in general. Laboratory analysis provides information concerning the specifics of any given sample. The expert's experience and training allow him to piece together all the information to complete the picture of the operation.

Figure 6.2 Opinion needs triangle.

6.3.1 Scene Information

The information gathered at the scene of a clandestine lab has a number of functions. First, it corroborates the prior information used to establish probable cause in obtaining the search warrant. Secondly, it guides the direction of the on-scene investigation. Thirdly, it helps the analytical chemist devise the analytical schemes he will use during the testing of the samples that are sent to the laboratory. Finally, it will be used as a basis for the expert's opinion concerning the workings of the operation.

The information gained during the initial scene walk-through sets the tone for the balance of the scene processing. The expert's initial impressions of what the operator was making and how he was making it determine what type of search is conducted. Drug labs may take one approach. A cautious and different path may be necessary with suspected explosive labs. If no lab is initially apparent, a different search tactic is taken.

The common thread in all searches is that the observations and the physical evidence guide the on-scene investigation. The desire to put someone in jail or detain a suspected terrorist should not be so great that the expert misinterprets or misrepresents the presence of common items to justify the presence of the police or military. Search warrants and site exploitation events for the wrong location have been performed based upon poor information. This situation should never deteriorate to the point where the forensic expert is looking for whatever evidence it takes to allow the law enforcement agency or military actors to save face. On the other hand, even if the operation is not readily apparent, the forensic expert should be as creative in his search techniques as the operators are in hiding and disguising the labs.

The analytical chemist uses the information provided by the scene chemist or expert to devise his analytical schemes. He reviews this information to determine what samples require examination and what type of testing is appropriate. The analytical chemist also uses the scene information to estimate the amount of final product the operation could produce. This information can also explain reaction by-products that are not normally encountered in the reaction mixtures from that type of lab.

The opinions rendered from scene information are forensic evidence and must stand up to peer review. Documentation of the observations made at scene is a critical component. Photographs, sketches and inventory lists all can be used to support the expert's opinions rendered in the report. An independent expert should be able to review the scene documentation and come to the same conclusion as the person generating the report. This is not to say that the independent expert may not have a differing opinion, but only that the documentation supports the report's conclusion when looked at by outside review.

The scene chemist must take care not to overstate his opinions concerning the scene. It is easy to get caught up in the frenzy of the moment and provide an opinion that is not completely supported by the facts. Producing a written report after he has had time to objectively evaluate all of the information concerning the operation is the wisest method of disseminating the opinions concerning the operation. The written report should contain information concerning his role in the scene processing as well as his opinion concerning his observations thereof. The criminal investigators, analytical chemists, and both prosecuting and defense attorneys use this information to guide their investigations or prepare their case for trial. The report should state more than a final conclusion of the expert opinion. There should be some narrative explanation of how the expert reached the conclusion.

In some instances, the expert's report will be the only expert evidence presented, and the trier of the fact should be presented with some explanation of how and why the expert came to his conclusion. A scene report containing a simple summary statement such as "The items found at 123 Oak Street were consistent with those found at a clandestine lab that manufactures a controlled substance" does not provide other parties involved in the investigation or prosecution with sufficient information to continue with their portion of the investigation.

The previous statement may be a valid conclusion. However, it is too generic and does not provide the reader with a sense of what was being manufactured or information to support the statement. A better statement that summarizes the observations made by the scene expert would be: "The items found at 123 Oak Street were consistent with those found at a clandestine lab that was manufacturing tetra-ethyl-death using the shake and bake method." This wording allows the scene expert to say that in his experience, the items that were observed at the scene were the same as those found in operations that produced a certain controlled substance using a particular manufacturing method. It also provides the option for alternative manufacturing theories.

A large quantity of information can be derived from the scene. This information can be used to render limited or generic opinions concerning the operation and its capabilities. It will also be evaluated at a later time to establish the particulars of the operation under investigation. Table 6.1 presents a relationship between the information that can be obtained at the scene and the opinions that can thus be generated.

6.3.2 Laboratory Analysis Information

The information from the scene provides the pieces of the clandestine lab puzzle and a generic outline of how they fit together. The laboratory analysis

Table 6.1 Information–Opinion Relationship

Information	Conclusions
Chemical inventory	• Used to establish the manufacturing method. • Used to establish the overall production capability of the operation.
Equipment inventory	• Used to establish the manufacturing method. • Used to establish the per batch capability.
Location of the items at the scene	• Used to establish the location of each portion of the process at the scene.
Original volumes or weights	• Used to establish the actual amount of final product or precursor chemicals seized. • Used to establish the amount of product that could potentially be produced with the available precursors.
Seized paperwork	• Used to provide a historical perspective of the operation. Some operators are detail oriented enough to keep records concerning the percentage yield of each batch. • Used to establish information concerning the source of precursor chemicals.
The chemical receipts	• Used to provide insight into the amount of chemicals that have been purchased over a period of time. • Used to provide insight as to what compounds were or could have been produced. • Used to provide insight concerning the knowledge level of the operator and the source of that knowledge.

provides detail to each piece. It fills in the holes and provides answers to any questions that are generated during the scene investigation. Laboratory analysis supports or refutes the opinions that are generated at the scene. In some instances, laboratory analysis generates additional questions, requiring supplemental information from the investigative sources before a complete opinion can be rendered.

Laboratory analysis can supply information concerning the identification of the controlled substances being produced as well as the precursor and reagent chemicals located at the lab site. It aids in the identification of by-products in the reaction mixtures and the waste products that may be used to establish a particular manufacturing method. Lab analysis produces the quantitation data that can be used to calculate the amount of controlled substance present in reaction mixtures and estimate its production values.

6.3.3 Experience and Training

All clandestine lab chemists are forensic chemists. However, not all forensic chemists are clandestine lab chemists. A clandestine lab chemist has the training and experience to separate the wheat from the chaff and render an opinion concerning the existence of a clandestine lab given a certain set of facts. Volumes of information can be generated from a clandestine lab case. The expert trained in clandestine lab matters is able to wade through the quagmire to determine which pieces of the puzzle are relevant and which pieces are superfluous.

To be an effective clandestine lab expert, a chemist must have a solid background in basic organic, inorganic and analytical chemistry. He also should be well schooled in the techniques underground chemists utilize to manufacture a wide range of controlled substances. Finally, he should have access to a variety of analytical databases so that he can cross-reference his analytical information to reduce it to its most logical scenario given all of the known facts. This last is a skill that is only learned over time. In some instances, it is an art based on science and experience.

There is presently no complete course of instruction formally developed to train a clandestine lab chemist in his trade. Clandestine lab chemists are like the forensic drug and explosive chemists of the past. Under ideal circumstances, they serve an apprenticeship under an experienced forensic chemist who mentors them in the application of the analytical techniques of chemistry and forensic science for the examination of clandestine lab evidence. Under less than ideal circumstances, the forensic chemist is given samples from a clandestine lab case and expected to do the analysis, answer questions and provide the opinions because he is the most (or in some cases, the only) chemist available. He is the resident expert by virtue of his education.

There are presently instructional programs that address segments of the forensic clandestine lab investigative process. The Drug Enforcement Administration (DEA) and the California Criminalistics Institute have programs designed specifically for drug analysis or generic clandestine lab response. Courses have been available to law enforcement and military personnel involved in the identification and seizure of homemade explosive laboratories. Some agencies have integrated bodies of instruction concerning the clandestine manufacture of drugs and explosives into the programs of instruction used to train new forensic chemists. These programs only provide the tools that loosely address the issue of the analysis and interpretation of evidence from clandestine labs. Even with this training. the chemist must rely on real-world experience to gain the knowledge necessary to make the tools gained in his basic training effective in his work.

6.4 What? How? How Much?

All of the "Who? What? When? Where? Why? How?" questions an expert needs to address can be boiled down into three basic questions. The answer to all of the other questions will fall into place if the expert at the scene investigation or the analytical chemist performing laboratory analysis can focus on addressing these questions. These three main questions are: *What is he making? How is he making it? How much can he make?*

Answering these questions is not the "be all and end all" of the forensic investigation. The answers generally lead to additional questions that need to be addressed. However, they point the forensic investigator in the direction the investigation should go.

6.4.1 What Is He Making?

What is he making? On the surface, the answer to this question is relatively easy to ascertain. The laboratory analysis of the final product quickly and definitively establishes what the operator was making. Laboratory analysis confirming the presence of a drug or an explosive makes a strong argument in favor of one type of manufacturing operation or another.

Establishing the identity of the final product in operational labs that do not contain an isolated finished product is just as simple. The presence of reaction mixtures or waste material usually will not deter the confirmation of a final product of the operation. Reaction mixtures and waste materials can each contain detectable levels of the finished product. The challenge for the chemist is to detect, isolate and identify the final product in a sample.

The analysis of waste materials is where the experience and training of the chemist begin to show. There may be only trace levels of a controlled substance. The analytical chemist should be able to recognize the potential final product from the reaction by-products within the mixture. If the proper compound or combination of compounds is present in a mixture, he should be able to devise an analytical scheme that can isolate and confirm the presence of the controlled substance with which they are associated.

Non-operational labs present a challenge. There are no reaction mixtures or waste products for the analytical chemist to examine. There is no instrumental data to hang his opinion on and say definitively what the final product of the operation was. These situations become a mental exercise for the expert. He uses the chemical and equipment information from the scene to identify the most probable product and method used. He evaluates the lists of chemicals that were found at the scene and categorizes them as precursors, reagents or solvents. He identifies the final products associated with each precursor and the synthesis routes used with each. He then does the same with each reagent chemical. The precursor possibilities are combined with the reagent possibilities, and a hypothesis of the type of operation is derived. To support the hypothesis, the type of equipment is factored into the equation along with any notes, receipts or additional paperwork to complete the puzzle. Section 6.5 contains a practical application that demonstrates how the chemical inventory can be used to piece together a manufacturing method to answer the question: *What is he making?*

Many chemicals can be used to produce more than one controlled substance. Some can be used for more than one synthesis route for the same controlled substance. With hundreds of different synthesis routes for the various controlled substances that are encountered in a clandestine lab, it is unlikely that the scene or analytical chemist will immediately recognize unfamiliar chemical combinations that are associated with obscure chemical reactions. He is less likely to put the two together in the chaos of a clandestine lab scene.

Matrixes containing the chemicals commonly encountered in clandestine labs can narrow the possibilities. If the chemist has access to this type of information, he can cross-reference the chemicals seized with the controlled substance they are associated with as well as the synthesis route. This information will assist him in objectively looking for a pattern that will indicate what controlled substance the operator was trying to produce.

"What else is he making?" is a corollary question that should be addressed. As a result of this mental exercise, the chemist should be able to determine what other controlled substances could be made with the combination of chemicals seized from the scene. It is not uncommon for the operator to be experimenting with other manufacturing methods. The expert should not focus on the most obvious final product. He should expand his evaluation to include or exclude all of the possibilities and any *"What if?"* questions that could be brought up during the peer review process. Many hypotheses fall by the wayside before a theory can emerge.

The expert must remember his limitations. He is an expert in the clandestine manufacture of controlled substances. Even if he has a PhD in chemistry with an emphasis in organic synthesis, he would be well advised not to speculate about any or all of the compounds that can potentially be produced using a given list of chemicals. The *"What if?"* questions in this situation can be endless and beyond the expertise of the chemist as well as the jurors. Just as the forensic drug chemist is an expert in the identification of cannabis but not in plant identification, the clandestine lab chemist is an expert in the clandestine manufacture of controlled substances, not in organic synthesis. He proposes what is the most likely product(s) that will result from the combination of chemicals and equipment that were seized given the totality of circumstances, not every possible combination that will result when factoring in irrelevant information.

6.4.2 How Is He Making It?

"How is he making it?" is the natural follow-up question. Determining what is being made in a clandestine lab is a relatively straightforward process. In

many cases, the final product is still unidentified. In other instances, the combination of chemicals is like a road map showing the way. The clandestine lab chemist must take the next step in the opinion process and determine either how the product was made in an after-the-fact analysis or how it would be made in a before-the-fact hypothesis.

The simplest method of answering this "How?" question is to ask the operator or look at his notes. Some operators can be very talkative. Others are quiet. In all cases, the operator's statements should be put into perspective, since their culture is one of deceit. However, their statements can be used as a guide for the scene investigation and to corroborate opinions generated from the physical evidence.

Clandestine lab operators do not generally have the education or training to cook without a recipe. The paperwork detailing the manufacturing method that is being utilized is often located somewhere at the scene. This paperwork, combined with the seized chemicals and equipment, demonstrates how the operator was manufacturing his product. However, some operators have committed the process they use to memory because it is so simple and they have been using it so long. Unfortunately, in these instances, there will be no recipe defining the manufacturing method, which will of necessity make other forms of corroboration more important.

The list of chemicals seized from the scene can give insight into the most likely manufacturing method used. Once the field of possible final products has been narrowed to the most likely candidates, the chemist can compare the list of chemicals required for individual synthesis routes with the chemicals seized from the scene. The synthesis route with the most complete list is the most likely manufacturing method being used. The operator may not have all of the components for the suspected route. However, the lack of a complete list of chemicals does not eliminate a synthesis route from consideration. Section 6.5 provides an example of this technique.

As with any opinion, an answer to the *"How is he making it?"* question should only be given after objective review of the physical evidence. However, there are manufacturing methods that are so commonly used, and the chemical combinations are encountered so frequently, that the on-scene chemist can usually provide a qualified opinion concerning the type of lab and the probable synthesis route. Much beyond that, he is probably treading in water he should not be in without time to reflect on the totality of the physical evidence.

Qualified opinions are necessary to guide the on-scene investigation. They are used to direct the search for items of physical evidence that will corroborate or supplement the evidence that has been located to that point of time. On-scene qualified opinions should never be used as defining statements associated with the expert opinion used in a criminal proceeding or used in justifying military action resulting in extreme consequences. The expert would simply be opening up his testimony for a cross-examination from Hell or professional sanctions if he does not rely on the scientific method to support his official opinion concerning the details of the operation in question.

Reaction mixtures and waste material can provide a wealth of information. Many of these liquids contain all of the information concerning the method used to produce the final product. The precursor and reagent chemical components of a reaction mixture, the reaction by-products in the final product or waste material, all can give information as to the method the operator was using to manufacture.

The mass spectrometer provides the clandestine lab chemist with a tool to identify all of the components within a reaction mixture. Appendix M is a compilation of mass spectral data of reaction by-products that are potentially found in reaction mixtures found in clandestine drug labs. These values are taken from the scientific literature and include the synthesis route(s) associated with each compound. The table contains the compound's five major ions and the synthesis routes with which each compound is associated. The chemist must remember that the ion sequence may differ on the instrument he is using. If possible, he should run the actual compound to obtain the actual mass spectrum for identification purposes.

The lack of primary standards for the reaction by-products complicates the identification process. The analytic chemist must rely on the analysis of reaction mixtures he has synthesized himself to obtain the mass spectral data of reaction by-products in various manufacturing methods. He should compare these spectra with the spectra in the literature to confirm the identity of the by-products encountered. Instrumental data from non-primary standards can be used for these identifications, since the identity of these components does not have to be established beyond a reasonable doubt.

The analytic chemist should attempt to reproduce the manufacturing process used by the operator to demonstrate that it actually works. The recipe being used may or may not produce the intended product. Sometimes, the reagents called for will not produce the desired effect. Other times, the operator does not have access to the chemicals listed in the recipe, and his lack of knowledge does not allow him to use the proper substitute. The analytic chemist should go through the steps outlined in the operator's recipe to determine whether or not it would function as designed. Understanding the theory of the reaction is one thing. Having direct knowledge as to whether or not the reaction will produce a controlled substance has a greater impact in an opinion. The best way to respond to the question "How do you know the operator's reaction will produce flubber?" is to respond: "I followed the directions found at the location, using samples of chemicals seized from the location, and the result was flubber."

There will be times when the chemical inventory and the laboratory analysis do not provide sufficient information to determine the synthesis route. These instances require the follow-up question: *Why is he using this method?* This question may be a mental exercise that does not have an answer. When all else fails, ask the operator or look at his notes. His level of cooperation may provide the expert with the insight he requires. There will be many instances when the expert has to accept the answer: "I do not know."

6.4.3 How Much …?

Depending on the size of the operation, the *"How is he making it?"* question can take a back seat to the *"How much is he making?"* question. The fact that the operator was making tetra-ethyl-death by the ABC method at times seems secondary to the quantity the operation could produce. The amount of finished product seized or that could be produced may or may not affect the type of charge or the sentence that is handed down if a conviction is obtained. The amount of destruction that the operators could produce is directly proportional to the amount of explosives that can be produced or seized. The operation's actual or projected production may have nothing to do with the manufacturing charges. The accused either is or is not manufacturing. The fact or opinion that the operation could potentially produce $10,000,000 worth of drugs or enough explosives to blow up the local police station is probably not an element of the crime. However, it may be used as demonstrative evidence to impress upon a jury the size and scope of the operation.

The three basic variations of the *How much?* questions are: *How much product is there? How much product could the operation produce per batch?* And, *How much product could the operation produce with the existing chemicals?* These questions may or may not come up at trial. However, some variation of each one will be asked of the expert at some point during the investigation or prosecution of the operator. Therefore, the expert should have the answers to each one.

6.4.3.1 How Much Product?

Of the three basic "how much" questions, *"How much product?"* is the most relevant. Many controlled substance statutes use a weight value to establish the severity of the offence. The wording of the statute will provide the analytical chemist with guidance in developing an analytical scheme to address the legal question of *How much is there?* The wording "… grams of substance…" may require a different analytical approach from the wording "… grams of substance containing ....." In either case, the analytical chemist should be able to tell how much of the controlled substance in question was in each sample analyzed.

In the United States, possession of explosives is different from possession of drugs from a criminal law perspective. Explosives have a variety of legitimate uses, and as such, their possession is generally legal or illegal. The illegal possession of explosives does not have the codified sliding scale of penalties exhibited in the drug statutes. Possession of explosives is a black and white issue, with possession being permissible more times than not. However, possession of an explosive can be an element of an underlying crime. For example, possession of an explosive that is a component of a destructive device, as defined by Title 26 United States Code, Chapter 53 (26 U.S.C. § 5845), is illegal.

How much explosive can be produced has more significance from an intelligence perspective. The amount of chemicals seized can be used to calculate the destructive potential of the final product. This has major significance in the realm of homemade fuel/oxidizer explosives. The destructive power of these mixtures will vary with the fuel/oxidizer balance used by the operator. Knowing "how much" fuel and oxidizer are on site, "how much" of each is used in the operator's mixture, and "how much" of the mixture is placed into each device allows explosives experts to calculate the various levels of destruction that could have been generated.

There are two basic methods of determining amount of substance. The direct method is applicable for situations in which the statutes use wording similar to "... substance containing ....." The indirect method is applicable to scenarios in which an accurate accounting of the amount of controlled substance is needed.

The direct method is straightforward. The analytical chemist measures the weight or volume of the substance prior to any analytical work. This establishes the weight at the time of seizure. For exhibits in which only a sample was received, the accurate documentation of the original weights or volumes is critical. Without documented weights or volumes, the court will rely on the only documented value available to them i.e. the weight or volume obtained by the analytical chemist during his analysis.

The indirect method ratios the calculated concentration of a sample into the original amount of substance. This calculation will provide an amount of substance at the time of seizure. Its use is appropriate when an accurate accounting of the amount of controlled substance is required. The concentration of a sample of the exhibit is determined, and that value is ratioed into the weight or volume of the original substance. Section 6.5 contains examples of how these calculations can be applied.

The resulting value may still be subject to interpretation. Issues concerning representative sampling done at the scene, the accuracy and precision of the test methods used to establish the concentration, and the original weight/volume information obtained from the scene will affect the final value.

Proper documentation during every phase of the process is essential. Without the supporting documentation, the analysis and the resulting calculations may end up being considered nothing more than speculation and hearsay. If it is not available for peer review, it may not be admissible at trial. However, it may still be useful as an investigative tool.

6.4.3.2 How Much per Batch?

Determining how much per batch is not a straightforward calculation. There are numerous variables that affect each batch's production. Equipment size, reaction type, recipe, actual versus theoretical yield, and the cook's experience all play a role in the operation's per batch production. The expert's training and experience are critical in interpreting the information and in factoring in the variables to establish a realistic estimation of the operation's production capabilities.

6.4.3.2.1 Equipment Limitations   The size of the equipment used in the operation is the major factor that establishes its per batch capability. The operation could have a limitless supply of chemicals operated by a PhD level chemist yet will still be limited by the size of the reaction flask. A 500 ml reaction flask will only produce a certain quantity of controlled substance during a given reaction cycle.

The "per batch" capacity of an operation that utilizes legitimate scientific equipment is a simple calculation. As a rule of thumb, the volume of the reaction mixture in a traditional round bottom reaction flask is two-thirds of its capacity (e.g., a 3000 ml reaction flask has approximately 2000 ml of usable volume). This allows uniform heat distribution and safe and efficient reflux or distillation. However, because of the operator's lack of technical expertise, they may fill the flask to the top or may only use 25% of the flask's capacity.

The operator's use of alternative equipment also negates any assumption of proper proportions. There is no rhyme or reason as to why or how full the makeshift reaction vessel is filled. In these situations, as well as the situation in which legitimate scientific equipment is used, the expert should rely on the operator's notes to provide guidance as to the per batch production, because operators do not usually deviate from their recipes.

Once the volume of the equipment has been determined, the ratio of chemicals used in the method is factored into the equation. Using the two-thirds capacity guideline, the reaction mixture maximum volume is established. The analytic chemist calculates the amount of precursor and reagent chemicals required to establish that volume. The calculated precursor amount is then used to calculate the amount of product that will be produced with this amount of precursor chemical. Section 6.5 contains an example of how this calculation can be applied.

6.4.3.2.2 Chemical Limitations The amount of precursor and reagent chemicals available can limit the per batch amounts. The operator cannot produce more product than the precursor chemicals he starts with allow, no matter what the reaction vessel size. By the same token, the amount of reagent chemicals present will limit the amount of precursor chemical that can be converted into the final product. These values have more relevance in the estimates of the operation's total production capability.

Two production estimates can be calculated for a fuel/oxidizer explosive mixture. The first is for the balanced reaction in which each weight of fuel has a proportionate weight amount of oxidizer. The other calculation incorporates the proportions utilized by the operator, as revealed in his notes or recipes or through interviews. It is suggested that both calculation be performed.

6.4.3.2.3 Reaction Limitation In calculating product yields, the expert must decide what value he wants to demonstrate: the actual or the theoretical. The maximum yield of a chemical reaction is theoretically 100% conversion of precursor to product; i.e., 1 mole precursor chemical will produce 1 mole product. The actual yield will always be less than the theoretical yield. This number will vary with the reaction, the recipe and the experience of the operator.

The expert must take into account the difference in molecular weight between the precursor chemical and the final product. The molecular weight of a substance is simply the weight of a single molecule of the substance. The ratio of the molecular weight of the final product and the precursor chemical involved provides a conversion factor that can be used to calculate the amount. Appendix N lists the conversion factors for commonly encountered controlled substances and their associated precursor chemicals. The conversion factor can be used to quickly calculate the weight of a final product from a known weight of precursor chemical assuming 100% conversion. Practical Application Section 6.5 contains examples of how to apply these principles.

The manufacturing methods used in clandestine labs are based on reactions that have been published in the scientific literature. These publications generally report the theoretic and actual yields for the reactions they are reporting on. Some reactions are efficient and will produce actual yields that approach 90%. The yields of other reactions are substantially less. The expert must remember that the published yields may not correspond to those of clandestine operations. The published yields are obtained under ideal conditions, and those of the operation under investigation are usually less than ideal.

The expert should rely upon the percentage yield of the reaction when estimating the amount of final product a given amount of precursor chemical would produce. These opinions should address three situations. First, he should address the perfect situation in which 100% of the precursor is converted into the final product. Second, he should address what the literature states the expected yield should be if the reaction were done under controlled laboratory conditions. Finally, he should address what the yield in a clandestine lab situation would be. The analytical chemist who performs the reaction, mimicking the operating conditions of the lab operation under investigation, can obtain the actual yield value. The analytical chemist may compare his values with those of the operator, who may have calculated production yields.

The only value the expert can produce with any degree of certainty is the 100% conversion value. This hard value is based upon the molar conversion of a specific amount of precursor to a specific amount of final product, taking into account the difference in molecular weights. The published yield values were obtained under controlled conditions that as a rule will not be experienced in a clandestine operation and thus, can only be used as a guide to estimate the production of a given amount of precursor chemical. The yield obtained by the analytical chemist when validating the method under investigation can be used to approximate the operation's yield. However, he should factor in his laboratory technique and the elimination of the variables introduced into the operation because of the operator's experience and training as well as the "lab" conditions of the operations.

When discussing the yield of an operation, the expert's opinion should simply state that a given amount of precursor chemical would theoretically produce a given amount of final product. He should overtly acknowledge the fact that the actual value will be lower because of the variables involved in the production of the product. He should also be prepared to describe how he arrived at the lower figures, either by the use of published data, from his own analytical experiments, through the use of notes from the operator himself, or a combination thereof. Being able to defend his opinion in a calm and organized fashion is crucial to his perceived reliability.

Multi-step reactions place additional variables into the equation that should also be accounted for. Each step of the manufacturing sequence has a characteristic yield that may or may not be the same as the previous step. In calculating the total production from a given operation, the expert needs to account for the yield estimates for each individual step in the sequence, and be able to describe the differences and why they exist if necessary.

6.4.3.3 How Much per Week?

There are a number of variables that affect an operation's production over a period of time. The synthesis route, the number of batches in the time frame, the cook's experience and the availability of chemicals all will affect this value. Even with these unknowns, there is information available that can be used to produce a historical perspective of the operation's production.

The seized paperwork is the best source of information concerning overall production. The relevant paperwork that can be used to establish production amounts includes sales ledgers, production logs, chemical receipts and recipes. Such information is generally available at the scene, again because of the typical operator's packrat nature. Receipts provide a purchasing pattern of the required precursor chemicals. Per batch and per week estimates can be extrapolated from these receipts and other information from the scene to obtain a historical production pattern or to project one into the future.

Some operators do the work for the expert. They have been known to document the per batch production, some to the extent of even calculating yield percentages (see Figure 6.3). Other paperwork found at the scene may document sale or distribution information. If the documents contain dates associated with amounts, a historical portrait of the operation may be obtained. The key to using this type of information is the ability to decipher the operator's shorthand or codes. This situation is an example of where the expert's knowledge of clandestine operations is essential. His ability to translate cryptic notes into understandable language gives investigators and attorneys involved in the case a better understanding of the various portions of the operation to which the notes refer.

The operator's recipes can also be extrapolated to provide a historical representation of the operation's production. However, in doing so, a number of assumptions must be made. Operators do not tend to deviate from the recipes they use. Therefore, per batch estimates can be calculated; the amount

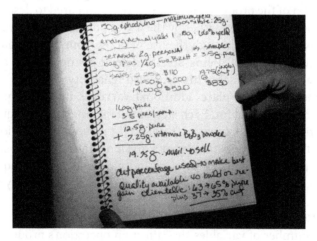

Figure 6.3 Handwritten notes.

of precursor chemical denoted can be used to estimate a time frame for their consumption. Any estimates beyond that will be considered speculation on the expert's part without additional information.

Assumptions come into play in this situation, and the numerous "*What if?*" questions can be asked. The expert can calculate the single batch production using the techniques listed earlier. The number of batches per day, per week, per ..., is subject to conjecture without additional information. These estimates can be made, but the variables used to provide the opinion should be made up front. For example, at 100% conversion, 1000 grams of ephedrine hydrochloride will produce 920 grams of methamphetamine hydrochloride. It is misleading for the expert to claim that the whole 920 grams could be produced at once if the operator's recipe called for 10 grams of ephedrine. All of the variables must be reasonably addressed. With the per batch information factored into the opinion, it would take 100 batches to convert all 1000 grams of ephedrine into methamphetamine. More information is needed to determine how long that would take, or another assumption would have to be made and presented to qualify the opinion.

It is in the best interest of the expert to be candid about the information used to produce his opinion. Playing word or number games in trial or deposition can compromise the expert's credibility or diminish his objectivity in the eyes of the jury. He must remember that his role is to provide all the information the trier of the fact needs to make an informed decision. Providing information concerning the assumptions used to make the opinion will reduce the "*What if?*" questions that can be posed by either counsel. Using the previous example, the expert would be wise to say that assuming 100% conversion, the operation could theoretically produce 920 grams of methamphetamine in 100 10-gram batches.

If more detail is requested, more information is required or more assumptions must be made. *How long does a single batch take?* and *How long between batches?* are not unreasonable questions. Their answers will affect the span of time necessary to convert the entire amount of known precursor into the finished product. The expert should not render an opinion concerning production time frames without qualifying his response by establishing the parameters that frame it.

6.4.3.3.1 Production with Available Chemicals *How much controlled substance could the operation produce at the time it was seized?* This question can be answered by using the chemicals that the suspect had on hand to estimate the amount of product that could be produced. To provide a total picture of the operation's production potential, the expert will need to calculate the amounts using the most abundant and least abundant chemicals. The most abundant chemical calculation will provide information concerning the operation's potential if the balance of the necessary chemicals is obtained. The least abundant chemical does provide information concerning the limitations on the amount of product the operation could produce at the time the operation was seized. The focus should still be the operation's potential product. The time requirement does not enter into these calculations. Obviously, depending on whether the expert is testifying for the prosecution or for the defense, there will be differing emphasis placed on cross-examination. It is always better to have all the information either way.

These calculations differ slightly from the "per week" estimates in that the expert takes into account the amount of reagent chemicals required. If the operator does not have the necessary reagent chemicals, the reaction cannot take place. For example, the amount of methamphetamine that can be produced from the 1000 grams of ephedrine from the previous section is zero if there are no reagent chemicals to facilitate the conversion. That is not to say that a clandestine lab does not exist. It only means that at the time of the seizure, the operation could not produce methamphetamine.

In establishing the chemical ratios required for these calculations, the expert should rely on the operator's notes or recipes. These will provide the most accurate information concerning the operation's production methods, which is used to estimate the operation's production potentials. Not having access to this information, the expert should fall back on ratios from clandestine operations using similar manufacturing techniques. If these sources are unavailable to the expert, the scientific literature should be consulted.

The expert should provide the 100% conversion value as well as an adjusted yield value using available percentage yield values. If these are unavailable, he should stick to the 100% yield value, acknowledging the fact that the actual value will be lower. As with all production estimates, the expert needs to acknowledge that the actual production will be less that the total conversion value.

6.5 Practical Application

The opinions provided by a trained clandestine lab chemist can be utilized during any portion of the investigation or prosecution of a clandestine lab operation. As previously stated, a clandestine lab chemist has knowledge beyond that of the traditional forensic chemist. Both the forensic chemist and the clandestine lab chemist must know the limitations of their training and experience and should not render opinions beyond that scope. The following are examples of the effects the opinions provided by chemists have upon various clandestine lab investigations or prosecutions.

6.5.1 Practical Example (Knowledge and Experience)

A clandestine lab investigator in the southwestern United States came to a senior forensic drug chemist for an opinion concerning the use of lead acetate in the production of phenylacetone. The chemist responded that lead acetate could not be used to produce phenylacetone. The reaction required sodium acetate. This was a true statement in that the synthesis route of choice in that region of the United States used sodium acetate in a reaction with phenylacetic acid. However, he was not a clandestine lab chemist and was not aware of an alternative reaction used in the Pacific Northwest that utilized lead acetate and phenylacetic acid to produce phenylacetone. The forensic chemist was an excellent bench chemist. However, he did not possess the *knowledge and experience* necessary to render the required opinion.

6.5.2 Practical Example (Knowledge and Experience)

Two experts were providing testimony concerning a clandestine lab that was located in a remote desert location. The prosecution's witness held a BS degree in chemistry with over 10 years of experience working with clandestine drug laboratories. The defense's expert held a PhD and was a respected former forensic laboratory director who had published extensively. However, he had never been directly involved with a clandestine drug lab case, and his publications had only mentioned that certain types of drugs were produced in clandestine drug labs. During the verdict portion of the bench trial, the judge commended the defense expert's service to forensic sciences. He then pronounced his testimony unbelievable. His opinion was valid. However, it was based on his reputation, not his actual *experience and training* in the area of clandestine labs.

6.5.3 Practical Application (Information Interpretation)

Volumes of technical information can be generated during a clandestine lab investigation. It is the job of the clandestine lab chemist to extract the relevant information and place it into some semblance of order. The following are examples of how a clandestine lab chemist brought calm to the chaos by using the physical evidence to answer some of the "who, what, when, where, why and how" questions that required answers.

The following chemicals were found at the scene of a suspected clandestine drug lab: acetic acid, acetic anhydride, aluminum powder, bromobenzene, hydrochloric acid, mercuric chloride, methylamine, nitroethane, phenylacetic acid and sodium acetate. The expert was asked to determine what was the most likely final product of the operation, based on the chemicals found at the scene. A sequential evaluation of the potential manufacturing routes allows the expert to establish the manufacturing route the operator most likely was implementing.

The evaluation of the chemicals begins by making a table with two columns. One column lists the chemicals found at the scene. The other column lists the potential products that could be manufactured and the chemical's role in the process (i.e., solvent, reagent or precursor). The chemist then should use a table that correlates chemicals to their potential end product. Drug Table 1 in Appendix R can be used for this purpose. The Chemistry Guide of the DEA Clandestine Laboratory Training manual is also a source of this information.

A pattern becomes apparent once all of the chemicals have been assigned a potential end product. At this point, the expert looks at individual synthesis routes to establish whether all, or a significant portion, of the chemicals are present. Chemicals for multiple synthesis routes may be present. Many times, there may be chemicals present that have nothing to do with the synthesis being used. The likely final product is the one with the most complete set of the required chemicals.

The information compiled in Table 6.2 indicates that the most probable end products from the list of chemicals are phenylacetone or methamphetamine. In one step, the expert narrowed the field of possible final products from all controlled substances that are commonly produced in clandestine labs to two.

Table 6.2   Practical Exercise 6.3 Drug Synthesis Opinion

Chemical	Product/Route
Acetic Acid	rA1 rMD1, 2, 5 pP4
Acetic Anhydride	pP1
Aluminum	rA1 rM1
Bromobenzene	pPC1, 2, 3
Hydrochloric Acid	rAll A, M, MD, PC
Mercuric Chloride	rM1
Methylamine	pMD3, 4 pM1,4,5
Nitroethane	pA2 pMD2 pP5
Phenylacetic Acid	pP1, 3, 4
Sodium Acetate	rP1
Legend	
A = Amphetamine	PC = Phencyclidine
M = Methamphetamine	# = Reaction route #
MD = MDA	p = Precursor
P = Phenylacetone	r = Reagent

Once the field of possible final products is narrowed to a manageable number, the expert compares the list of known chemicals with the list of precursor and reagent chemicals required for each of the various synthesis routes (Drug Table 2 of Appendix S). In this example, the list of chemicals supports two different synthesis routes. One route suggests the production of phenylacetone using a phenylacetic acid, sodium acetate and acetic anhydride. The other suggested synthesis route produces methamphetamine using phenylacetone, methylamine, aluminum and mercuric chloride. Each method supports the existence of the other. At the time of this seizure, the manufacturing method of choice for the production of methamphetamine was a two-step process, which used phenylacetone as an intermediate product.

6.5.4 Practical Application (Information Interpretation)

The same process can be used to propose the explosives that could be manufactured from the chemicals seized at a given location from a list that had been acquired during the course of an investigation. Opinions based upon the list of chemicals seized at a suspected clandestine explosive manufacturing operation can be corroborated through the analysis of final product, reaction mixtures and waste products. Opinions based upon only a list of chemicals present more of a challenge.

The challenge has to do with the interjection of bias into the evaluation process. As with drug manufacture, most, if not all, of the chemicals used in the clandestine manufacture of explosives have legitimate commercial, industrial, household or agricultural applications. However, when context is added to the evaluation equation, a list of chemicals that are available at the hardware store may take on a nefarious nature.

A list containing the following chemicals was taken from an individual with connections to a terrorist organization. The list contained the following: stump remover, battery acid and toluol. The questions to the forensic chemist are: what are these chemicals, and what explosive end product could be made?

Identifying the chemicals in the first step in the process. Stump remover is comprised of potassium nitrate (KNO_3). Battery acid is the common name for sulfuric acid (H_2SO_4). Toluol is the common name for toluene, which is a used as a paint thinner.

As with the previous drug manufacturing example, the first step is to determine which chemical can be used in the manufacture of explosives. Using Explosives Table 1 of Appendix T, we determine that potassium nitrate is associated with black powder, flash powder and pyrotechnics. Sulfuric acid and toluene are associated. Potassium nitrate is the oxidizer in black powder and black powder substitutes. Sulfuric acid is used as a reagent in the nitration of organic compounds into explosives. Toluene is the precursor chemical used to manufacture TNT, trinitrotoluene.

On the surface, these chemicals appear to be unrelated and cannot be used to manufacture an explosive. Although potassium nitrate is an oxidizer used in fuel/oxidizer explosives, commonly used fuels are not part of the list of ingredients. The nitration of organic materials, toluene in this case, into explosives requires nitric acid in concert with sulfuric acid. Although sulfuric acid and toluene are on the list, the essential precursor required for the nitration appears to be missing. As such, initially, it appears that the ingredients on the list will not produce an explosive or an explosive mixture.

However, when the ingredient list is compared with the basic requirements of a fuel/oxidizer explosive, a different conclusion may be reached. Potassium nitrate is the oxidizer for black powder and numerous other low-explosive and pyrotechnic mixtures. Toluene, an organic material and the major component of gasoline, is more than an adequate fuel and in the proper ratio with potassium nitrate will produce a fuel/oxidizer explosive.

Experienced clandestine lab operators live by the rule: if you cannot buy an ingredient, make it. Explosive manufacturers make nitric acid if it is unavailable through other means. Just as drug manufacturers make hydrogen chloride with salt and battery acid, explosive manufacturers can make nitric acid using a nitrate salt and sulfuric acid. Once the nitric acid is made, it can be combined with the toluene in the presence of sulfuric acid to make TNT.

The previous explanation is based on the assumption that the chemicals in the list were going to be used to manufacture an explosive. As such, contextual bias was injected into the thought process used to generate the opinions. The clandestine lab expert needs to ensure that the recipient of his opinions understands that additional information may be required to increase the level of certainty associated with his opinion.

6.5.5 Practical Application (Data Interpretation)

The comprehensive analysis of a reaction mixture from an operational clandestine lab produces a volume of information concerning the method the operator was utilizing to manufacture the controlled substance involved. The thorough examination of the data from any given analytical technique may be all that is required to profile the synthesis route being used. The following is an example of how the mass spectral analysis of a reaction mixture sample can be used to determine the synthesis route.

The mass spectral analysis of a clandestine lab reaction mixture produced six significant peaks in addition to the detected controlled substance, methamphetamine (Table 6.3). Each of the compounds was tentatively identified

Table 6.3   Reaction Mixture Components

Compound	Peak 1	Peak 2	Peak 3	Peak 4	Peak 5	Mole. Wt.	Drug/Synthesis Route
Ephedrine	58	69	79	41	59	165	M2, M3
Phenyl-2-propanone	91	134	92	43	65	134	Numerous A and M routes
1,2 Dimethyl-3-phenylaziridine	146	105	42	132	91	147	M2, M3
1-Benzyl-3-methylnaphthalene	232	217	108	215	202	232	M3, A1
1,3-Dimethyl-2-phenylnaphthalene	232	215	217	108	202	232	M3, A1

by a search of the five most prominent ions in its mass spectrum (Appendix M). The table also indicates which manufacturing methods are associated with each compound. The pattern that emerges from the evaluation of the potential manufacturing routes indicates that the operator was converting ephedrine to methamphetamine using the hydriodic acid reduction technique.

6.5.6 Practical Application (Production Estimates)

How much controlled substance the operation could produce is a question that will always be asked at some point during the investigation or prosecution. The production amount may or may not be an element of the crime, but it may be significant during the prosecution or the sentencing phase if a conviction is obtained.

The expert should routinely calculate the operation's estimated production as one of his opinions. The information to determine these production estimates is readily available if the lab scene was documented properly. The relevance of the value that is calculated may be debatable. Is the amount of controlled substance that could be produced with the chemicals on hand the benchmark figure? Or, is the amount of finished product that could be produced if the operator had all the chemicals necessary to completely use the chemicals found at the site the appropriate value? This philosophical difference in opinion necessitates that the expert calculates a range. The following are examples of calculations used to determine the production of various controlled substances.

The expert needs to know the amount of phenylacetone that can be produced from 1000 grams of phenylacetic acid. He first calculates the reaction's conversion factor (n) by dividing the molecular weight of the phenylacetone (the product) by the molecular weight of the phenylacetic acid (the reactant).

Appendix N contains the conversion factors for chemicals and drugs most commonly encountered in clandestine labs.

Conversion Factor (n)	= Molecular weight Product/Molecular weight Precursor
	= Molecular weight phenylacetone/Molecular weight phenylacetic acid
	= 134/136 = **0.98**

The weight of the phenylacetic acid (precursor chemical) is multiplied by the conversion factor to produce the weight of the phenylacetone (final product) at 100% conversion.

Weight Product$_{Theoretical}$	= n * Weight Precursor
Weight Phenylacetone$_{Theoretical}$	= n * Weight Phenylacetic Acid
	= 0.98 * 1000 gm = **980 g**

In some instances, the precursor is in a solution. The lower concentration must be accounted for in the production calculation. For example, methylamine is commonly found as a 40% (weight/volume) aqueous solution. The following example illustrates the modifications necessary to account for a diluted solution of 1000 grams of a methylamine solution used as a precursor chemical.

Weight Product$_{Theoretical}$	= n * Volume$_{Precursor}$ * Dilution factor
Weight Meth. HCl$_{Theoretical}$	= n * Volume$_{Methylamine}$ * Dilution factor
	= 5.96 * 1000 ml * 0.40 g/ml = **2384 g**

6.5.7 Practical Application (Production Estimates, Multi-Step)

Multi-step reactions contain an intermediate that must be accounted for. Each step of the reaction has a conversion factor that figures into the final calculation. In these instances, the calculation boils down to a sequence of single-step calculations that use the weight of the previous calculation as the starting point for the next in the sequence. The intermediate acts as the product in one calculation and the precursor in the next. The calculated weight of the intermediate is used as the precursor weight for the second step of the process.

The benzyl cyanide synthesis of phenylacetone and subsequent conversion to methamphetamine HCl can be used to demonstrate the calculation sequence. The conversion factors for both steps can be calculated as in the previous examples or taken from a table of recalculated values.

n_1	= Molecular weight phenylacetone/Molecular weight benzyl cyanide
	= 134/117 = **1.14**
n_2	= Molecular weight methamphetamine HCl/Molecular weight phenylacetone
	= 185/134 = **1.38**

The weight of the phenylacetone intermediate is calculated as an independent step. Using the standard 1000 grams of benzyl cyanide as a starting point, the calculation is as follows.

Weight Phenylacetone$_{Theoretical}$	= n * Weight Benzyl Cyanide
	= 1.14 * 1000 g = **1140 g**
Weight Meth. HCl$_{Theoretical}$	= n_2 * Weight Phenylacetone$_{Theoretical}$
	= 1.38 * 1140 g = **1573 g**

6.5.8 Practical Application (per Batch Production Estimates)

The per batch estimate can be a significant point of debate. The potential of an operation to produce 10 kilograms of controlled substance loses its significance if it can only be produced in 10 gram batches due to limitations placed upon it by the size of the available equipment. The following is an example of estimating the per batch production of an operation using the equipment as a limiting factor.

The expert is asked to calculate the amount of methamphetamine that could be produced using a 1000 ml single-neck round bottom reaction flask. Without additional information, the expert makes the following assumptions. First, he uses a common chemical ratio for the methamphetamine reaction, which is 4 liters of acid, 1 kilogram precursor and 500 grams of an additional reagent. He also assumes a reaction mixture volume of 2/3 the total volume of the reaction flask. His calculations are as follows.

Volume$_{Reaction\ Mixture}$	= Volume$_{Flask}$ * 66%
	= 1000 ml * 0.66 = **660 ml**
Reaction Ratio	= Precursor Amount/Acid Amount
	= 1000 g/4000 ml = **0.25 g/ml**
Weight Precursor$_{Reaction\ Mixture}$	= Reaction Ratio * Volume$_{Reaction\ Mixture}$
	= 0.25 g/ml * 660 ml = **165 g**
Weight Product $_{Theoretical}$	= n * Weight Precursor$_{Reaction\ Mixture}$
	= 0.92 * 165 g = **152 g**

6.6 Summary

The clandestine lab investigator must answer the *Who, What, When, Where, Why* and *How* questions concerning the operation. The forensic clandestine lab investigator will be most concerned with specific questions of *What, Where* and *How*.

The *What* questions consist of:

- *What were they making?*
- *What chemicals and equipment were used in the operation?*
- *What production methods were used?*

The *Where* questions consist of:

- *Where was the lab located?*
- *Where were specific parts of the lab located?*
- *Where was the finished product or waste material located?*

The *How?* questions consist of:

- *How was the operator making the controlled substance?*
- *How much finished controlled substance was there?*
- *How much finished controlled substance could the operator make?*

To answer these questions, the expert requires information from a variety of sources in order to form a strong objective opinion. The information can come from the scene of the operation, from the laboratory analysis of samples taken from the scene or from the expert's specialized training and experience in the area of the clandestine manufacture of controlled substances. The information from three sources is combined to formulate the total picture of the clandestine operation.

Information to answer the "*What is he making?*" question can be obtained from the seized notes and recipes, the lab operator himself, the laboratory analysis of samples from the scene or the chemical inventories. Information to answer the "*How is he making it?*" question can be obtained from seized notes and recipes, the chemical inventory, the equipment inventory or the laboratory analysis of samples from the scene. Finally, information to answer the "*How much ...?*" questions can be obtained from the seized notes and recipes, the laboratory analysis of samples from the scene or the chemical inventories.

The opinions provided concerning the existence of a clandestine lab should be neutral and based upon the known facts. As with all forensic evidence, it should be presented in an objective fashion that allows the judge or jury to make an informed decision based on objective information.

The expert opinions provided in the investigation and prosecutions of clandestine labs are key in determining the direction of the investigation or the subsequent trial. The expert opinion can be used to imply whatever the receiver of the information deems reasonable. The prosecutor may imply that the operation was the largest ever seized. The defense may, on the other hand, reduce the significance of the same set of facts to diminish the seizure to an insignificant occurrence. How the opinion is used is outside the expert's control, which necessitates that the opinion be fact based, free of bias and presented with qualitative uncertainty statements when appropriate.

The forensic expert must remember that his purpose is to evaluate the evidence in an objective manner and provide his opinions in an understandable fashion. He is not in court to establish guilt or innocence. The purpose of an expert opinion is to assist those who are charged with doing just that, by providing the information they require to make an informed decision.

Presenting Results

7

Abstract

The effective dissemination of the expert's opinion is just as important as the information contained within the opinion. The expert may have found the true meaning of life, the key to the fountain of youth or the answer to any of the other mysteries of life. However, if the expert does not have effective communication skills, his vast amount of knowledge will fall upon deaf ears or be misconstrued and given a meaning diametrically opposed to what was intended.

This chapter provides an overview of the methods used to relay expert opinions to their intended target audience. A significant portion of the chapter is dedicated to testimony and presentation of opinions in a judicial forum. However, presenting information in scientific reports that conform to a structural norm will also be covered. Additionally, oral presentations in a non-judicial forum will be addressed.

7.1 Introduction

The analysis of evidence submitted to forensic laboratories has often been the focus of relevant articles published in the scientific literature. New and innovative methods to detect and identify drugs, explosives and the chemicals associated with their manufacture are published continuously. Technology perpetually increases instrumentation sensitivity and selectivity, allowing forensic examiners to detect and identify drugs and explosives at levels unheard of by Paul Kirk. The accepted use of standardized examination methodology within the forensic laboratory community has provided a means for the exchange of data leading to the effective peer review of analytical results and the associated opinions. The standard use of good laboratory practices monitored by internal auditing and confirmed through third party audits has increased the reliability of the data generated by forensic laboratories.

The accuracy and precision of the data generated through laboratory examination and opinions associated with the meaning of the results are of little consequence if that information cannot be effectively communicated to the end user. The laboratory can be staffed with the most knowledgeable scientists, utilizing the most sensitive instrumentation available. However, all this is useless if the written report is undecipherable or the oral presentations leave the audience with more questions than answers.

DOI: 10.4324/9781003111771-7

Presenting the analytical results is the ultimate goal of laboratory examinations. These results must accurately reflect the interpretation of the data generated from the analysis of the samples. Just as importantly, the written report or associated oral presentation should not misrepresent the significance or meaning of the presence of the items identified or their relevance to other items associated with the investigation.

7.2 Written Report

The official report is the final component of the forensic investigation process. This report summarizes the examination of the evidence into a single concise document. All of the information from the crime scene, analytical data from the laboratory examination and the examiner's professional opinions should be reflected in this report. The final report should provide the reader with a road map of the trip the exhibit took through the forensic examination process. Ideally, this report should be able to stand on its own and not require courtroom explanations from the examiner.

Every laboratory has its own examination report format. Most of these formats are based on the criteria set forth by the American Society of Testing Materials (ASTM), the International Standards Organization (ISO) or one of the various scientific working groups (SWGs). Although visual layout may vary, every report should include:

- Examining laboratory identity
- Case file number
- Name of the individual requesting the examination(s)
- Examiner's name
- A list and description of the exhibit(s) submitted for examination
- Description of the examination(s) performed
- Results of the examination
- Chain of custody information

As previously stated, there should be sufficient information in the final report to make the testimony of the examiner unnecessary. Simple one- and two-word responses in the administrative sections are usually adequate. The descriptions of the examinations that were conducted and the analytical results do require more detail. Some report formats have separated these sections. Other formats include a description of the testing process in the results narrative. In either case, the report's reader should be able to discern what controlled substance was identified and the testing process used to make that determination.

The following are two examples of styles of reporting examination results.

Example One

Items   1. White powder
            2. Plant material

Exam   Drugs

Results   1. Contained cocaine, a narcotic drug. Substance weight 1.32 grams. A usable quantity.
            2. Contained marijuana. Substance weight 6.29 grams. A usable quantity.

Example 2

Items   1. Item 1 contained a paper packet containing a white powder.
            2. Item 2 contained a plastic bag containing green leafy plant material.

Results   1. The examination of Item 1 using wet chemical tests, microcrystal tests, gas chromatography and infrared spectroscopy concludes that Item 1 contained a usable quantity of cocaine. The total substance weighing was 1.32 grams, which is considered a usable quantity. Cocaine is defined as a narcotic drug under ARS 13-3401.20.
            2. The examination of Item 2 using microscopic and wet chemical techniques concludes that Item 2 contained marijuana. The total substance weighing was 6.29 grams, which is considered a usable quantity. Marijuana is defined as a narcotic drug under ARS 13-3401.20.

The first example provides the basic information in a Spartan format. The report quickly identifies the contents of each exhibit, the gross quantity of the substance, its classification under the governing statutes and a case law required opinion concerning the amount of substance seized. However, information concerning how the examiner reached his conclusions is not presented. This omission may lead to an unnecessary and time-consuming court appearance.

By contrast, Example 2 includes information concerning how the examiner reached his conclusions. The key pieces of information, i.e., the identity, weight and classification of the controlled substance, do not jump out at the reader as in Example 1. However, the information concerning the basis for the examiner's conclusions is included. This addition may lead to more stipulations by the opposing attorney, thus reducing the number of court appearances required by the examiner.

Figure 7.1 presents a compromise in report writing style. It provides individual sections in which the examiner can provide the details required to satisfy the ASTM and ISO formatting suggestions as well as placing the information in a concise, easy-to-read format.

Opinions must be supported by the analytical data as well as the totality of circumstances associated with the exhibit under evaluation. The examiner must avoid the perception of bias when rendering an opinion concerning the

Scientific Examination Report

Agency Name	Maricopa County Sheriff's Office	Case #	09-01-021
Agency #:	09-24567		
Officer	Eccles #2543		
Date:	27 January 2009		

Exhibits

Exhibit 1 contained a paper packet containing a white powder.

Exhibit 2 contained a plastic bag containing green leafy plant material.

Exhibit 3 contained a sample vial containing a black granular substance

Examinations

The examination of Exhibit 1 used a combination of wet chemical tests, microcrystal tests, gas chromatography and infrared spectroscopy.

The examination of Exhibit 2 used a combination of microscopic examination and wet chemical techniques.

The examination of Exhibit 3 used a combination of microscopic examination, infrared spectroscopy, X-ray fluorescence, X-ray diffraction and gas chromatography mass spectroscopy.

Results

Exhibit 1 contained cocaine, a narcotic drug as defined in under ARS 13-3401.20. The total substance weighing was 1.32 grams, which is considered a usable quantity.

Exhibit 2 contained marijuana, a narcotic drug as defined under ARS 13-3401.20. The total substance weighing was 6.29 grams, which is considered a usable quantity.

Exhibit 3 contained potassium nitrate and sulfur. The combination of these materials and its smooth rock shaped morphology can be found in black powder, an explosive as defined under ARS 13-3101.A.3.

Chain of Custody

Received:	John Smith, 24 January 2009	**Examiner**	Donnell Christian
Disposition:	Mary Jones, 26 January 2009		

Figure 7.1 Written report example.

significance or use of a specific chemical. This is especially relevant when dealing with the clandestine manufacture of drugs and explosives using chemicals with legitimate commercial, agricultural or household applications. Context should be part of any opinion statement.

Consider the following scenarios when placing contextual statements into the reported opinion concerning the evidence collected from an exploited site consisting of a four-room structure in which ammonium nitrate and aluminum powder were identified under the following conditions:

- Condition 1: Ammonium nitrate and aluminum powder were located in separate rooms.
  - Report: Ammonium nitrate was identified in Exhibit 1. Aluminum was identified in Exhibit 2. Ammonium nitrate and aluminum have numerous commercial, industrial, agricultural and household uses.
- Condition 2: Ammonium nitrate and aluminum powder were located in separate containers in the same room.
  - Report: Ammonium nitrate was identified in Exhibit 1. Aluminum was identified in Exhibit 2. Ammonium nitrate and aluminum have numerous commercial, industrial, agricultural and household uses. The combination of ammonium nitrate and aluminum may be found in fuel/oxidizer explosives.
- Condition 3: Ammonium nitrate and aluminum powder were located in the same container.
  - Report: Ammonium nitrate and aluminum were identified in Exhibit 1. The combination of ammonium nitrate and aluminum may be found in fuel/oxidizer explosives.

The previous example demonstrates that context matters. Absent additional information, in Condition 1, an association between the ammonium nitrate and the aluminum cannot be made. Therefore, an opinion concerning their use as a fuel/oxidizer explosive would be interjecting a bias on the part of the examiner. A statement concerning their legitimate use is warranted.

An association between the two chemicals in Condition 2 can be established. The proximity to each other merits a statement concerning their use as an explosive mixture. However, the chemicals being in separate containers requires a legitimate use statement to be incorporated into the opinion.

Condition 3 does not require a legitimate use statement. The combination of ammonium nitrate and aluminum powder is a fuel/oxidizer mixture that will explode. There is no documented legitimate commercial, agricultural or household use for the mixture of ammonium nitrate and aluminum other than as an explosive.

7.3 Oral Presentations

Oral presentations are another means of disseminating examination results and the associated opinions. Some agencies have strict policies concerning the verbal dissemination of laboratory reports, to the extent of prohibiting the practice except under subpoenaed testimony in a deposition or trial setting. Other agencies do not encourage or discourage conveying report results orally as long as the communication is documented and the information circulated stays within the bounds of the facts of the case and opinions that can be supported by those facts.

Depending on the forum, this means of communication has its benefits and disadvantages. An advantage of oral presentations is that they offer a venue to clarify confusion or misunderstandings resulting from language used in the official written report. Technical jargon can be translated into common vernacular. Descriptions of the testing process and the underlying data supporting the identification of chemicals can be provided. The rationale used to formulate the opinions associated with the analytical results can be expressed, questioned and debated.

By contrast, oral presentations have disadvantages. Misinterpretations of opinions can result for a number of reasons. The expert may interject personal (biased) opinions referencing the meaning or significance of the observed analytical results that are not based upon demonstrable evidence. Responses to leading questions, if not properly qualified, can lead to an unintended interpretation of the data and opinions. Questioning during the adversarial litigation or deposition phase of a criminal trial can result in more confusion for the jury if the presentation is not properly prepared prior to testimony.

Oral presentation can be generically divided into verbal reports/summary briefs and testimony. Verbal reports are means to disseminate information to one or more individuals, usually the individual who requested the examination or the end user of the information. Testimonial presentations are generally conducted in a judicial environment in which an objective "trier of the fact" uses the expert's opinion, coupled with responses to questions from each side, to render a judgment.

The target audiences may differ, but the basic premise of disseminating facts based on analytical data and rendering opinions based upon those facts remains a constant. Analytical results must be based upon valid data, and opinions must be supported by facts. The following will discuss approaches to providing oral presentations in each situation. Although presented separately, the concepts discussed in the individual sections can be used in either situation.

7.3.1 Verbal Presentations

7.3.1.1 *Verbal Reports*

Verbal reports and summary briefs both relay examination information and opinions to a target audience. The presentation timeline is generally different. These differing timelines lead to differing levels of preparation prior to disseminating report information.

Verbal reports are generally requested because of the expedited need for examination results. The report information may be required as a component of an affidavit submitted to justify a search warrant related to the clandestine manufacture of drugs or explosives. The results of the examination of items from an exploited site may be required to prepare a targeting package in a military operation. An authorized end user of the information may require examination results sooner than they would receive them during the normal course of business.

The dissemination of examination results via a verbal report is subject to the same dissemination criteria as a written report. All information must receive administrative and technical review prior to dissemination to ensure that the results meet the organization's standards for accuracy and the reporting language is within policy guidelines. As such, verbal reports should not be issued until they have received all of the approvals required by a written report as well as the appropriate authorization for verbal dissemination.

The only difference between a verbal and a written report should be the mechanism of conveying the information. Although a verbal report provides the opportunity for the information receiver to ask questions, the expert should be leery of providing information that is not part of the written report. Information outside of the approved report language has not been subjected to administrative or technical review. Although there may be no malice intended, providing more information than what has been approved may lead to unintended consequences. An unguarded response to a request for additional information or rendering an opinion based on a slight change in circumstances can be misinterpreted or at worst, manipulated into an unrealistic meaning to serve as rationalization for an unjustifiable action.

The expert should document the dissemination of verbal report results. Not because it is an organizational policy. Not because it is an accreditation requirement. Documentation of dissemination of verbal report results should be done because it is a good business practice designed to preserve the expert's integrity. Too many times, an expert has been misquoted, or his words have been taken out of context. The expert's integrity can easily be called into question without documentation of who was provided with what information, along with the time and place the conversation occurred.

Without his integrity, an expert's opinion has little value. Therefore, every time an expert provides a verbal report, he should document the time, date and location of the conversation(s), the person or persons who were involved in the conversation, and a summary of the information that was provided or not provided.

7.3.1.2 Summary Briefs

Summary briefs provide reportable information to one or more individuals who are generally not part of the report dissemination list. They are generally provided to command level personnel after the written report has been issued. Summary briefs are usually delivered in an abbreviated presentation format, where the expert presents the analytical results and the associated opinions. This format gives the expert the ability to present foundational information that is used as the foundation for his opinion in a way that a written or verbal report cannot.

Summary briefs differ from verbal reports in that there is a question and answer component: a situation that should be avoided when delivering a verbal report. However, because a summary brief's audience is at the command level, they may require clarification of any ambiguities in the report due to reporting language or lack of context. These questions may be slightly outside the scope of the original report and require the expert to provide an opinion(s) concerning the relevance of the analytical results under different scenarios. Additional facts may have come to light, following which the decision makers require an expert opinion on how the new information will impact the expert's original opinion.

In this situation, the expert must be cautious and present opinions based on new information. He has an obligation to render opinions founded on fact-based reality and not tell commanders what they want to hear because of their position in the chain of command. The expert should not get caught up in the Helsinki syndrome and mold his opinions to meet the operational goals of the mission. The best way to ensure operational success is to keep opinions within a fact-based reality.

7.3.2 Testimony

Presentation of analytical results and opinions to a jury is just as important as issuing a written report but receives little notice. If the information that has been gathered during the course of the investigation and laboratory analysis is not relayed to the trier of the fact (i.e., the judge or the jury) in an understandable fashion, it may essentially fall on deaf ears and may even be ignored altogether.

The presentation of forensic evidence to a jury involves conveying technical information to a group of people who may have little to no knowledge of the subjects under discussion. The expert's task is to present his information in such a way as to inform without insulting the broad range of personalities who find themselves sitting on a jury. Educational levels will certainly vary, but more likely is the fact that few will have any knowledge about the clandestine production of drugs or explosives. The expert's testimony, therefore, plays a key role in educating a judge or jury about what a clandestine lab is and how the evidence does or does not indicate that one exists in the particular case before them.

This section will focus on the format that courtroom presentation of forensic evidence in a clandestine lab case should take to be most productive. The discussions will be directed toward forensic chemists. However, these principles can be applied to anyone involved in presenting physical evidence and can be used in the summary brief process.

The expert's testimony is an essential element in the prosecution of a clandestine lab. Investigators gather the pieces of the clandestine lab puzzle, collecting the facts that establish who are the participants in the illegal activity and delineating the items of evidence that were seized. The expert's explanation puts the pieces of the puzzle together. His description of how the chemicals and equipment can be combined to manufacture a controlled substance is critical in establishing that a crime has even been committed or terrorist activity is taking place. If the expert does not effectively relay this information to the jury, a conviction may be unnecessarily difficult to obtain.

There are two situations in which a forensic expert may be called to testify in a clandestine lab case. The first situation is where the expert was an active participant in the lab seizure or performed laboratory analyses on the evidentiary samples. In the other situation, the forensic expert has acted simply as an independent expert who evaluated the information concerning a suspected clandestine lab operation. Although each situation is handled in a similar manner, there are still subtle differences.

7.3.2.1 Case Preparation

Trial preparation for a clandestine drug lab case begins long before the trial or deposition subpoena is issued. Conversations with the investigator who is preparing an affidavit for a search warrant are very important. Every comment or opinion the expert gives to the investigator can potentially have evidentiary value. His expert opinions are used in the affidavit to justify the search warrant or guide the investigation in one direction or another. If he provides incorrect or faulty information, the search warrant could potentially be invalidated, or the investigation could go into an ineffective or inappropriate direction.

The expert providing opinions to investigators or attorneys concerning the potential existence of a clandestine lab should document his conversations, noting with whom he spoke, any facts that were presented and the opinion(s) he provided. The technical nature of the information the expert provides to a non-technical individual has the potential to be inaccurately reflected in documents that are submitted by others to the court. Therefore, it is essential for the expert to maintain files documenting and hopefully clarifying these conversations.

What the scene expert does and says at the clandestine lab scene itself should also be documented. As the on-scene technical advisor, his words and actions guide the investigation. Even if he is not the person who actually finds or seizes a particular item of physical evidence, his opinion is the basis for the investigator's actions. Everything the scene expert does and says at the scene has potential evidentiary value. The expert determines which samples are taken, which items should be disposed of due to contamination, and what controlled substances were potentially being manufactured.

A good photographic record supplemented with comprehensive notes taken at or shortly after the search will be invaluable come trial time. These records will allow the expert to remember details of that particular lab operation and the sampling procedures used, on which he based his opinion of the existence of a clandestine laboratory operation. Photographs and notes will also help the expert remember why certain items were sampled and others were not.

Generically, the laboratory analysis and testimony concerning samples from a clandestine lab are no different from the analysis necessary to identify drugs, explosives or chemical identification in general. A reaction mixture is, in essence, a liquid in which the expert is trying to identify any controlled substance; he also can use this to identify precursors, by-products, diluents, reagents and solvents as well. Many of the same techniques used to identify the drugs and explosives are simultaneously used to identify these other components.

The generic chemist's testimony concerning the analytical results is as straightforward as the analysis. The expert describes how and when he received the sample, describes his examination procedures and finally provides his results, all of which should be supported by his working notes. A standard set of questions is generally encountered, and a great deal of trial preparation is not required for the experienced chemist.

The clandestine lab expert takes the analysis of clandestine lab samples beyond the basic identification of a controlled substance. The opinions required to establish the elements of a clandestine manufacturing operation mandate the identification of the other components of the various mixtures found at the scene. The expert should devise his analytical scheme in anticipation of presenting his results to a jury. Opinions that are generated concerning the operation or a specific exhibit must be substantiated by the analytical chemist's laboratory data to establish the facts beyond a reasonable doubt. If the analytical expert is going to talk about a specific chemical within a mixture, he should have the analytical data to support its existence. For example, if the analytical expert is going to talk about the synthesis route used to produce the methamphetamine detected in the reaction mixture he analyzed, he should have also identified the components in the mixture that support his conclusion.

The clandestine lab chemist's trial role takes a significant deviation from the traditional forensic chemist's role by providing opinions concerning the "how" and "why" of the operation. Traditionally, the forensic drug chemist's testimony does not include significant speculation concerning the condition or potential use of the exhibit he examined. The clandestine lab chemist's testimony can potentially be filled with opinions concerning the various aspects of clandestine lab chemistry. With the information from the laboratory analysis and the scene information, the expert should be able to form opinions concerning the following: the synthesis route being used, the exact step in the synthesis at the time of seizure, other synthesis routes that may have been used as well, estimation of the production of each synthesis route using the chemicals and equipment on hand, and finally, determination of the total amount of finished product. Each one of these opinions needs to be supported by some type of objective physical evidence.

The chemist's courtroom presentation of a clandestine lab case is the other half of his job and in some instances, may be the most important. His testimony ties all the pieces of information together, and this must be done in an understandable fashion. His presentation of the technical information is a key factor in establishing the cause and effect relationships between what appear to be ordinary items and the manufacturing of drugs or explosives. A conviction may be difficult to obtain if the expert cannot demonstrate how it all fits together, no matter how much evidence is presented.

The chemist's courtroom presentation can be broken into two distinct phases: the pretrial conference and the actual testimony. This discussion will be from the perspective of a generic expert witness. Although the prosecution presents most of the expert testimony in clandestine lab cases, the defense does have the opportunity to present its own expert's opinion of the significance of the physical evidence. Whether the expert is testifying for the prosecution or the defense should be a moot point, since in either case, the expert should be objectively evaluating the evidence and only after careful consideration relaying his opinion. Although the steps in the process for either side are essentially the same, it is obviously in the interpretation of the evidence and formation of an opinion that differences may arise.

7.3.2.2 Pretrial Conference

A pretrial conference with the attorney should be scheduled as soon as the expert knows he will be testifying in a clandestine lab case. Ideally, the attorney handling the case will be knowledgeable about clandestine lab prosecutions. In reality, prosecutors handling these cases are often inexperienced and are not at all knowledgeable of the associated intricacies. Because of the special nature of clandestine lab cases, defense attorneys may not be any better prepared. Therefore, the chemist's first job is that of a teacher.

During this pretrial conference, and all subsequent meetings, the expert should educate the attorney about clandestine drug labs in general; educate the attorney about the specifics of this clandestine drug lab; explain what indicates that a clandestine drug lab exists in this instance; determine what items are missing; determine the sampling procedures that were used; and finally, explain the chemical disposal process (if used).

7.3.2.2.1 Education about Clandestine Drug Labs in General   The expert should explain to the attorney exactly what a clandestine lab is and delineate the many forms it may take. The attorney must truly understand that clandestine labs come in many shapes and sizes. The expert must explain what they are and just as importantly, what they are not. The attorney must understand that clandestine drug labs can range from a simple "crack" conversion operation performed in a kitchen to an elaborate chemical synthesis using exotic chemicals and expensive equipment. He must be informed that the common household equipment and the proper combination of chemicals that can be found in his own house can be turned into drugs or explosives. However, the totality of circumstances surrounding the case will determine whether a clandestine lab exists or not.

Auxiliary issues concerning clandestine labs should also be addressed in the initial interviews. Information concerning the direct and indirect hazards related to the operation should be relayed. The prosecutor must realize that the significance of the case goes beyond the manufacturing of the operation's final product. He must understand the jeopardy to the health and welfare of everyone who had contact with the operation. This may not be a significant point in proving the facts of the case. However, it may place a different perspective on the overall significance to the priority of this type of case.

7.3.2.2.2 Education Concerning the Generalities of This Clandestine Lab   Once the attorney knows and understands what a clandestine lab is, the expert can explain what type of operation is being dealt with in this particular instance. The expert should begin with generally explaining the synthesis route he suspects was used in the operation and which chemicals and equipment are needed. He should walk the attorney through the process, providing a step-by-step explanation of how all the items fit together. This big picture overview should be done in a generic way to paint a picture of how the operation would be conducted in a perfect world.

7.3.2.2.3 What Indicates a Clandestine Drug Lab Exists in This Instance   Once the attorney understands the process being used, the expert can explain the specifics concerning which items of evidence support the opinion about this operation. At this point, the expert shifts his focus from the ideal situation to the case at hand. He may explain how the reaction flask and heating mantle in an ideal scenario is an ordinary mason jar and pressure cooker in the scenario in the present case. During this portion of the interview, he explains how each piece of physical evidence fits together to complete the manufacturing operation puzzle.

7.3.2.2.4 What Items Are Missing   Crime scene and site exploitation puzzles are always missing pieces. Clandestine lab scenes are no different. The expert must inform the attorney what items were not found at the scene, what their significance to the process was, and whether their absence affects the opinion about this operation. If there is a problem with justifying the existence of a clandestine lab through the physical evidence (or lack thereof), the attorney should be made aware of it.

The expert is not there to prove the attorney's case by manipulating insufficient physical evidence to give the illusion that something exists. He is there as an objective seeker of the scientific truth. If the case is legitimate, the physical evidence is there. The expert identifies what pieces of the puzzle are missing and what could possibly have been substituted. Again, if the evidence is not there, the expert should make the attorney aware of what is missing and what is needed to fill the hole.

7.3.2.2.5 What Sampling Procedures Were Used   The attorney needs to understand the sampling scheme used at the scene and the thought process behind it. He must understand that it is unrealistic to sample every container at the scene. Therefore, the expert must explain the purpose behind the sampling process and how this scientific method has brought calm to chaos through the condensing and consolidating of the mass of physical evidence into manageable packages. This process does everything possible to preserve the integrity of the scene and to document all of the physical evidence that was located. The attorney must understand the thought process behind the determination of which items were sampled and how each selected sample serves as a piece of the puzzle. The attorney must be able to trust and rely upon the training and experience of the scene chemist.

7.3.2.2.6 Explain Chemical Disposal (If Used)   The toxic and hazardous nature of many of the chemicals and equipment seized from a clandestine lab require disposal according to protocol by following guidelines mandated by an established set of local, state or federal regulations. The lack of proper storage facilities has led many jurisdictions to opt for more immediate disposal after proper sampling and documentation has been done. The attorney needs to clearly understand that this protocol exists for the safety of the personnel at the scene, because of lack of proper storage facilities for the seized chemicals, and to avoid contaminating the courtroom by bringing hazardous material in during trial. It must also be stressed that no evidence was destroyed. It is therefore critical that all evidence be documented and photographed and only then sampled; photographed and original volume must also be noted before material is disposed of according to the appropriate regulations.

7.3.2.2.7 Outline Testimony Trial attorneys follow a basic rule concerning questioning: do not ask a question unless you know the answer. There is no reason why the expert should be surprised by a question from the attorney whose client he is representing. The expert should work with the attorney to create a known line of questioning for the direct examination. It should flow smoothly, with no surprises and with all relevant questions being addressed. The practical example section presents a scenario of shaky testimony resulting from incomplete testimony preparation.

During this part of the pretrial process, the expert and the attorney should develop an understanding concerning their respective limitations in each other's field of expertise. Each should have enough knowledge of the other's job to be helpful, yet at the same time understand and accept that such knowledge can be dangerous in less experienced hands. As the number of cases each attorney and expert presents at trial increase, their knowledge and comfort level will increase. However, each should know and understand, separately or together, his own limitations and particular role in the process. The expert should present his knowledge of the relevant statutes concerning the manufacture of controlled substances but simultaneously not attempt to make legal assumptions without consulting the attorney. The attorney should never make assumptions concerning certain pieces of physical evidence as scientific fact without consulting the expert.

For example, while attorneys are masters of the English language as it relates to the law, their expertise dwindles to the apprentice level when English is applied to the technical arena of forensic science. In this instance, the chemist's knowledge of the proper terms should be utilized to create a script of questions that will present the evidence in a logical sequence that uses the appropriate terminology. This provides the expert with knowledge of what the question is, and the attorney with the knowledge of what the response will be.

A word of caution is necessary for trial. It is important for both sides to stick to the script. Slightly changing the way a question is phrased may elicit a response that is not expected. At the same time, an unexpected response may result in additional questions to explain what should have been apparent if the original answer had been presented as planned. Again, it is always dangerous to play with the unknown, as well as giving opposing counsel grounds to challenge the testimony that really counts. The strength of expert testimony may very well lie simply in how conversant in his subject the judge is or how the jury perceives the expert to be.

How the exhibits will be presented should be discussed. It is not wise for either party to see the exhibits for the first time in the courtroom. The expert and the attorney should discuss which exhibits demonstrate the points they want to make. At the same time, they should develop a logical presentation sequence, which will relate each item's use in the operation to the jury. This type of preparation provides a flowing and understandable format for the testimony that does not surprise either party.

7.3.2.2.8 Discuss Visual Aids   The use of simple, concise visual aids should be discussed with the attorney. Used properly, visual aids can take a complicated process and reduce it to simpler terms. The combination of oral explanations with a visual reinforcement provides a more interesting testimony format as well as enhancing the retention of the information presented. For example, visual aids can be used to demonstrate what items are necessary to manufacture a controlled substance and compare these with the actual evidence exhibits. They can also demonstrate how the equipment exhibits fit together to make the reaction apparatus.

7.3.2.3 Testimony

The testimony of the chemist or forensic expert can make or break a case that is presented to a jury. Their appearance, demeanor and presentation all affect how the jury interprets the information they receive. Lack of testimony skills or an improper attitude or presentation can substantially reduce the credibility of even the most qualified expert. During his presentation, the expert should present a professional appearance, tell the truth, address answers directly to the jury and answer only the questions presented.

7.3.2.3.1 Direct Testimony   Clandestine lab forensic testimony is divided into three basic phases. The level of expertise and the number of people involved in the forensic portion of the investigation will dictate the number of people required to testify. The three phases of testimony include the scene processing, the analytical examination and the expert opinions.

In many situations, the scene expert acts only as a technical advisor during the seizure of a clandestine lab. He is not actually the person responsible for the seizure of the physical evidence and may not be called to testify. However, he is the person who provides the expert opinions and the technical knowledge used to guide the investigator's decision to seize or not seize a given item. As such, the scene expert should be able to articulate the thought process he used during the processing of the scene if he is called to testify. He should also be able to articulate his role in the chain of custody of the items and samples seized into evidence.

The analytical chemist's testimony is similar to giving testimony concerning the identification of drugs, explosives or related chemicals that have been submitted to a forensic laboratory for analysis. The voir dire process, the testimony establishing the chain of custody and the results of the laboratory analysis are essentially the same. The level of a chemist's expertise in the clandestine manufacture of controlled substances will determine how far beyond the identification of the sample components his testimony may be allowed to extend. In some instances, he may be able to articulate how the laboratory results can be used to outline or depict the role of the samples in the manufacturing process. In other instances, his lack of knowledge would dictate that someone who specializes in clandestine manufacturing techniques would be more qualified to address those issues.

Expert testimony concerning how seized items could have been used to manufacture drugs or explosives should be the dynamic part of the presentation. This is the pinnacle of the expert testimony and should be done in such a way as to relate technical information in a non-technical format that is not condescending. This testimony should fit all of the pieces of the puzzle together. The expert or forensic expert links the testimony concerning the scene processing to the analytical results, hopefully painting a picture of how the manufacturing process worked. The expert should interact with the appropriate exhibits and prepared visual aids in a manner that will demonstrate to the jury how the exhibits relate to each other.

There are a few basic principles that apply universally to expert testimony. The witness's appearance, how he answers questions, how he addresses the jury and how he interacts with exhibits are the same as those when providing testimony concerning laboratory results.

The forensic expert should present a professional appearance. He should dress in business attire that is appropriate for the region of the country and the court in which he is testifying. While a sport coat, tie and cowboy boots may be acceptable in some state court situations in the western United States, this would be distracting in federal court in Manhattan. He should address the attorney representing each side with the same respect. The expert must clearly present himself as being there to establish the scientific facts of the case, and not merely as a hired gun for one side or the other.

The expert should direct his answers to the jury. The testimony is for their benefit. Answers to questions should be presented in terms that the layperson can understand. Obtuse technical language should be avoided. However, when technical jargon is required, it should be explained in a non-condescending manner.

The expert must adjust his testimony style to facilitate the job of the court reporter and hence, the official written record. During his testimony, he must refer to evidence items by their exhibit number. Even though he is holding the exhibit, and the jury can see what is being described, he must always be aware that there is a written record being taken. The written transcript must reflect his actions as well as his words. The reader of the transcript would not understand "Insert this here" but would comprehend "Place Exhibit #13 into the outlet on Exhibit #24." Therefore, the expert should always describe what he is doing with the evidence items so that it will accurately be reflected in the written record.

The attorney's style, the personality of the expert and what the court will allow all dictate the format used to testify about clandestine labs in general, as well as the specifics of the particular case. Testimony formats range from questions with rambling narrative answers to questions with specific answer responses. The expert and attorney should work together to establish a basic approach so they will both be comfortable with this portion of the testimony. If the expert is not comfortable giving narrative answers in front of the jury, his testimony may be more effective with a specific answer format.

In the narrative answer format, the attorney asks an open-ended question. The expert then provides a narrative answer covering as much of the topic as appropriate. When necessary, he asks the court's permission to use the exhibits or prepared visual aids to demonstrate his answer to the jury. This method of testimony can be entertaining as well as informative. Such information may have more jury impact because of the deviation from the normal dry question and answer format.

The specific answer format uses specific questions with specific answers to present the information to the jury. This format is drier and can be more time consuming than the narrative answer approach. The jury may get bored, lose interest in the testimony and end up not retaining key information concerning the facts of the case. Still, there are always instances where specific questions with a yes/no response should be used to introduce large amounts of boring material. Alternating narrative answers with a quick yes/no format is also a far more effective way of keeping the jury's attention during lengthy testimony.

There may be occasions where regardless of how prepared the expert and attorney are for a narrative answer format, the opposing counsel may object, and the court rule that only a specific answer format be used. It is therefore essential that the attorney is knowledgeable of the particular type of clandestine drug lab being tried so that appropriate questions may be substituted at a moment's notice.

The points the opposing counsel would normally address should be covered during the direct examination. This is done to avoid an appearance of deception. Examples of these points are: What are the legitimate uses for the chemicals and equipment that were seized from the scene? Why were the chemicals and equipment disposed of? What other products could result from the combination of chemicals that were seized? Addressing these issues keeps the attorney focused on objectivity and transparency. Addressing questionable issues during direct examination also takes the wind out of the opposing counsel's sails by addressing them in an up-front manner.

There are situations in which although not all of the chemicals or equipment necessary for a particular synthesis route were found at the scene, there is still enough evidence to indicate that a particular synthesis was being used. In these instances, the expert's testimony should address this issue during the direct examination. The opposing counsel will contend that the expert cannot make the determination of the existence of a clandestine drug lab, since not all the necessary items were found at the scene. If the issue is addressed immediately during direct examination, and the expert is confident with his opinion, he is less likely to get caught in the "What if?" game the opposing counsel will try to play in order to establish "reasonable doubt."

7.3.2.3.2 Cross-Examination   The cross-examination methods used to question experts vary widely. The options can range from "No questions" to a wide range of hypothetical questions designed to cloud the issues raised during the direct examination. There should be no problem with the cross-examination if the expert is properly prepared for direct examination.

During cross-examination, the expert should keep in mind the following: be sure to understand the question, answer only the question, do not argue with the attorney, know your limits and be truthful.

Understanding the question and answering only the question that is asked are closely related issues. Attorneys who specialize in litigation are often wordsmiths who ask questions a particular way for a reason. They carefully word questions in a specific manner to elicit a specific response. The question may be intentionally vague and unclear to obtain a particular response from the witness in an effort to undermine their credibility. The expert should not outfox himself by playing word games with a professional wordsmith. Answering what he thinks the question is or what he thinks the attorney wants to hear will only lead to problems and misunderstandings. If the expert does not understand the question or does not like the way it is worded, he should ask the attorney to repeat or rephrase it. However, if the expert thinks the response to a particular question would be intentionally misleading, he should preface his response with a brief explanation. All of this can be cleared up during redirect examination. However, the chemist's credibility may be compromised and a certain amount of doubt created.

Do not argue with the attorneys. Opposing counsels have been known to badger and harass expert witnesses. This combative approach to cross-examination is done to shake the witness's confidence or fluster the witness into giving inaccurate answers to questions. The opposing counsel should be treated in the same courteous manner as the attorney the expert is representing; no appearance of favoritism should be visible on the part of the expert. If the expert loses composure, a certain amount of credibility may also be lost. A calm, composed response to an offensive line of questioning will give the jury the impression that the expert is being abused by the opposing counsel, potentially giving the testimony more credibility.

The expert should know the limits of his expertise. Even if he possesses knowledge beyond what the common juror would, the expert would be well advised to limit his testimony to the areas in which he is strongest. A degree in chemistry does not make the expert an expert in all areas of science. If the answer to a question is beyond his expertise, he should simply say so. There is no sin in defining the scope of his knowledge. Stating facts without direct knowledge, embellishing the truth to make a point, or showing off his expertise could all end up doing more harm than good.

Telling the truth to the best of one's knowledge is the best defense against an attack on your credibility. Experts work long and hard to build a reputation as an expert; one bluff or exaggeration can forever tarnish their credibility.

7.3.2.3.3 Independent Expert There are times when a qualified expert or forensic expert will be called to provide testimony concerning a suspected clandestine lab with which they had no direct involvement. An expert or forensic expert familiar with clandestine labs can give expert testimony concerning his interpretation of the facts of the case. The independent expert takes similar steps in preparing for this type of case as if he had had direct involvement in the case.

The pretrial conferences with the attorney cover essentially the same areas as the pretrial conferences when the expert was an active participant in the lab seizure. His review of the case documentation may bring up additional questions that were not addressed during the original investigation. The independent expert may also ask for additional examinations to be performed or additional items of information to be provided.

The independent expert's testimony is almost solely opinion. As such, they are given a little more latitude in the content of their testimony. They must stick to the facts of the case, but as an expert giving an opinion, they are allowed to make some assumptions based on those facts, being careful not to exaggerate the significance of a point. The independent expert should be conservative in their opinions; keeping in mind that the goal is the truth.

Many of the items seized in clandestine drug labs have legitimate uses, and the independent expert should be willing to admit to such uses. If the facts as presented indicate that the items were being used to manufacture a controlled substance, they should be willing to say so. By the same token, they should be willing to accept the legitimate reasons for the combination of chemicals and equipment if the totality of circumstances dictates it. The independent expert should not be a hired gun. They are there to present all of the options to the court to allow it to make a fully informed decision.

7.3.2.4 Visual Aids

The use of visual aids is an important component of the trial presentation of a clandestine lab case. Visual aids allow the expert to demonstrate the relationship among items that helped form the opinion that the items were used to manufacture a controlled substance. Studies show that an audience will retain approximately 55% of what they see as opposed to 10% of what they hear. Thus, the use of appropriate visual aids should enhance a point the expert is trying to make. Such a deviation from the normal courtroom presentations will also break up the dull question and answer format of a trial, leaving a greater impression on the jury.

The expert can use visual aids as a memory refresher when talking about a process or a particular set of exhibits. A well-prepared set of visual aids can be used as an outline for a narrative testimony. The expert can use the key words and items the jury must remember to remind him of what he wants to say.

To be most effective, visual aids should be simple, easy to read, easy to understand and colorful.

7.3.2.4.1 Simple   Visual aids should be simple. They should be prepared using the one to one rule, i.e., one central idea per visual aid. If a visual aid is too busy or confusing, the point may be lost in the clutter, thus resulting in the jury ignoring it. Figure 7.2 is an example of a busy visual aid. It presents a lot of relevant technical information concerning conventional serology testing in one package. All of this may be informative to the jury. However, what single piece of information does the expert want the jury to retain? Does he want to correlate the number of positive tests between the suspect, victim and evidentiary sheet? Does he want the jury to know the frequency of occurrence each set of tests represents? Does he want the jury to remember that the stain on the evidentiary sheet was the same as the blood of the victim and not the suspect?

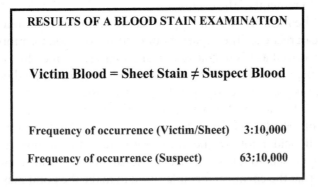

RESULTS OF A BLOOD STAIN EXAMINATION

Sample	ABO	PGM	EsD	GLO	EAP	ADA	AK
Suspect	O	1+1-	1	2	B	1	1
Victim	O	1+1-	1	1	B	1	2-1
Sheet	O	1+1-	1	1	B	1	2-1

Frequency of occurrence (victim)	3:10,000
Frequency of occurrence (suspect)	63:10,000

Figure 7.2 Busy visual aid.

RESULTS OF A BLOOD STAIN EXAMINATION

Victim Blood = Sheet Stain ≠ Suspect Blood

Frequency of occurrence (Victim/Sheet)	3:10,000
Frequency of occurrence (Suspect)	63:10,000

Figure 7.3 Less busy visual aid.

In designing the visual aid, the expert should ask himself: "What single piece of information do I want the jury to retain?" From there, the visual aid designs itself. Complete sentences should be avoided unless absolutely necessary. Too many unnecessary words create a busy visual aid. Key words, phrases or ideas that emphasize the expert's opinion and will stick in the jury's mind should be used. Figure 7.3 is an example showing how the information from Figure 7.2 can be simplified to demonstrate the relationship between the victim's blood, the stain on the evidentiary sheet and the suspect's blood. The visual aid simply states the expert's opinion that the victim's blood and the stain are the same. It supplements that conclusion by adding statistical information concerning the frequency of occurrence that would produce the test results. The information concerning the results of the individual tests used to make the conclusion is unnecessary detail that may detract from the central point being made.

7.3.2.4.2 Easy to Read The expert should put himself in the jury's position and ask if everything on the visual aid can be seen. Points that cannot be seen clearly will not receive the visual reinforcement desired and will lose their impact. Easy-to-understand terms and symbols should be used. Again, key words and phrases in large bold print produce a good visual impact.

7.3.2.4.3 Easy to Understand Technical symbols and abbreviations should also be avoided unless they are explained and will result in ease of comprehension. The expert should also stay away from the use of chemical structures and formulas. Most jurors do not have an extensive science background. However, the expert can use the jury's basic knowledge of chemistry to remember a chemical. An example would be hydriodic acid. For the non-expert, the word is hard enough to pronounce, much less spell. If the expert explains the use of the abbreviation "HI," he may provide the jury with an easier way to remember the chemical.

Many chemicals have some type of abbreviation or common name that the jurors can relate to. The expert should utilize common terminology in his visual aids as well as his narrative explanations. The same philosophy holds true with diagrams of reaction apparatus; the diagrams should be as simple and generic as possible yet still get the point across.

Figure 7.4 is a simplified version of the reaction mechanism for the reduction of ephedrine or pseudoephedrine into methamphetamine utilizing hydriodic acid. It follows the simplicity guide in that it shows that a group of chemicals reacts to form a controlled substance. The shortcoming of this example is that the chemical formulas used to make the point may be intimidating to a layperson. Some jurors may subliminally block any information that uses this format, even though it is presented in its simplest form.

Figure 7.5 presents a simple list as an alternative to the technical diagram in the previous figure. The list presents all the chemicals required to

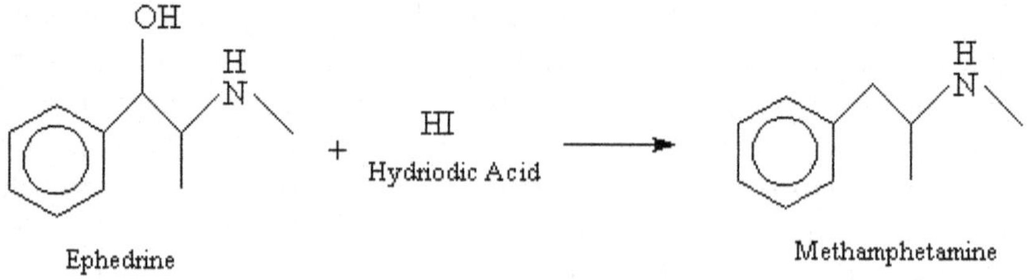

Ephedrine + HI Hydriodic Acid → Methamphetamine

M3

Figure 7.4 Symbol visual aid: Meth HI reaction mechanism.

# 2	Ephedrine
# 17	HI
	Phosphorus
# 8	Lye
# 37	Freon
# 22	HCl

Figure 7.5 Simple list visual aid.

manufacture methamphetamine with the HI reduction technique using the chemical's common household name. Presented properly, this list can be used to demonstrate the same concepts relayed in Figure 7.3 in a visual manner that will not cause some of the jurors to tune out.

Figure 7.5 is also an example of a multi-purpose visual aid that can be used to violate the concept of one idea per visual aid. A minimum of three concepts can be addressed using this example. 1) The right column lists the chemicals used to reduce ephedrine into methamphetamine. 2) The left column is the corresponding court exhibit number for the chemicals that were found at the scene. 3) The lack of the exhibit number for phosphorus acknowledges that the chemical was not at the scene and gives the expert the opportunity to comment on the relevance of that piece of information.

7.3.2.4.4 Colorful   The use of color can be an effective tool, especially if a single visual aid will be used to make a number of different points. While each color represents separate ideas or concepts within a single visual aid, the one idea per one visual aid concept is maintained. An example of this would be to group the list of chemicals in Figure 7.5 by their place in the manufacturing process. The chemicals required for each step of the process could have a distinct color associated with them. (Ephedrine, HI and phosphorus would be in red (step 1), lye would be written in blue (step 2), etc.) In a different color, the exhibit number could be placed next to the corresponding chemical, indicating which chemicals were present at the lab scene. (Numbers in the left column would be written in black.)

This situation allows the expert to use the visual for more than one purpose. First, it provides a complete list of the operation's chemical requirements. Second, it lays out the sequence in which the chemicals are used. Finally, it directly relates the chemicals required to manufacture the controlled substance in question to the items that were actually located at the scene. The different colors distinguish the different concepts involved. Therefore, this single visual aid can be used to emphasize three separate but interrelated ideas. These two additional ideas can be presented on the original visual aid, but they are differentiated by different colors.

7.3.2.5 Types of Visual Aids

The types of visual aids that may be used in court are photographs, slides, flip charts or the evidence exhibits. More often than not, some combination of these is used during the trial.

7.3.2.5.1 Photographs   Photographs are one of the most commonly used types of visual aids. They allow the expert the ability to display the suspected lab site in its original condition. Photographs demonstrate the items that were disposed of or otherwise unavailable for presentation at trial. They confirm the presence of the original containers evidentiary samples were taken from. They can be used in court in lieu of the bulky seized items. The jury can easily handle them during testimony and review them during deliberation. They can be marked to emphasize specific aspects of the scene the photograph represents. When placed in the proper sequence, they can be used to prompt narrative testimony, thus telling and enhancing the story of the expert's opinion of what occurred at the location.

The use of photographs has a few disadvantages. Their small size prevents the jury from seeing what the expert is talking about during his testimony unless the photograph is poster size or is projected onto a screen. It is difficult to demonstrate how two exhibits physically fit together using photographs. Photographs being handled by the jury during testimony will distract from the actual testimony.

Photographs should be a minimum of 8″ x 10″. This size enables the jury to minimally see the details in the photograph as the expert describes them. Preferably, the significant portion of the photograph should be marked for later jury review.

Cropping superfluous portions out of photographs has advantages and disadvantages. Removing extra content surrounding the object under examination allows the viewer to focus on the main point of the observation. However, it can be argued that cropping an image can distort its significance because relevant information has been removed, which alters the context of the photograph, giving the depicted information a distorted significance. Presenting the cropped and uncropped images is the best way to alleviate these arguments.

7.3.2.5.2 Projected Images   Slides are a relic of the past and have been replaced by digital images projected on to a screen using a computer and a multimedia projector. Projected images have many of the same advantages as photographs. Their projection onto a screen presents a larger picture for the jury to see during the expert's testimony. This allows the expert to point out specific items of interest to be emphasized during his testimony.

Projected images have disadvantages. The room lights may have to be dimmed, thus obscuring the jury's view of the expert. If the lights are not sufficiently dimmed, the jury will not be able to see the image. Projected images are hard to review in the jury room without the proper equipment, although digital images can easily be reviewed using laptop computers. As with photographs, it is hard to demonstrate how two exhibits physically fit together using projected images. Photographs can be taped together or placed together on a bulletin board to demonstrate a relationship. Projected images cannot. Projected images cannot be written on during testimony to stress a point or indicate an exhibit number or relationship to another exhibit.

7.3.2.5.3 Flip Charts   Flip charts can be some of the most versatile visual aid media available to the expert. They can be made in advance using simple, easy-to-understand lists or diagrams. The jury easily sees them during testimony. They can show inter-relationships between exhibits. One chart can be used for both general and specific explanations. They can be written on during testimony to stress points. Faint marks that are unobservable to the jury can be placed on them to refresh the expert's memory. This tactic has been deemed acceptable as long as all of the marks are written over during the testimony in such a way as to be completely visible to the court. Flip charts can be taken into the jury room during deliberation.

7.3.2.5.4 Multimedia Presentation   The use of multimedia presentations for technical presentations has become an expectation due to the exponential increase in the availability and use of electronic communication media. Computers have provided experts with a powerful tool that can be used in making courtroom presentations. Software packages like PowerPoint™ and Presentation™ give the expert the ability to present a choreographed multimedia show to the jury. Photographs can be inserted. Color can be effectively utilized. Diagrams can be presented and altered as the testimony progresses. Bells and whistles can literally be added if the expert thinks it will assist in making his point. However, the expert must be sure that the glitz used to dazzle the jury does not distract from the facts of the case.

7.3.2.5.5 Evidence Exhibits   Using the actual items seized from a clandestine drug lab during the chemist's testimony can have significant impact. The jury can see the actual items that were seized from the lab. They can make a visual connection between the expert's verbal descriptions and individual exhibits, thus reinforcing the information presented. This type of presentation educates the jury in an entertaining manner, hopefully motivating them come to agree with the expert's opinion concerning how the combination of exhibits can be used to clandestinely manufacture drugs or explosives.

Unfortunately, there are issues that may prohibit using the actual items. The hazardous nature of the chemical residues on the equipment used in the manufacturing process requires the disposal of many of the actual items at the scene, making them unavailable for physical presentation in court. If the items were not disposed of, they may still pose a potential chemical hazard, again making them unavailable. In some instances, the size of the items may prohibit their use in the courtroom.

7.3.2.5.6 Combination of Visual Aids   In the actual court presentation, the expert will use some combination of visual aids. The expert should determine which types of visual aids he prefers and devise a basic presentation concerning how and why this particular set of circumstances constitutes a clandestine lab. The presentation should be flexible enough to include or exclude any type of evidence that is available, since each case will have a different set of evidentiary items to work with.

Once the basic explanation is established, with a little practice, courtroom presentations of clandestine manufacturing evidence should become second nature. Facts concerning individual cases and exhibits used will change, but overall principles remain the same. Working with the attorney he is testifying for, the expert should be able to mold his presentation to fit the facts of the case.

7.3.2.5.7 Court Exhibits   The expert should not get attached to his visual aids. As part of the trial process, many jurisdictions require that visual aids be placed into evidence as part of the trial record. That is, the visual aid remains with the court until the case is adjudicated, including through the appeal process. The photographs, slides, charts, graphs and computer files used during the expert's presentation all stay in the courtroom once his testimony is completed. They may or may not be returned once the adjudication process has been completed. However, it would be a safe assumption that the visual aids used in the trial will not be returned to the expert at the conclusion of the litigation process. Therefore, it would be a wise course of action to make copies beforehand if he so chooses.

7.4 Practical Application

7.4.1 Practical Example: Terminology

The following is an example of the need for understanding the definition of the terms that are used during the testimony. The original exchange was between a defense attorney and the state's expert during a controlled substance trial. This exchange could easily have happened between an attorney and his own expert without pretrial preparation.

Attorney: Mr. Chemist, to avoid confusion I think it is important that we utilize proper terminology during our discussion. Would you agree?

Chemist: Yes sir.

Attorney: Mr. Chemist did you perform a qualitative analysis on Exhibit A?

Chemist: Yes sir.

Attorney: What percentage of Exhibit A contained a controlled substance?

Chemist: I did not perform that examination.

Attorney: Did you perform a qualitative analysis on Exhibit A?

Chemist: Yes sir.

Attorney: And what percentage of exhibit A contained a controlled substance?

Chemist: I did not perform that examination.

Attorney: Mr. Expert you said you performed a qualitative analysis on Exhibit A?

Chemist: Yes sir.

Attorney: Then what percentage of Exhibit A contained a controlled substance?

Chemist: I did not perform that examination.

Attorney: Mr. Chemist, why won't you answer my question?

Chemist: Sir, I am trying to use proper terminology as you requested.

This exchange demonstrates that the attorney clearly did not understand the difference between qualitative (what is it?) analysis and quantitative (how much is there?) analysis. The chemist's responses clearly created an adversarial atmosphere by specifically answering the question posed. He could have provided the information the attorney desired in a less combative style, placing him in a better light with the jury.

7.4.2 Practical Example: Pretrial Preparation

Lawyers have a simple rule when it comes to direct examination of witnesses. Never ask a question you do not know the answer to. Pretrial preparation should include reviewing the questions that will be asked as well as the responses that will be provided. In some instances, a certain amount of wordsmithing may be required to ensure that the information relayed to the jury is as factually correct as possible. The following is an example of an awkward moment created by not taking the time to prepare.

The expert analyzed a 10-bale representative sample of a 200-bale seizure of marijuana. The pretrial preparation consisted of brief introductions. The following is an excerpt of the resulting testimony.

Attorney: Mr. Chemist, I show you Exhibit A. Do you recognize it?

Chemist: No sir, I do not. (The expert was presented with a 50-pound, plastic-wrapped bale of plant material. He looked and did not find the identifying marks he placed on it at the time of analysis. He had expected to be presented with one of the 10 bales he had actually analyzed.)

Attorney: ** Silence. ** (The attorney had expected a YES, which would lead into his next question.)

Attorney: Do you recognize anything on Exhibit A?

Chemist: Yes sir, I recognize the case number.

Attorney: Where do you recognize that case number?

Chemist: From a submission I analyzed in July of this year.

Attorney: What did that submission consist of?

Chemist: Ten plastic-wrapped bales of plant material.

Attorney: Did those exhibits resemble Exhibit A?

Chemist: Yes sir.

Attorney: Did you analyze the exhibits submitted to you in July of this year?

The testimony continued. The results of the analysis of the representative samples were eventually allowed into the record.

With a short pretrial conference, this embarrassing exchange could have been avoided and the direct testimony condensed into a few flowing questions. Do you recognize the exhibit and how? Did you analyze the exhibit? What are the results of your analysis? However, the lack of proper preparation resulted in both parties being surprised, resulting in additional questions and answers to establish the same facts.

7.5 Summary

The presentation of forensic evidence is probably the most neglected part of the expert's job. Nowhere is it more important than in the presentation of a clandestine lab used to manufacture drugs or explosives, for the case often hinges on the expert's opinion. The significance of the information can be lost to the reader if the initial reporting is obscured with technical terms and jargon. If the expert makes a poor presentation, his information may not be deemed credible, and the trier of the fact will not consider it during its deliberation.

The expert's education of everyone involved in the investigation and prosecution of a clandestine lab case is essential to the success of the operation. All parties must know what made this particular situation a clandestine manufacturing operation so that proper questions are asked. Instigators and command staff should be briefed on the significance of the laboratory examination results and how those pieces fit into the clandestine lab puzzle. Pretrial meetings are essential for the expert and attorney to devise a script to present all of the information to the court without any surprise questions to the expert or unanticipated answers to the attorney.

The proper use of visual aids will make an expert's presentation more effective, because the jury will retain five times as much of what they see as opposed to what they hear. The effective presentation of forensic evidence in court is a skill the expert must develop to the same extent as his analytical technique. If the expert is not proficient in courtroom presentation, the most sophisticated evidence in the world may literally fall on deaf ears.

Explosives Labs

8

8.1 Introduction

The emphasis in the original book was placed on the investigation of clandestine drug labs, with explosives and their manufacture discussed as corollary. That coincided with the author's experience in investigating hundreds of clandestine lab incidents. His experience at that time was that approximately 1% of the operations he responded to involved the actual manufacturing of explosives. While this may seem to be insignificant in the overall scheme of things, it proved to be the proverbial tip of the iceberg. Most of the operations involving explosives were indeed detected as a result of a clandestine drug lab investigation or found by emergency medical personnel when responding to a call for assistance.

The events of September 11, 2001 created a heightened awareness of the potential for terrorist acts here in the United States. The terrorist bombings at the Murrah Federal Building in Oklahoma City in 1995 and the initial bombing of the World Trade Center in 1993 involved homemade explosive mixtures. At the other end of the spectrum, young pranksters deposit small but effective homemade explosive devices into the mailbox of a neighbor or one selected at random. Somewhere in the middle rest the events that played out at the Columbine High School in Littleton, Colorado, where a group of teenagers brought numerous homemade explosive devices to school to assist in their carnage. The size of the target audience may differ, as well as the profile of the terrorist, but the effect is the same.

Ideological differences have created chasms not overtly seen since the crusades. Inter- and intra-religious differences have become a global problem. Acts of violence have replaced civil discourse as a means of addressing political differences that have been historically settled at the ballot box. Terrorism has become a common tool to settle disagreements under the philosophy that might makes right. All too often, explosives are used as a means to an end.

8.2 Explosives Lab Operators and Manufacturers

The motivating forces driving the operator of an explosive lab usually differ from those of the drug lab operator and vary widely. There is one group of explosive lab operators who direct the final product to an end user (or victim) of a terrorist act. Another group utilizes hazardous materials in booby traps to protect their drug manufacturing operation from law enforcement and rival drug manufacturers. Finally, there is the hobbyist, who manufactures explosives and explosive mixtures without criminal intent but for the entertainment value they receive when they detonate their destructive mixtures.

An Assistant United States Attorney who was prosecuting a militia group in Arizona in 1998 summarized the motivating factors of the individuals involved in an explosives manufacturing incident. The individuals were charged with manufacturing dangerous devices with the intent to deposit them at federal and local courthouses and law enforcement offices. When asked what the motivating factors of the last two suspects in the case were, the attorney responded, "One is on a mission from God. The other one is just stupid."

As inflammatory as this comment was, it did summarize two of the motivating factors for clandestinely making explosives. The first portion of the statement indicates that one individual had an ideology or a statement he wanted to express through a violent act. Manufacturing the explosives and the explosive device was the route he chose to take. The other part of the statement is consistent with the hobbyist who does not realize the potential damage that could be done to himself or innocent people.

The manufacture of explosives is not an illegal activity per se in the United States. However, the federal government heavily regulates it. The widespread legitimate use of explosives has recently led to laws controlling their manufacture and regulations governing their use. These regulations were formulated to prevent accidents and eliminate incidents that might jeopardize public safety.

8.3 Regulations

Individual jurisdictions may impose additional restriction on the manufacture of explosives. These laws are dictated with public health and safety in mind. The lack of stiff criminal penalties provides no incentive to initiate an investigation into the clandestine manufacture of explosives unless it is in conjunction with a major felony.

While the manufacturing of explosives may not be illegal, what the end product is used for may be. For example, combining an oxidizer, a fuel and a sensitizing agent of flash powder may not be illegal in itself. However, when the components are placed into a sealed pipe, the combination becomes a destructive device (a bomb), which is considered a deadly weapon. Coupling the destructive device with the components necessary to create a booby trap increases the intensity of the criminal act. Rigging the booby trap to function further demonstrates the premeditation.

The manufacturing of explosives itself can be used to demonstrate the intent to commit other illegal activities. It can also be used as an aggravating factor to increase the seriousness of a criminal act. For example, the premeditated act of manufacturing an explosive, i.e., using it to construct a destructive device, which is to be used as a booby trap could be used as an aggravating circumstance during trial. These premeditated acts could be construed as an assault with a deadly weapon, which would be further aggravated if law enforcement personnel encountered the device.

Fireworks, on the other hand, have a different distinction. Many of their components are classified as explosives in the broad sense by the U.S. government. However, many local jurisdictions have classified them as contraband substances, thus making their manufacture and subsequent possession illegal. Such local jurisdictions feel that the general public does not have a legitimate need to possess fireworks. As with drugs, exceptions are well defined. Properly licensed commercial operations can manufacture, possess and use fireworks under well-defined circumstances. What would the Fourth of July be without a fireworks display? However, in many jurisdictions, the general public is criminally prohibited from all such activities except as spectators.

The three categories of materials that need to be considered with regard to such clandestine labs are explosives, fireworks and pyrotechnics. The differences in each of their definitions determine whether their possession or manufacture is illegal or simply regulated. Properly configured, the components of any one of these groups can be incorporated into a destructive device. This act is illegal in all jurisdictions, but for different reasons.

The U.S. government defines explosives, fireworks and pyrotechnic compositions in Title 27, Code of Federal Regulations Section 55.11 (27 CFR 55.11). Explosives are defined as any chemical compound, mixture or device whose primary or common purpose is to function by explosion. The term includes, but is not limited to, dynamite and other high explosives, black powder, propellant powder, initiating explosives, detonators, safety fuses, squibs, detonating cord, igniter cord and igniters. The list of explosive materials is contained in 27 CFR 55.23 and in Appendix E of this book.

Fireworks are defined as any composition or device designed to produce a visual or an audible effect by combustion, deflagration or detonation, and which meets the definition of "consumer fireworks" or "display fireworks" as defined by this section. Finally, pyrotechnic compositions are defined as a chemical mixture that upon burning and without explosion produces visible, brilliant displays, bright lights or sounds.

By contrast, state and local jurisdictions may address explosives and fireworks under different sections of their legal code. The broad brush that declares the manufacture of drugs illegal on all levels is not generally applicable when it comes to explosives and fireworks. For example, Arizona addresses the possession of explosives in Title 13 (Criminal Code), and the possession and manufacture of fireworks are addressed in Title 36 (Public Health and Safety). Therefore, it is imperative that all federal, state and local laws be considered during the investigation of explosive manufacturing situations.

8.4 Scene-Processing Procedures

The criminal status of the manufacturing of explosives is not an issue when it comes to the procedures used to process the scene of a clandestine manufacturing operation. The procedures are the same as with any clandestine lab, the only difference being that the safety issue becomes much more apparent. Personnel processing the scene must keep in the forefront of their mind the fact that the sole purpose of the operation is to produce a substance that by definition is a hazardous material (an explosive). The three general hazards associated with clandestine labs listed in Chapter 2 increase the complication of the processing process by at least one order of magnitude.

The operators have little chemical training. They do not understand the hazardous potential of the chemicals they are working with. In many instances, the chemicals involved in these operations are on the extreme end of the hazard scale. The acids are some of the most corrosive and possess oxidizing characteristics. The oxidizers are extremely reactive, some to the point of becoming shock or friction sensitive when combined with the right or wrong component.

The lack of understanding of proper laboratory technique increases this general hazard. Quality control is not significant in the world of the clandestine chemist, and certainly not where it should be: in an arena where the purity of the final product is critical. Impurities lead to an increased or decreased sensitivity of the compound or mixture, which in turn, leads to unpredictability in the explosive characteristics. The simple fact that the pH is too high or too low may change a relatively stable explosive into one that detonates with the slightest provocation.

As in drug labs, the operator's lack of chemical knowledge leads to the improper storage of chemicals. Ethers exposed to the atmosphere for extended periods of time form explosive peroxides. The accidental explosive potential of picric acid increases dramatically when it is allowed to dry completely or is stored in a container with a metallic lid.

Finding an unlabeled container is the scariest situation of all. The operator himself may or may not know what it contains. Accidental detonation of unlabeled containers or unknown substances leads to the detection of many explosive manufacturing operations. It cannot be stressed too often that extreme caution should always be exercised when handling these containers, because the simple act of moving them may cause them to explode.

Curious juveniles operate many clandestine explosive labs. As in clandestine drug labs, their source of information is underground literature or the Internet. The reliability of these recipes is suspect at best. For example, many of the recipes encountered on the Internet have been known to be missing one or more of the steps in the manufacturing process. The closet explosive chemist does not have the technical background to detect missing or additional steps in a chemical reaction. This lack of technical knowledge can lead to the production of a final product that either does not work at all or is extremely sensitive and explodes with the slightest provocation.

The makeshift nature of clandestine explosive operations increases the potential for disaster. Most reactions are performed under less than ideal conditions using equipment not intended for explosives manufacturing. Sparks, friction or incompatibility between the chemicals and the reaction vessels all can potentially lead to an accident during the manufacturing process or while the emergency responders are trying to identify and abate the hazards.

The sequence of events used to process a clandestine explosives lab scene is the same as for a clandestine drug lab. Pre-raid planning ensures that all of the resources required to safely process the scene are available. The scenario is discussed and the assignments are handed out at the briefing. A trained entry team secures the location as quickly as possible and reports its observations. The evaluation/abatement team identifies and neutralizes any obvious hazards and provides an additional perspective for the search team. The search team processes the site for physical evidence and prepares the site for the disposal company.

The most significant difference between the processing of a clandestine explosive lab and a clandestine drug lab is the interaction between the evaluation team and the search team. In many instances, these functions are combined. The potential for encountering explosive compounds exists throughout the search phase of the operation simply because explosives are the final product. Therefore, the bomb technicians, who generally take a passive role during the search of the scene, become active participants as individual items are examined and evaluated.

The documentation of the scene of a clandestine explosive lab is just as important as the documentation necessary for a clandestine drug lab. Even if the manufacturing of explosives is not illegal, the activities that are associated with it may be a serious felony. The manufacturing operation may be construed as the overt act in a homicide or terrorism conspiracy case. Therefore, proper documentation of the scene is essential.

The disposal component of an explosive manufacturing operation takes on a different light. The explosive nature of the end products and many of the precursor chemicals used in the manufacturing process limits disposal options. Many commercial chemical disposal companies will not take explosives or chemicals with explosive potential. If they do, the cost of disposal may be prohibitive.

In these incidences the local bomb squad often has the authority to perform the disposal operation. This squad has the expertise to safely dispose of the explosive components, which may be completely consumed through combustion or detonation. Even then, regulations concerning the environmental impact of the act must be considered when utilizing this method of disposal.

8.5 Summary

Explosives were invented to cause damage and destruction. The act of clandestinely manufacturing them is not generally illegal in and of itself. Clandestine production often demonstrates a criminal intent for the explosives' end use and can be directly linked to heinous acts of violence and terrorism. The concept is the same, whether it is the mischievous act of blowing up a neighbor's mailbox or the wholesale destruction of an office building, murdering countless innocent occupants. Therefore, if a substance with a legitimate function was manufactured for a criminal purpose, it should be treated with the same fervor.

The concepts associated with the processing of a clandestine explosives lab scene are the same as those used in clandestine drug labs. The hazardous nature of the end product demands extra vigilance when it comes to adhering to safety protocols. The dangerous nature of the end product and the operating conditions used to produce it dramatically increase the potential for disaster through an accidental explosion.

Appendix A
SWGDRUG
Glossary of Terms

www.swgdrug.org/Documents/SWGDRUG%20Recommendations%20Version%208_FINAL_ForPosting_092919.pdf
This glossary of terms and definitions has been developed and adopted by the Scientific Working Group for the Analysis of Seized Drugs (SWGDRUG) core committee from a variety of sources. In some instances, the core committee modified existing definitions or created definitions where none could be found in standard references.

Term	Definition
accuracy	Closeness of agreement between a test result or measurement result and the true value
analyst	a designated person who: • examines and analyzes seized drugs or related materials, or directs such examinations to be done, • independently has access to unsealed evidence in order to remove samples from the evidentiary material for examination and, • as a consequence of such examinations, signs reports for court or other purposes
analyte	The component of a system to be analyzed
audit systematic	Independent and documented process for obtaining audit evidence and evaluating it objectively to determine the extent to which audit criteria are fulfilled
bias	The difference between the expectation of the test results and an accepted reference value
blank	Specimen or sample not containing the analyte or other interfering substances
by-product	A secondary or incidental product of a manufacturing process
calibration	Operation that, under specified conditions, in a first step, establishes a relation between the quantity values with measurement uncertainties provided by measurement standards and corresponding indications with associated measurement uncertainties and, in a second step, uses this information to establish a relation for obtaining a measurement result from an indication

(*Continued*)

Term	Definition
capacity	The amount of finished product that could be produced, either in one batch or over a defined period of time, and given a set list of variables
catalyst	A substance whose presence initiates or changes the rate of a chemical reaction, but does not itself enter into the reaction
certified reference material (CRM)	Reference material characterized by a metrologically valid procedure for one or more specified properties, accompanied by a certificate that provides the value of the specified property, its associated uncertainty, and a statement of metrological traceability
chain of custody	Procedures and documents that account for the integrity of a specimen or sample by tracking its handling and storage from its point of collection to its final disposition
clandestine	Secret and concealed, often for illicit reasons
combined standard uncertainty	Standard uncertainty of the result of a measurement when that result is obtained from the values of a number of other quantities, equal to the positive square root of a sum of terms, the terms being the variances or covariances of these other quantities weighted according to how the measurement result varies with changes in these quantities
control	Material of established origin that is used to evaluate the performance of a test or comparison
deficiency of analysis	Any erroneous analytical result or interpretation, or any unapproved deviation from an established policy or procedure in an analysis
detection limit	The lowest concentration of analyte in a sample that can be detected, but not necessarily quantitated under the stated conditions of the test
expanded uncertainty (U)	Quantity defining an interval about a result of a measurement that may be expected to encompass a large fraction of the distribution of values that could reasonably be attributed to the measurand
false negative	Test result that states that an analyte is absent, when, in fact, it is present above the established limit of detection for the analyte in question
false positive	Test result that states that an analyte is present, when, in fact, it is not present or, is present in an amount less than a threshold or designated cut-off concentration
finished product	A manufactured product ready for use
intermediate	Substance that is manufactured for and consumed in or used for chemical processing to be transformed into another substance
limit of quantitation	The lowest concentration of an analyte that can be determined with acceptable precision (repeatability) and accuracy under the stated conditions of the test (EURACHEM)A.2.25 linearity defines the ability of the method to obtain test results proportional to the concentration of analyte
pharmaceutical identifiers	Physical characteristics of tablets, capsules or packaging indicating the identity, manufacturer or quantity of substances present
population	The totality of items or units of material under consideration

(Continued)

Term	Definition
precision	Closeness of agreement between independent test/measurement results obtained under stipulated conditions
precursor	A chemical that is transformed into another compound, as in the course of a chemical reaction, and therefore precedes that compound in the synthetic pathway
procedure	Specified way to carry out an activity or process
proficiency testing	Ongoing process in which a series of proficiency specimens or samples, the characteristics of which are not known to the participants, are sent to laboratories on a regular basis
qualitative analysis	Analysis in which substances are identified or classified on the basis of their chemical or physical properties, such as chemical reactivity, solubility, molecular weight, melting point, radiative properties (emission, absorption), mass spectra, nuclear half-life, etc.
quality assurance	Part of quality management focused on providing confidence that quality requirements will be fulfilled (ISO 9000:2005 (E))
quality management	Coordinated activities to direct and control an organization with regard to quality
quality manual	Document specifying the quality management system of an organization
quantitative analysis	Analyses in which the amount or concentration of an analyte may be determined (estimated) and expressed as a numerical value in appropriate units
random sample	The sample so selected that any portion of the population has an equal (or known) chance of being chosen
reagent	A chemical used to react with another chemical, often to confirm or deny the presence of the second chemical
recovery	Term used in analytical and preparative chemistry to denote the fraction of the total quantity of a substance recoverable following a chemical procedure
reference material (RM)	Material, sufficiently homogeneous and stable with respect to one or more specified properties, which has been established to be fit for its intended use in a measurement process
repeatability (of results of measurements)	Closeness of the agreement between the results of successive measurements of the same measurand carried out subject to all of the following conditions: - the same measurement procedure; - the same observer; - the same measuring instrument, used under the same conditions; - the same location; - repetition over a short period of time
reproducibility (of results of measurements)	Closeness of the agreement between the results of measurements of the same measurand, where the measurements are carried out under changed conditions such as: - principle or method of measurement; - observer; - measuring instrument; - location; - conditions of use; - time

(Continued)

Term	Definition
robustness	Measures a method's capacity to remain unaffected by small, but deliberate variations in method parameters and provides an indication of its reliability during normal usage
ruggedness	The degree of reproducibility of test results obtained by the analysis of the same samples under a variety of conditions, such as different laboratories, analysts, instruments, lots of reagents, elapsed assay times, assay temperatures, or days
sample	Subset of a population made up of one or more sampling units
sampling	Act of drawing or constituting a sample
sampling plan	A specific plan which states the sample size(s) to be used and the associated criteria for accepting the lot
sampling procedure	Operational requirements and/or instructions relating to the use of a particular sampling plan; i.e., the planned method of selection, withdrawal and preparation of sample(s) from a lot to yield knowledge of the characteristic(s) of the
sampling scheme	A combination of sampling plans with rules for changing from one plan to another
selectivity (in analysis)	1. (Qualitative): The extent to which other substances interfere with the determination of a substance according to a given procedure 2. (Quantitative): A term used in conjunction with another substantive (e.g., constant, coefficient, index, factor, number) for the quantitative characterization of interferences A.2.51 standard uncertainty of the result of a measurement expressed as a standard deviation
traceability	Ability to trace the history, application or location of that which is under consideration
trueness	Closeness of agreement between the expectation of a test result or a measurement result and a true value
uncertainty (measurement)	Parameter, associated with the measurement result, or test result, that characterizes the dispersion of the values that could reasonably be attributed to the particular quantity subject to measurement or characteristic subject to test
validation	Confirmation, through the provision of objective evidence, that the requirements for a specific intended use or application have been fulfilled
verification	Confirmation, through the provision of objective evidence, that specified requirements have been fulfilled
yield, expected	The quantity of material or the percentage of theoretical yield anticipated at any appropriate phase of production based on previous laboratory, pilot scale, or manufacturing data
yield, theoretical	The quantity that would be produced at any appropriate phase of production based upon the quantity of material to be used, in the absence of any loss or error in actual production

Appendix B
TWGFEX
Glossary of Terms

www.nist.gov/system/files/documents/2018/09/21/twgfex_glossary_of_terms.pdf

This glossary of terms was developed and adopted by the Technical Working Group for Fire and Explosives (TWGFEX) Core Committee from a variety of sources.

ANFO	A mixture of ammonium nitrate and fuel oil.
Base Charge	The main high explosive charge in a blasting cap.
Binary Explosive	Two substances which are not explosive until they are mixed.
Black Powder	A low explosive traditionally consisting of potassium nitrate, sulfur and charcoal. Sodium nitrate may be found in place of potassium nitrate.
Black Powder Substitutes	Modified black powder formulations such as but not limited to: Pyrodex, Black Canyon, Golden Powder, Clean Shot, and Clear Shot.
Blasting Agent	A high explosive with low-sensitivity usually based on ammonium nitrate and not containing additional high explosive(s).
Blasting Cap	A metal tube containing a primary high explosive capable of initiating most explosives.
Bomb	A device containing an explosive, incendiary, or chemical material designed to explode.
Booby Trap	A concealed or camouflaged device designed to injure or kill personnel.
Booster	A cap sensitive high explosive used to initiate other less sensitive high explosives.
Brisance	The shattering power associated with high explosives.
C4	A white pliable military plastic explosive containing primarily Cyclonite (RDX).
Cannon Fuse	A coated, thread-wrapped cord filled with black powder designed to initiate flame-sensitive explosives.
Combustion	Any type of exothermic oxidation reaction, including, but not limited to burning, deflagration and/or detonation.
Deflagration	An exothermic reaction that occurs particle to particle at subsonic speed.
Detasheet (Det Sheet)	A plastic explosive in sheet form containing PETN, HMX or RDX.

(Continued)

Detonation	An exothermic reaction that propagates a shockwave through an explosive at supersonic speed (greater than 3300 ft/sec)
Detonation Cord (Det-Cord)	A plastic/fiber wrapped cord containing a core of PETN or RDX.
Detonator	A device used for detonating many types of high explosives.
Double Base	A smokeless powder which contains both nitroglycerine and nitrocellulose.
Dud	An explosive device which has undergone a complete arming and firing cycle but has failed to explode. It should be noted that this is a very dangerous situation.
Dynamite	Originally a mixture of nitroglycerine and an absorbent filler now used to designate an entire class of high explosives.
Electric Match	A metal wire coated with a pyrotechnic mixture designed to produce a small burst of flame designed to initiate a low explosive.
Electric Squib	A metal wire surrounded by a pyrotechnic mixture and encased within a metal tube which produces a small jet of flame designed to initiate a low explosive.
Explosion	A rapid expansion of gases resulting from a chemical or physical action that produces a pressure wave.
Explosive	A chemical substance or mixture capable of producing an explosion.
Explosive Compound	A single chemical compound capable of causing an explosion.
Explosive Mixture	A mixture of chemical compounds capable of causing an explosion.
Explosive Train	A series of combustible or explosive components arranged in order of decreasing sensitivity designed to initiate explosives.
Firing Train	See "Explosive Train."
Flex-X	See "Det Sheet."
Frag	Any item(s) produced and cast away from an explosion.
Fuel	Any substance capable of reacting with oxygen or oxygen-carriers (oxidizers).
Fuse	A fiber wrapped cord of black powder used to initiate blasting caps or low explosives.
Fuze	A mechanical, chemical, or electrical device designed to initiate an explosive train.
Gunpowder	See "Black Powder."
High Explosive	Generally a chemical substance or mixture capable of detonation.
HMX	Octagen, a high explosive formed as a by-product during the manufacture of RDX.
Hoax	A "dummy" device intended to appear as a bomb but not containing an explosive.
Hobby Fuse	See "Cannon Fuse."

(Continued)

Improvised Explosive Device (IED)	A non-commercially produced device designed to explode.
Incendiary	A compound, metal or mixture capable of producing intense heat.
Inert	A simulated explosive or device that contains no explosive, pyrotechnic, or chemical/biological agent.
Initiator	The part of an explosive train which starts the reaction.
Low Explosive	Generally a chemical compound or mixture that can deflagrate without the addition of atmospheric oxygen.
Main Charge	The main or final explosive in an explosive train.
Munitions	Any and all military explosives.
Munroe Effect	The focusing of the force produced by an explosion resulting in an increased pressure wave.
Ordnance	See "Munitions."
Oxidizer	A chemical compound which supplies the oxygen in a chemical reaction.
PETN	Pentaerythritoltetranitrate, a high explosive used in many applications.
Plastic Bonded Explosives (PBX)	A high explosive in a pliable plastic matrix, i.e., C4, Det Flex.
Plastic Explosives	Common term for PBX.
Primary High Explosive	A high explosive sensitive to heat, shock, spark, and/or friction.
Primer	See "Initiator."
Primer Cap	A small metal device containing an impact sensitive primary high explosive commonly found in ammunition or used in initiators.
Pyrotechnic Fuse	See "Cannon Fuse."
Pyrotechnic Mixtures	An oxidizer/fuel mixture which produces bright or colored lights, heat, fogs, or acoustic effects.
RDX	Cyclonite, high explosive used in PBX and other applications.
Report	A loud sound produced by an explosion.
Safety Fuse	A water-proof coated, thread-wrapped cord filled with black powder designed to be used to initiate a non-electric blasting cap.
Secondary High Explosive	A less sensitive high explosive initiated by another explosive.
Semtex	Plastic explosive primarily containing Pentaerythritoltetranitrate (PETN) made in Czech Republic.
Shaped Charge	An explosive device which is designed to direct or focus explosive energy into a narrow jet. The created plasma has a synergistic effect increasing the heat and energy on the target area.
Shock Tube	Hollow plastic tube coated with a thin coating of HMX and powdered aluminum used in non-electric firing systems.

(Continued)

Shrapnel	Objects which are attached to the outside or included inside a device to increase the blast damage and/or injure/kill personnel. The device/container walls themselves can also function in this manner.
Single Base	A smokeless powder which contains nitrocellulose but does not contain nitroglycerine or nitroguanidine.
Smokeless Powder	A low explosive used in ammunition as a propellant, which can be single, double, or triple based.
Triple Base	A smokeless powder which contains nitrocellulose, nitroglycerine, and nitroguanidine.

Appendix C
Equipment

1 Authentic Variations

Scientific Equipment Encountered at Clandestine Labs

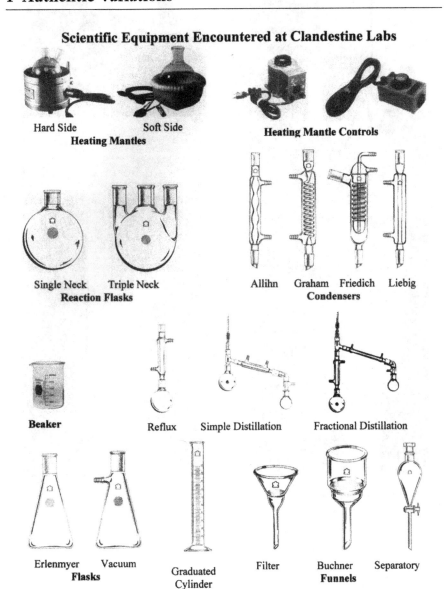

Hard Side   Soft Side
Heating Mantles   **Heating Mantle Controls**

Single Neck   Triple Neck   Allihn   Graham Friedich   Liebig
Reaction Flasks   **Condensers**

Beaker   Reflux   Simple Distillation   Fractional Distillation

Erlenmyer   Vacuum   Filter   Buchner   Separatory
Flasks   Graduated   **Funnels**
                          Cylinder

2 Reflux Variations

3 Distillation Variations

4 Hydrogenator Variation

5 Vacuum Filtration Variations

6 Extraction Equipment Variations

7 Makeshift Ventilation

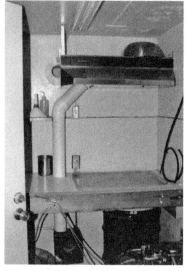

Appendix D
Chemical Lists

1 Legitimate Use Table

Chemical	Legitimate Uses
Acetaldehyde*	Manufacture of perfumes, flavors, plastics; silver mirrors
Acetic acid*	Food preservative; organic synthesis
Acetic anhydride*	Organic synthesis; dehydrating agent
Acetone	General solvent
Acetonitrile*	Solvent; organic synthesis
Allylbenzene*	None listed
Allylchloride*	Organic synthesis
4 Allyl 1,2 methylenedioxybenzene*	None listed
Aluminum powder	Paint additive, pyrotechnics, explosives, metal alloys
Aluminum chloride*	Disinfectant, deodorant, wood preservative
Ammonia gas*	Refrigerant; manufacture of nitric acid, fertilizer and explosives
Ammonium chloride*	Batteries, electroplating, textiles
Ammonium formate*	Inorganic metal analysis
Ammonium hydroxide*	None listed
Ammonium nitrate	Fertilizer, explosives, matches, pyrotechnics
Aniline*	Manufacture of dyes, perfumes, varnishes, organic synthesis
Aspirin	Over-the-counter medicine
Barium chlorate	Pyrotechnics, explosives, matches, dye processes
Barium nitrate	Pyrotechnics, explosives, vacuum tube manufacture
Benzaldehyde*	Manufacture of dyes, perfumes, flavors
Benzene*	Manufacture of pharmaceuticals, dyes; industrial solvent
Benzyl chloride*	Manufacture of dyes, pharmaceuticals, perfumes, resins
Benzyl cyanide*	None listed
Bromobenzene*	Manufacture of Grignard reagent
Bromoethane*	None listed
Carbon dioxide	Beverage carbonation, fire extinguisher, dry ice
Carbon tetrachloride*	Dry cleaning, fire extinguisher, general solvent
Chloro-2-propanone*	Intermediate for perfumes, drugs, insecticides; photography

(Continued)

Chemical	Legitimate Uses
Copper oxide	Ceramic and glass pigment, glass polishing
Copper sulfate	Fungicide, photography, food additive, paint pigment
Cyclohexanone*	Industrial solvent
Dichloroethane*	None listed
Ephedrine*	Bronchodilator
Ergotamine tartrate*	Treatment of migraine headaches
Erythritol	Sugar substitute
Ethyl acetate*	Manufacture of photo film, perfume, gun powder; dry cleaning
Ethylene glycol	Antifreeze, polymer precursor
Formamide*	Organic synthesis
Formic acid*	Tanning, electroplating, wool dying
Glycerin	Food preservative, antimicrobial, soap production
Guanidine	Plastic production
Hydrazine	Reducing agent, rocket fuel
Hydriodic acid*	Manufacture of pharmaceuticals, disinfectants
Hydrobromic acid*	Analytical reagent
Hydrochloric acid	Pool chemical, masonry and metal cleaner, mining
Hydrogen	Welding
Hydrogen peroxide*	Rocket fuel, bleaching agent
Hydroxylamine HCl*	Photography, antioxidant for soaps
Iodine*	Manufacture of germicides and antiseptics; catalyst
Lithium metal*	Catalyst, metal alloys, batteries
Lithium aluminum hydride*	Reducing agent
Lithium hydroxide	Photographic developers
Magnesium metal	Metal alloys, pyrotechnics, Grignard reagent
Mannitol	Sweetener, food additive, pharmaceutical applications
Mercuric chloride	Preservative; photography, mining, steel etching
Mercury*	Thermometers, switches, lighting, mining, dentistry
Methanol	Solvent, antifreeze, gas additive, camping fuel
Methyl acrylate*	Manufacture of plastics and textiles
Methylamine*	Tanning; organic synthesis
Nitric acid	Manufacture of fertilizers, dyes, explosives and a variety of organic chemicals
Nitroethane	Fuel additive, solvent
Perchloric acid*	Metal plating; explosives
Pentaerythritol	Fire retardant, paints and coating, organic synthesis
Phenylacetic acid*	Manufacture of perfumes
Phenylacetonitrile*	Organic synthesis
Phenylmagnesium bromide*	Organic synthesis

(Continued)

Chemical	Legitimate Uses
Phosphoric acid*	Manufacture of fertilizers, detergents; food add.; cleaning solvent
Phosphorus red*	Pyrotechnics; matches; fertilizers; pesticides
Phosphorus pentachloride*	Catalyst
Piperidine*	Organic synthesis
Potassium carbonate	Manufacture of soap, glass, pottery; engraving; tanning
Potassium chlorate	Explosives, pyrotechnics, matches, dye industry
Potassium chromate	Leather tanning, rust proofing metal
Potassium cyanide	Mining, electroplating, steel hardening
Potassium dichromate	Leather tanning, dyes, paints, pyrotechnics, matches, waterproofing fabrics
Potassium hydroxide	Paint/varnish remover, photo engraving, printing inks
Potassium iodide	Oxidizer for chemical analysis
Potassium nitrate	Tempering steel, tobacco curing, glass manufacture, explosives
Potassium perchlorate	Explosives, pyrotechnics, photography
Propylene glycol	Food additive, antifreeze, polymer production
Pyridine*	Solvent, organic synthesis
Raney nickel*	Catalyst
Sodium metal*	Manufacture of sodium compounds, sodium vapor lamps
Sodium acetate	Photography, food additive
Sodium bicarbonate	Baking soda, fire extinguishers, cleaning compounds
Sodium bisulfite	Disinfectant; textile bleaching; food additive
Sodium hydroxide	Drain cleaner
Sodium nitrate*	Glass manufacturing, pottery, fertilizer, meat and tobacco preparation
Strontium nitrate	Pyrotechnics, road flares, matches
Sulfuric acid	Battery acid; drain cleaner; manufacture of fertilizers, explosives
Tartaric acid	Food additive; photography; tanning; ceramics
Thorium nitrate*	Catalyst
Toluene	Paint solvent, industrial chemical, printing and leather tanning processes
Trifluoro acetic anhydride*	None listed
Urea	Fertilizers, pharmaceutical applications
Zinc metal	Metal alloys, batteries, mining, printing plates, household utensils

*No legitimate home or hobby use

Appendix E
List of Explosive Materials

Pursuant to 18 U.S.C. 841(d) and 27 CFR 555.23, the Department of Justice must publish and revise at least annually in the **Federal Register** a list of explosives determined to be within the coverage of 18 U.S.C. 841 *et seq.* The list covers not only explosives, but also blasting agents and detonators, all of which are defined as "explosive materials" in 18 U.S.C. 841(c).

Date approved: December 27, 2019.

Marvin G. Richardson, Associate Deputy Director.

Bureau of Alcohol, Tobacco, Firearms and Explosives, Department of Justice

www.federalregister.gov/documents/2020/01/02/2019-28316/commerce-in-explosives-2019-annual-list-of-explosive-materials

A

Acetylides of heavy metals.
Aluminum containing polymeric propellant.
Aluminum ophorite explosive.
Amatex.
Amatol.
Ammonal.
Ammonium nitrate explosive mixtures (cap sensitive).
Ammonium nitrate explosive mixtures (non-cap sensitive).
Ammonium perchlorate having particle size less than 15 microns.
Ammonium perchlorate explosive mixtures (excluding ammonium perchlorate composite propellant (APCP)).
Ammonium picrate [picrate of ammonia, Explosive D].
Ammonium salt lattice with isomorphously substituted inorganic salts.
* ANFO [ammonium nitrate-fuel oil].
Aromatic nitro-compound explosive mixtures.
Azide explosives.

B

Baranol.
Baratol.
BEAF [1, 2-bis (2, 2-difluoro-2-nitroacetoxyethane)].
Black powder.
Black powder based explosive mixtures.
Black powder substitutes.
*Blasting agents, nitro-carbo-nitrates, including non-cap sensitive slurry
    and water gel explosives.
Blasting caps.
Blasting gelatin.
Blasting powder.
BTNEC [bis (trinitroethyl) carbonate].
BTNEN [bis (trinitroethyl) nitramine].
BTTN [1,2,4 butanetriol trinitrate].
Bulk salutes.
Butyl tetryl.

C

Calcium nitrate explosive mixture.
Cellulose hexanitrate explosive mixture.
Chlorate explosive mixtures.
Composition A and variations.
Composition B and variations.
Composition C and variations.
Copper acetylide.
Cyanuric triazide.
Cyclonite [RDX].
Cyclotetramethylenetetranitramine [HMX].
Cyclotol.
Cyclotrimethylenetrinitramine [RDX].

D

DATB [diaminotrinitrobenzene].
DDNP [diazodinitrophenol].
DEGDN [diethyleneglycol dinitrate].

Detonating cord.
Detonators.
Dimethylol dimethyl methane dinitrate composition.
Dinitroethyleneurea.
Dinitroglycerine [glycerol dinitrate].
Dinitrophenol.
Dinitrophenolates.
Dinitrophenyl hydrazine.
Dinitroresorcinol.
Dinitrotoluene-sodium nitrate explosive mixtures.
DIPAM [dipicramide; diaminohexanitrobiphenyl].
Dipicryl sulfide [hexanitrodiphenyl sulfide].
Dipicryl sulfone.
Dipicrylamine.
Display fireworks.
DNPA [2,2-dinitropropyl acrylate].
DNPD [dinitropentano nitrile].
Dynamite.

E

EDDN [ethylene diamine dinitrate].
EDNA [ethylenedinitramine].
Ednatol.
EDNP [ethyl 4,4-dinitropentanoate].
EGDN [ethylene glycol dinitrate].
Erythritol tetranitrate explosives.
Esters of nitro-substituted alcohols.
Ethyl-tetryl.
Explosive conitrates.
Explosive gelatins.
Explosive liquids.
Explosive mixtures containing oxygen-releasing inorganic salts and hydrocarbons.
Explosive mixtures containing oxygen-releasing inorganic salts and nitro bodies.
Explosive mixtures containing oxygen-releasing inorganic salts and water insoluble fuels.
Explosive mixtures containing oxygen-releasing inorganic salts and water soluble fuels.

Explosive mixtures containing sensitized nitromethane.
Explosive mixtures containing tetranitromethane (nitroform).
Explosive nitro compounds of aromatic hydrocarbons.
Explosive organic nitrate mixtures.
Explosive powders.

F

Flash powder.
Fulminate of mercury.
Fulminate of silver.
Fulminating gold.
Fulminating mercury.
Fulminating platinum.
Fulminating silver.

G

Gelatinized nitrocellulose.
Gem-dinitro aliphatic explosive mixtures.
Guanyl nitrosamino guanyl tetrazene.
Guanyl nitrosamino guanylidene hydrazine.
Guncotton.

H

Heavy metal azides.
Hexanite.
Hexanitrodiphenylamine.
Hexanitrostilbene.
Hexogen [RDX].
Hexogene or octogene and a nitrated N-methylaniline.
Hexolites.
HMTD [hexamethylenetriperoxidediamine].
HMX [cyclo-1,3,5,7-tetramethylene 2,4,6,8-tetranitramine; Octogen].
Hydrazinium nitrate/hydrazine/aluminum explosive system.
Hydrazoic acid.

I

Igniter cord.
Igniters.
Initiating tube systems.

K

KDNBF [potassium dinitrobenzo-furoxane].

L

Lead azide.
Lead mannite.
Lead mononitroresorcinate.
Lead picrate.
Lead salts, explosive.
Lead styphnate [styphnate of lead, lead trinitroresorcinate].
Liquid nitrated polyol and trimethylolethane.
Liquid oxygen explosives.

M

Magnesium ophorite explosives.
Mannitol hexanitrate.
MDNP [methyl 4,4-dinitropentanoate].
MEAN [monoethanolamine nitrate].
Mercuric fulminate.
Mercury oxalate.
Mercury tartrate.
Metriol trinitrate.
Minol-2 [40% TNT, 40% ammonium nitrate, 20% aluminum].
MMAN [monomethylamine nitrate]; methylamine nitrate.
Mononitrotoluene-nitroglycerin mixture.
Monopropellants.

N

NIBTN [nitroisobutametriol trinitrate].
Nitrate explosive mixtures.
Nitrate sensitized with gelled nitroparaffin.
Nitrated carbohydrate explosive.
Nitrated glucoside explosive.
Nitrated polyhydric alcohol explosives.
Nitric acid and a nitro aromatic compound explosive.
Nitric acid and carboxylic fuel explosive.
Nitric acid explosive mixtures.
Nitro aromatic explosive mixtures.
Nitro compounds of furane explosive mixtures.
Nitrocellulose explosive.
Nitroderivative of urea explosive mixture.
Nitrogelatin explosive.
Nitrogen trichloride.
Nitrogen tri-iodide.
Nitroglycerine [NG, RNG, nitro, glyceryl trinitrate, trinitroglycerine].
Nitroglycide.
Nitroglycol [ethylene glycol dinitrate, EGDN].
Nitroguanidine explosives.
Nitronium perchlorate propellant mixtures.
Nitroparaffins Explosive Grade and ammonium nitrate mixtures.
Nitrostarch.
Nitro-substituted carboxylic acids.
Nitrotriazolone [3-nitro-1,2,4-triazol-5-one].
Nitrourea.

O

Octogen [HMX].
Octol [75 percent HMX, 25 percent TNT].
Organic amine nitrates.
Organic nitramines.

P

PBX [plastic bonded explosives].
Pellet powder.

Penthrinite composition.
Pentolite.
Perchlorate explosive mixtures.
Peroxide based explosive mixtures.
PETN [nitropentaerythrite, pentaerythrite tetranitrate, pentaerythritol tetranitrate].
Picramic acid and its salts.
Picramide.
Picrate explosives.
Picrate of potassium explosive mixtures.
Picratol.
Picric acid (manufactured as an explosive).
Picryl chloride.
Picryl fluoride.
PLX [95% nitromethane, 5% ethylenediamine].
Polynitro aliphatic compounds.
Polyolpolynitrate-nitrocellulose explosive gels.
Potassium chlorate and lead sulfocyanate explosive.
Potassium nitrate explosive mixtures.
Potassium nitroaminotetrazole.
Pyrotechnic compositions.
Pyrotechnic fuses.
PYX [2,6-bis(picrylamino)] 3,5-dinitropyridine.

R

RDX [cyclonite, hexogen, T4, cyclo-1,3,5,-trimethylene-2,4,6,-trinitramine; hexahydro-1,3,5-trinitro-S-triazine].

S

Safety fuse.
Salts of organic amino sulfonic acid explosive mixture.
Salutes (bulk).
Silver acetylide.
Silver azide.
Silver fulminate.
Silver oxalate explosive mixtures.
Silver styphnate.
Silver tartrate explosive mixtures.

Silver tetrazene.

Slurried explosive mixtures of water, inorganic oxidizing salt, gelling agent, fuel, and sensitizer (cap sensitive).

Smokeless powder.

Sodatol.

Sodium amatol.

Sodium azide explosive mixture.

Sodium dinitro-ortho-cresolate.

Sodium nitrate explosive mixtures.

Sodium nitrate-potassium nitrate explosive mixture.

Sodium picramate.

Squibs.

Styphnic acid explosives.

T

Tacot [tetranitro-2,3,5,6-dibenzo-1,3a,4,6a tetrazapentalene].

TATB [triaminotrinitrobenzene].

TATP [triacetonetriperoxide].

TEGDN [triethylene glycol dinitrate].

Tetranitrocarbazole.

Tetrazene [tetracene, tetrazine, 1(5-tetrazolyl)-4-guanyl tetrazene hydrate].

Tetrazole explosives.

Tetryl [2,4,6 tetranitro-N-methylaniline].

Tetrytol.

Thickened inorganic oxidizer salt slurried explosive mixture.

TMETN [trimethylolethane trinitrate].

TNEF [trinitroethyl formal].

TNEOC [trinitroethylorthocarbonate].

TNEOF [trinitroethylorthoformate].

TNT [trinitrotoluene, trotyl, trilite, triton].

Torpex.

Tridite.

Trimethylol ethyl methane trinitrate composition.

Trimethylolthane trinitrate-nitrocellulose.

Trimonite.

Trinitroanisole.

Trinitrobenzene.

Trinitrobenzenesulfonic acid [picryl sulfonic acid].

Trinitrobenzoic acid.

Trinitrocresol.
Trinitrofluorenone.
Trinitro-meta-cresol.
Trinitronaphthalene.
Trinitrophenetol.
Trinitrophloroglucinol.
Trinitroresorcinol.
Tritonal.

U

Urea nitrate.

W

Water-bearing explosives having salts of oxidizing acids and nitrogen bases, sulfates, or sulfamates (cap sensitive).
Water-in-oil emulsion explosive compositions.

X

Xanthomonas hydrophilic colloid explosive mixture.

Appendix F
Hazards

1 Chemical Hazard Table

Chemical	Hazards[j] NFPA Rating	Flash Point (deg. F)	Explosive Limit Lower/Upper	Incompatibilities
Acetaldehyde	3/4/2	−6	4%/60%	Strong oxidizers, acids, bases, alcohols, ammonia and amines, phenols, ketones, HCN, H_2S. (Note: Prolonged contact with air may cause formation of peroxides that may explode and burst containers; easily undergoes polymerization.)
Acetic Acid	3/2/0	104	4%/19.9%	Metals, acetic anhydride, alcohols, amines, ammonium nitrate, chlorine trifluoride, nitric acid, permanganates, peroxides, sodium hydroxide, sodium peroxide, hydrogen peroxides, acetaldehyde, caustics (e.g., ammonia, ammonium hydroxide, calcium hydroxide, potassium hydroxide, sodium hydroxide), acid anhydrides, chlorosulfonic acid, oleum, chromium trioxide, potassium hydroxide, carbonates, bromine pentafluoride, perchloric acid, chromic anhydride, potassium-tert-butoxide, calcium salts, ethyleneimine, attacks some forms of plastics, rubbers and coatings, 2-aminoethanol, ethylene diamine, phosphorus trichloride, chromic acid anhydride, phosphorus isocyanate, diallyl methyl carbinol + ozone, nitric acid + acetone, xylene, sodium salts.
Acetic Anhydride	3/2/1	NA	2.9%/10.3%	Strong oxidizing agents, strong reducing agents, bases, alcohols, metal powders, moisture.
Acetone	1/3/0	−4	2.5%/12.8%	Strong oxidizing agents, strong acids, perchlorates, aliphatic amines, chromyl chloride, hexachloromelamine, chromic anhydride, chloroform + alkali, potassium tert-butoxide.

(*Continued*)

Chemical	Hazards[i] NFPA Rating	Flash Point (deg. F)	Explosive Limit Lower/Upper	Incompatibilities
Acetonitrile	2/3/0	−4	4.4%/16%	Oxidizing agents, reducing agents, acids, bases, alkali metals, fluorine, nitric acid, perchlorates, sulfuric acid, chlorosulfonic acid, oleum, dinitrogen tetraoxide, sulfites, indium, moisture, attacks some forms of plastics, rubbers and coatings, nitrating agents, N-fluoro compounds (e.g., perfluorourea + acetonitrile), lanthanide perchlorates, iron (III) perchlorate, 2-cyano-2-propyl nitrate, trichlorosilane, diphenyl sulfoxide.
Allylbenzene	NA	177	NA/NA	NA
Allylchloride	2/3/1	30	2.9%/11.1%	Explosion hazard when exposed to acids or oxidizing agents, explosive reaction with alkyl aluminum chlorides + aromatic hydrocarbons (e.g., benzene or toluene). Violently exothermic polymerization reaction with Lewis acids (e.g., aluminum chloride, boron trifluoride and sulfuric acid). Incompatible with ethylene imine, ethylenediamine, chlorosulfonic acid, oleum, sodium hydroxide and nitric acid.
4 Allyl 1,2 Methylenedioxybenzene	NA	NA	NA/NA	NA
Aluminum (powder)	1/3/1 W	NA	NA	Strong oxidizers and acids, halogenated hydrocarbons. (Note: Corrodes in contact with acids and other metals. Ignition may occur if powders are mixed with halogens, carbon disulfide or methyl chloride.)
Aluminum Chloride	3/0/2	NA	NA/NA	Water, organic materials. Aluminum chloride reacts violently with water, producing hydrochloric acid and heat.
Ammonia gas	3/1/0	NA	15%/28%	Strong oxidizers, acids, halogens, salts of silver and zinc. (Note: Corrosive to copper and galvanized surfaces.)

(Continued)

Chemical	Hazards[i] NFPA Rating	Flash Point (deg. F)	Explosive Limit Lower/Upper	Incompatibilities
Ammonium Chloride	2/0/0	NA	NA/NA	Acids, alkalis and their associated carbonates. Substance reacts with lead and silver salts to form a fulminating compound. Substance reacts with ammonium compounds, bromine pentafluoride, bromine trifluoride, hydrogen cyanide, iodine heptafluoride, nitrates and potassium chlorate.
Ammonium Formate	2/0/0	NA	NA/NA	Strong oxidizing agents.
Ammonium Hydroxide	3/1/0	NA	16%/27%	Acrolein, acrylic acid, chlorosulfonic acid, dimethyl sulfate, fluorine, gold + aqua regia, hydrochloric acid, hydrofluoric acid, iodine, nitric acid, oleum, propiolactone, propylene oxide, silver nitrate, silver oxide, silver oxide + ethyl alcohol, nitromethane, silver permanganate, sulfuric acid, halogens. Forms explosive compounds with many heavy metals and halide salts.
Ammonium Nitrate	2/0/3 OXY	>200	NA	Reducing agents, powdered metals, strong acids.
Aniline	3/2/0	158	1.3%/11%	Strong acids and strong oxidizers, albumin, solutions of iron, zinc, aluminum, toluene diisocyanate, and alkalis. Ignites spontaneously in the presence of red fuming nitric acid and with sodium.
Benzaldehyde	2/2/0	NA	1.4%/8.5%	Performic acid and other oxidizing materials. An explosion occurred after mixing sodium hydrosulfite, aluminum powder, potassium carbonate and benzaldehyde.
Benzene	2/3/0	12	1.2%/7.8%	Strong oxidizers, many fluorides and perchlorates, nitric acid.
Benzyl Chloride	3/2/2	153	1.1%/14%	Oxidizers, acids, copper, aluminum, magnesium, iron, zinc, tin. (Note: Can polymerize when in contact with all common metals except nickel and lead. Hydrolyzes in H_2O to benzyl alcohol.)
Benzyl Cyanide pdc	2/1/0	223	NA/NA	Strong acids, strong bases, strong oxidizing agents, strong reducing agents, sodium hypochlorite.
Bromobenzene pdc	2/2/0	NA	0.5%/2.5%	Bromobutane + sodium, strong oxidizing agents, alkali metals.

(Continued)

Chemical	Hazards[i] NFPA Rating	Flash Point (deg. F)	Explosive Limit Lower/Upper	Incompatibilities
Bromoethane	3/1/0	NA	10%/16%	Risk of fire and explosion on contact with aluminum, zinc or magnesium.
Carbon Dioxide	NA	NA	NA/NA	Dusts of various metals, such as magnesium, zirconium, titanium, aluminum, chromium and manganese, are ignitable and explosive when suspended in carbon dioxide. Forms carbonic acid in water.
Carbontetrachloride	3/0/0	NA	NA/NA	Aluminum, bromine trifluoride, calcium hypochlorite, dimethyl formamide, ethylene oxide, fluorine, lithium, magnesium, potassium, potassium-tert-butoxide, silver perchlorate, sodium, uranium, chlorine trifluoride, dinitrogen tetraoxide, methanol.
Cellulose	0/0/0	NA	NA/NA	Water, bromine pentafluoride, sodium nitrate, fluorine, strong oxidizers.
Chloro-2-propanone	3/2/0	102	NA/NA	Strong acids, strong bases, strong oxidizing agents, strong reducing agents.
Citric Acid	1/0/0	NA	NA/NA	Oxidizers, sulfuric acid, nitric acid, strong bases.
Copper Sulfate	2/0/0	NA	NA/NA	Moisture, air, steel, finely powdered metals, hydroxylamine, magnesium, hydrazine, nitromethane.
Copper Oxide	2/0/0	NA	NA/NA	Aluminum, boron, cesium acetylene carbide, hydrazine, magnesium, phospham, potassium, rubidium acetylene carbide, sodium, titanium and zirconium. Forms explosive acetylides with acetylene in caustic solutions. Exposure to moist air at >212 °F can result in spontaneous combustion.
Cyclohexanone	1/2/0	111	1.1%/9.4%	Oxidizing agents, strong acids, amines, nitric acid, plastics, rubber, sulfuric acid, aliphatic amines, lead, red metals, resins.

(Continued)

Chemical	Hazards[i] NFPA Rating	Flash Point (deg. F)	Explosive Limit Lower/Upper	Incompatibilities
Dichloroethane	2/4/2	>233	13%/23%	Strong oxidizing agents, liquid oxygen, nitric acid, potassium, lithium, sodium, caustics (e.g. ammonia, ammonium hydroxide, calcium hydroxide, potassium hydroxide, sodium hydroxide), potassium-tert-butoxide, sodium potassium alloys, powdered aluminum, active metals (such as potassium and magnesium), nitrogen tetroxide, N-methyl-N-nitrosourea + potassium hydroxide, powdered magnesium.
Ephedrine	1/0/0	NA	NA/NA	Oxidizing agents, direct light.
Erythritol	0/0/0	NA	NA/NA	Strong oxidizers, strong bases, oxidizers.
Ethyl Acetate	1/3/0	24	2.0%/9.0%	Chlorosulfonic acid, lithium aluminum hydride + 2-chloromethylfuran, lithium tetrahydroaluminate, oleum, potassium t-butoxide. Substance coming in contact with nitrates or strong acids/oxidizers/alkalis may cause fire.
Formamide	2/1/0	310	2.7%/19%	Strong oxidizing agents, acids, bases, aluminum.
Formic Acid	3/2/0	>233	18%/57%	Strong oxidizing agents, strong bases, finely powdered metals, permanganates, sulfuric acid, hydrogen peroxides, nitromethane, furfuryl alcohol, hydrated thallium nitrate.
Glycerol	1/1/0	320	NA/NA	Strong oxidizers (e.g., chromium trioxide, potassium chlorate, potassium permanganate). (Note: Hygroscopic (i.e., absorbs moisture from the air).)
Hexamine	2/2/1	482	NA/NA	Strong oxidizing agents, strong acids.
Hydrazine	4/2/1	99	2.9%/98%	Oxidizers, hydrogen peroxide, nitric acid, metallic oxides, acids. (Note: Can ignite SPONTANEOUSLY on contact with oxidizers or porous materials such as earth, wood and cloth.)

(*Continued*)

Chemical	Hazards[i] NFPA Rating	Flash Point (deg. F)	Explosive Limit Lower/Upper	Incompatibilities
Hydriodic Acid (HI)	3/0/0	NA	NA/NA	Explodes on contact with ethyl hydroperoxide, ignites on contact with magnesium, perchloric acid, potassium + heat, potassium chlorate + heat, and oxidants, violent reaction with $HClO_4$ + Mg, metals, potentially violent reaction with phosphorus.
Hydrobromic Acid (HBr)	3/0/0	NA	NA/NA	Strong oxidizers, strong caustics, moisture, copper, brass, zinc (Note: Hydrobromic acid is highly corrosive to most metals.)
Hydrochloric Acid (HCl)	3/0/0	NA	NA/NA	Hydroxides, amines, alkalis, copper, brass, zinc. (Note: Hydrochloric acid is highly corrosive to most metals.)
Hydrogen	0/4/0	gas	4%/75%	Oxidizing agents, some metals, alkaline material, halogens.
Hydrogen Peroxide	4/0/1	NA	40%/100%	Strong oxidizing agents, strong reducing agents, acetic acid, acetic anhydride, alcohols, brass, copper, copper alloys, finely powdered metals, galvanized iron, hydrazine, iron, magnesium, nitric acid, sodium carbonate, potassium permanganate, cyanides (e.g., potassium cyanide, sodium cyanide), ethers (e.g., dioxane, furfuran, tetrahydrofuran (THF)), urea, chlorosulfonic acid, alkalis, lead, nitrogen compounds, triethylamine, silver, nickel, palladium, organic matter, charcoal, sodium borate, aniline, platinum, formic acid, cyclopentadiene, activated carbon, tert-butyl alcohol, hydrogen selenide, manganese dioxide, mercurous chloride, rust, ketones, carboxylic acids, glycerine, sodium fluoride, sodium pyrophosphate, soluble fuels (acetone, ethanol, glycerol), wood, asbestos, hexavalent chromium compounds, salts of iron, copper, chromium, vanadium, tungsten, molybdenum and platinum. (Note: Contact with combustible material may result in SPONTANEOUS combustion.)
Hydroxylamine HCl	2/0/3	305	NA/NA	Strong oxidizing agents, heat plus sodium acetate or ether, carbonyl compounds, copper sulfate, zinc and phosphorus chlorides.

(Continued)

Chemical	Hazards[i] NFPA Rating	Flash Point (deg. F)	Explosive Limit Lower/Upper	Incompatibilities
Iodine	NA	NA	NA/NA	Incompatible with ammonia, powdered metals, alkali metals and strong reducing agents. Reaction can be violent or explosive with acetaldehyde and acetylene. Reacts with ammonium hydroxide to form shock-sensitive iodides on drying.
Isosafrole*	1/1/0	120	NA/NA	Oxidizing agents.
Lead Acetate	1/0/0	NA	NA/NA	Strong acids, oxidizers.
Lead Nitrate	NA	NA	NA/NA	Strong reducing agents, organic materials, powdered metals.
Lithium	3/2/2	NA	NA/NA	Moisture, acids, oxidizers, oxygen, nitrogen, carbon dioxide.
Lithium Aluminum Hydroxide	3/0/2	NA	NA/NA	Water.
Magnesium Turnings	0/1/1	NA	NA/NA	Oxygen, moisture, chlorinated solvents, methanol, hydrogen peroxide, oxidizing agents, sulfur compounds, metal oxides, metal cyanides, metal oxide salts, fluorine, carbonates, halogens, phosphates.
Manganous Carbonate	1/0/1	NA	NA/NA	Contact with acids may generate carbon dioxide gas. Oxidizes toxic sulfur dioxide to the more toxic sulfur trioxide and causes violent decomposition of hydrogen peroxide.
Manganous Chloride	1/0/1	NA	NA/NA	Strong reducing agents, hydrogen peroxide, potassium, sodium and zinc.
Mercuric Chloride pdc	4/0/1	NA	NA/NA	Reacts violently with potassium and sodium. Incompatible with many compounds: formates, sulfites, phosphates, albumin, ammonia, gelatin, carbonates, hypophosphites, sulfides, alkalis, alkaloid salts, lime water, antimony and arsenic, bromides, borax, reduced iron, copper, iron, lead, tannic acid and vegetable astringents.
Mercury	3/0/0	NA	NA/NA	Acetylenes, ammonia, ethylene oxide, chlorine dioxide, azides, metal oxides, methyl silane, lithium, rubidium, oxygen, strong oxidants, metal carbonyls.

(Continued)

Chemical	Hazards[i] NFPA Rating	Flash Point (deg. F)	Explosive Limit Lower/Upper	Incompatibilities
Methanol	3/3/1	54	6%/36%	Strong oxidizing agents such as nitrates, perchlorates or sulfuric acid. Will attack some forms of plastics, rubber and coatings. May react with metallic aluminum and generate hydrogen gas.
Methylamine	3/3/0	39	4.9%/20.8%	Nitromethane, acids, oxidizing agents, chlorine, hypochlorite, halogenated agents, mercury, copper, copper alloys, zinc, zinc alloys, aluminum, perchlorates.
Methyl Ethyl Ketone	1/3/0	16	1.4%/11.4%	Strong oxidizers, amines, ammonia, inorganic acids, caustics, isocyanates, pyridines.
Methyl Ethyl Ketone Peroxide	2/2/4	75	NA/NA	Organic materials, heat, flames, sunlight, trace contaminants. (Note: A strong oxidizing agent. Pure MEKP is shock sensitive. Commercial product is diluted with 40% dimethyl phthalate, cyclohexane peroxide or diallyl phthalate to reduce sensitivity to shock.)
Methylformamide	1/1/0	NA	NA/NA	Strong oxidizing agents, acids, bases, acid chlorides.
Nitric Acid	3/0/2 OXY	NA	NA/NA	Combustible materials, metallic powders, hydrogen sulfide, carbides, alcohols. (Note: Reacts with water to produce heat. Corrosive to metals.)
Nitroethane	1/3/3	82	3.4%/NA	Amines; strong acids, alkalis and oxidizers; hydrocarbons; combustibles; metal oxides.
Norpseudoephedrine	1/0/0	NA	NA/NA	Oxidizing agents, direct light.
Palladium Sulfate	NA	NA	NA/NA	Strong oxidizing agents. Protect from freezing.
Perchloric Acid	3/0/3	102	NA/NA	Incompatible with numerous materials, including combustible materials, organic chemicals, strong dehydrating agents, reducing and oxidizing agents. Reacts violently with benzene, calcium hydride, wood, acetic acid, charcoal, olefins, ethanol, sulfur and sulfuric acid. Do Not Use Perchloric Acid in a Hood Designed for Other Purposes.
Phenylacetic Acid	2/0/0	168	NA/NA	NA

(Continued)

Chemical	Hazards[i] NFPA Rating	Flash Point (deg. F)	Explosive Limit Lower/Upper	Incompatibilities
Phenylmagnesium Bromide	NA	NA	NA/NA	Water
Phenylacetone	NA	NA	NA/NA	NA
Phosphorus (red)	0/2/2	NA	NA/NA	Halogens, halides, sulfur, oxidizing materials and alkalis (forms phosphine).
Phosphorus Pentachloride	3/0/2	NA	NA/NA	Reacts violently with water. Alcohols, amines, aluminum, sodium, potassium, acids.
Piperonal	1/0/0	NA	NA/NA	Strong oxidizing agents.
Piperidine	3/3/0	NA	NA/NA	Acids; acid chlorides; acid anhydrides; carbon dioxide; strong oxidizing agents; dicyanofurazan; N-nitrosoacetanilide; 1-perchlorylpiperidine
Platinum	NA	NA	NA/NA	Aluminum, acetone, arsenic, ethane, hydrazine, hydrogen peroxide, lithium, phosphorus, selenium, tellurium, various fluorides.
Platinum Chloride	NA	1076	NA/NA	Strong oxidizing agents.
Platinum Oxide	2/0/0	NA	NA/NA	Oxidizing agents.
Potassium Carbonate	2/0/1	NA	NA/NA	Acids, chlorine trifluoride, magnesium. An explosion occurred after mixing sodium hydrosulfite, aluminum powder, potassium carbonate and benzaldehyde.
Potassium Chlorate	2/0/3 OXY	NA	NA/NA	Extremely reactive or incompatible with reducing agents, combustible materials, organic materials.
Potassium Cyanide	4/0/0	NA	NA/NA	Violent reactions can occur with oxidizing agents such as nitric acid, nitrates and peroxides. Contact with acids liberates extremely toxic and flammable hydrogen cyanide gas. Hydrogen cyanide may form by a reaction with carbon dioxide and moisture when this material is in prolonged contact with air in a closed system.

(Continued)

Chemical	Hazards¹ NFPA Rating	Flash Point (deg. F)	Explosive Limit Lower/Upper	Incompatibilities
Potassium Hydroxide	3/0/2	NA	NA/NA	Contact with water, acids, flammable liquids and organic halogen compounds, especially trichloroethylene, may cause fire or explosion. Contact with nitromethane and other similar nitro compounds causes formation of shock-sensitive salts. Contact with metals such as aluminum, tin and zinc causes formation of flammable hydrogen gas.
Potassium Nitrate	1/0/2 OX	NA	NA/NA	Strong reducing agents, powdered metals, strong acids, organic materials.
Potassium Perchlorate	1/0/1 OX	NA	NA/NA	Strong reducing agents, powdered metals, strong acids, organic materials, forms shock-sensitive mixtures with certain other materials, alcohols.
Pumice	NA	NA	NA/NA	NA
Pyridine	2/3/0	68	1.8%/12.4%	Strong oxidizers, strong acids.
Raney Nickel*	NA	NA	NA/NA	NA
Sodium	3/3/3	NA	NA/NA	Water, oxygen, carbon dioxide, carbon tetrachloride, halogens, acetylene, metal halides, ammonium salts, oxides, oxidizing agents, acids, alcohols, chlorinated organic compounds, many other substances.
Sodium Acetate	3/0/1	NA	NA/NA	Nitric acid, fluoride, potassium nitrate, strong oxidizers and diketene.
Sodium Azide	4/0/0	NA	NA/NA	Acids, metals, water. (Note: Over a period of time, sodium azide may react with copper, lead, brass or solder in plumbing systems to form an accumulation of the HIGHLY EXPLOSIVE compounds of lead azide and copper azide.)
Sodium Bisulfate*	2/0/0	NA	NA/NA	Strong bases, strong oxidizing agents, strong reducing agents.
Sodium Chlorate	1/0/2 OX	NA	NA/NA	

(Continued)

Chemical	Hazards[i] NFPA Rating	Flash Point (deg. F)	Explosive Limit Lower/Upper	Incompatibilities
Sodium Hydroxide	3/0/2	NA	NA/NA	Metals, acids, nitro compounds, halogenated organics (e.g., dibromoethane, hexachlorobenzene, methyl chloride, trichloroethylene), nitromethane, flammable liquids.
Sodium Nitrate	2/1/3 OX	NA	NA/NA	A mixture of sodium nitrate and sodium hypophosphite constitutes a powerful explosive (Mellor 8, Supp. 1:154 1964). Sodium nitrate and aluminum powder mixtures have been reported to be explosive (Fire, 1935, 28, 30). The nitrate appears to be incompatible with barium thiocyanate, antimony, arsenic trioxide/iron(II) sulfate, boron phosphide, calcium-sodium alloy, magnesium, metal amidosulfates, metal cyanides, powdered charcoal, peroxyformic acid, phenol/trifluoroacetic acid, sodium, sodium nitrite/sodium sulfide, sodium phosphinate, sodium thiosulfate, tris(cyclopentadienyl)cerium and even wood.
Sodium Sulfate	1/0/0	NA	NA/NA	Strong oxidizing agents, aluminum, magnesium, potassium, mercury, lead, calcium, silver, barium, ammonium ions, strontium.
Sucrose (Sugar)	1/0/0	NA	NA/NA	Oxidizers, sulfuric acid, nitric acid.
Sulfuric Acid	3/0/3	NA	NA/NA	Water, potassium chlorate, potassium perchlorate, potassium permanganate, sodium, lithium, bases, organic material, halogens, metal acetylides, oxides and hydrides, metals (yields hydrogen gas), strong oxidizing and reducing agents and many other reactive substances.
Toluene	2/3/0	39	1.2/7	Oxidizing agents, acids. Reacts violently with strong oxidizing agents.
Thionyl Chloride	4/0/2	NA	NA/NA	Water, ammonia, chloryl perchlorate, dimethyl sulfoxide, linseed oil, quinoline, sodium, 2,4-hexadiyn-1-6-diol, o-nitrobenzoyl acetic acid and o-nitrophenylacetic acid.
Thorium Nitrate	NA	NA	NA/NA	Strong oxidizers.
Urea	0/0/0	NA	NA/NA	Incompatible with halogens, hydrogen peroxide, chlorinated hydrocarbons, fluorine, nitric acid, oxidizing agents and sulfuric acid.

1.1 Key to NFPA Ratings

HEALTH (H) HAZARD – The left quadrant. BLUE.	FIRE (F) HAZARD – The top quadrant. RED.	REACTIVITY (R) HAZARD – The right quadrant. YELLOW.	IMPORTANT MESSAGES – The bottom quadrant. WHITE.
4. DANGER: May be fatal on short exposure. Specialized protective equipment required.	4. DANGER: Flammable gas or extremely flammable liquid.	4. DANGER: Explosive material at room temperature.	W- Avoid use of water.
3. WARNING: Corrosive or toxic. Avoid skin contact or inhalation.	3. WARNING: Flammable liquid. Flash point below 100 °F.	3. DANGER: May be explosive if shocked, heated under confinement, or mixed with water.	COR– Corrosive.
2. WARNING: May be harmful if inhaled or absorbed.	2. CAUTION: Combustible liquid. Flash point of 100 °F to 200 °F.	2. WARNING: Unstable, or may react if heated or mixed with water.	LAS – Laser Electrical Hazard.
1. CAUTION: May cause irritation.	1. Combustible if heated.	1. CAUTION: May react if heated or mixed with water	AZK – Alkali.
0. No unusual hazard.	0. Not combustible.	0. Stable. Not reactive when mixed with water.	ACID – Acid.
			Radiation
			OXY – Oxidizing Chemicals.
			RED – Dangerous reducing agent –metal hydride.

2 Health Hazards

Chemical	Effects	Target Organs	Immediately Dangerous To Life or Health (IDLH) (ppm)	National Institute of Occupational Safety and Health (NIOSH) Relative Exposure Limit (REL) (ppm)	Occupational Safety and Health Administration (OSHA) Relative Exposure Limit (PEL) (ppm)
Acetaldehyde	Causes severe eye irritation. Vapors may cause eye irritation. May cause transient corneal injury. Lachrymator (substance which increases the flow of tears). May cause skin irritation. May cause skin sensitization, an allergic reaction, which becomes evident upon re-exposure to this material. May cause gastrointestinal irritation with nausea, vomiting and diarrhea. May be harmful if swallowed. May cause central nervous system depression. Causes respiratory tract irritation. May cause narcotic effects in high concentration. Exposure produces central nervous system depression. Vapors may cause dizziness or suffocation. Can produce delayed pulmonary edema. Inhalation of large amounts may cause respiratory stimulation, followed by respiratory depression, convulsions and possible death due to respiratory paralysis.	Eyes, skin, respiratory system, kidneys, central nervous system, reproductive system	2000	None Listed	200

(Continued)

Chemical	Effects	Target Organs	Immediately Dangerous To Life or Health (IDLH) (ppm)	National Institute of Occupational Safety and Health (NIOSH) Relative Exposure Limit (REL) (ppm)	Occupational Safety and Health Administration (OSHA) Relative Exposure Limit (PEL) (ppm)
Acetic Acid	Causes severe eye irritation. Contact with liquid or vapor causes severe burns and possible irreversible eye damage. Causes skin burns. May be harmful if absorbed through the skin. Contact with the skin may cause blackening and hyperkeratosis of the skin of the hands. May cause severe and permanent damage to the digestive tract. Causes severe pain, nausea, vomiting, diarrhea and shock. May cause polyuria, oliguria and anuria. Rapidly absorbed from the gastrointestinal tract. Effects may be delayed. Causes chemical burns to the respiratory tract. Exposure may lead to bronchitis, pharyngitis and dental erosion. May be absorbed through the lungs.	Eyes, skin, respiratory system, teeth	50	10	10
Acetic Anhydride	In case of contact, immediately flush eyes with plenty of water for at least 15 minutes. Get medical aid immediately. In case of contact, immediately flush skin with plenty of water for at least 15 minutes while removing contaminated clothing and shoes. Get medical aid immediately. Wash clothing before reuse. If swallowed, do NOT induce vomiting. Get medical aid immediately. If victim is fully conscious, give a cupful of water. Never give anything by mouth to an unconscious person. If inhaled, remove to fresh air. If not breathing, give artificial respiration. If breathing is difficult, give oxygen. Get medical aid.	Central nervous system, eyes, skin, mucous membranes	200	5	5

(Continued)

Chemical	Effects	Target Organs	Immediately Dangerous To Life or Health (IDLH) (ppm)	National Institute of Occupational Safety and Health (NIOSH) Relative Exposure Limit (REL) (ppm)	Occupational Safety and Health Administration (OSHA) Relative Exposure Limit (PEL) (ppm)
Acetone	Flush eyes with plenty of water for at least 15 minutes, occasionally lifting the upper and lower eyelids. Get medical aid immediately. Flush skin with plenty of soap and water for at least 15 minutes while removing contaminated clothing and shoes. Get medical aid if irritation develops or persists. Wash clothing before reuse. Do NOT induce vomiting. If victim is conscious and alert, give 2–4 cupfuls of milk or water. Never give anything by mouth to an unconscious person. Get medical aid immediately. Remove from exposure to fresh air immediately. If not breathing, give artificial respiration. If breathing is difficult, give oxygen. Get medical aid. Do NOT use mouth-to-mouth resuscitation. If breathing has ceased apply artificial respiration using oxygen and a suitable mechanical device such as a bag and a mask.	Eyes, skin, respiratory system, central nervous system	2500	250	1000

(*Continued*)

Chemical	Effects	Target Organs	Immediately Dangerous To Life or Health (IDLH) (ppm)	National Institute of Occupational Safety and Health (NIOSH) Relative Exposure Limit (REL) (ppm)	Occupational Safety and Health Administration (OSHA) Relative Exposure Limit (PEL) (ppm)
Acetonitrile*	Causes eye irritation. Lachrymator (substance which increases the flow of tears). May produce superficial reversible injury. Causes mild skin irritation. Harmful if absorbed through the skin. May be metabolized to cyanide, which in turn acts by inhibiting cytochrome oxidase, impairing cellular respiration. May cause gastrointestinal irritation with nausea, vomiting and diarrhea. May cause effects similar to those for inhalation exposure. May cause tissue anoxia, characterized by weakness, headache, dizziness, confusion, cyanosis (bluish skin due to deficient oxygenation of the blood), weak and irregular heart beat, collapse, unconsciousness, convulsions, coma and death. May cause central nervous system depression. Metabolism may release cyanide, which may result in headache, dizziness, weakness, collapse, unconsciousness and possible death. Aspiration may lead to pulmonary edema. Vapors may cause dizziness or suffocation. Causes upper respiratory tract irritation. May be metabolized to cyanide, which in turn acts by inhibiting cytochrome oxidase, impairing cellular respiration. May cause tissue anoxia, characterized by weakness, headache, dizziness, confusion, cyanosis (bluish discoloration of skin due to deficient oxygenation of the blood), weak and irregular heart beat, collapse, unconsciousness, convulsions, coma and death.	Respiratory system, cardiovascular system, central nervous system, liver, kidneys	500	20	40

(Continued)

Chemical	Effects	Target Organs	Immediately Dangerous To Life or Health (IDLH) (ppm)	National Institute of Occupational Safety and Health (NIOSH) Relative Exposure Limit (REL) (ppm)	Occupational Safety and Health Administration (OSHA) Relative Exposure Limit (PEL) (ppm)
Allylbenzene*	Immediately flush eyes with plenty of water for at least 15 minutes, occasionally lifting the upper and lower eyelids. Get medical aid. Flush skin with plenty of soap and water for at least 15 minutes while removing contaminated clothing and shoes. Get medical aid if irritation develops or persists. Wash clothing before reuse. If victim is conscious and alert, give 2–4 cupfuls of milk or water. Never give anything by mouth to an unconscious person. Get medical aid. Remove from exposure to fresh air immediately. If not breathing, give artificial respiration. If breathing is difficult, give oxygen. Get medical aid if cough or other symptoms appear.	None	None Listed	None Listed	None Listed
Allylchloride*	Irritation eyes, skin, nose, mucous membrane; pulmonary edema; in animals: liver, kidney injury.	Eyes, skin, respiratory system, liver, kidneys	250	1	1
4 Allyl 1,2 Methylenedioxybenzene	None Listed	None Listed	None Listed	None Listed	None Listed
Aluminum (powder)	None Listed	Eyes, skin, respiratory system	None Listed	10 mg/m^3	10 mg/m^3
Aluminum Chloride*	Causes severe eye burns. Causes skin burns. Causes gastrointestinal tract burns. May cause corrosion and permanent tissue destruction of the esophagus and digestive tract. Causes delayed lung injury. Causes severe irritation of upper respiratory tract with coughing, burns, breathing difficulty, and possible coma.	Eyes, skin, mucous membranes	None Listed	10 mg/m^3	15 mg/m^3

(Continued)

Chemical	Effects	Target Organs	Immediately Dangerous To Life or Health (IDLH) (ppm)	National Institute of Occupational Safety and Health (NIOSH) Relative Exposure Limit (REL) (ppm)	Occupational Safety and Health Administration (OSHA) Relative Exposure Limit (PEL) (ppm)
Ammonia gas*	Irritation eyes, nose, throat; dyspnea (breathing difficulty), wheezing, chest pain; pulmonary edema; pink frothy sputum; skin burns, vesiculation; liquid: frostbite.	Eyes, skin, respiratory system	300	25	50
Ammonium Chloride*	Irritation eyes, skin, respiratory system; cough, dyspnea (breathing difficulty), pulmonary sensitization.	Eyes, skin, respiratory system	None Listed	10 mg/m^3	None Listed
Ammonium Formate*	Causes eye irritation. Causes skin irritation. May cause gastrointestinal irritation with nausea, vomiting and diarrhea. The toxicological properties of this substance have not been fully investigated. May cause respiratory tract irritation.	None Listed	None Listed	None Listed	None Listed
Ammonium Hydroxide*	Contact with liquid or vapor causes severe burns and possible irreversible eye damage. Causes severe skin irritation. Causes skin burns. May cause deep, penetrating ulcers of the skin. Contact with the skin may cause staining, inflammation and thickening of the skin. Harmful if swallowed. May cause severe and permanent damage to the digestive tract. Causes gastrointestinal tract burns. Causes throat constriction, vomiting, convulsions and shock. Effects may be delayed. Causes severe irritation of upper respiratory tract with coughing, burns, breathing difficulty and possible coma.	Eyes, skin, mucous membranes	None Listed	None Listed	None Listed
Ammonium Nitrate		Eyes, skin, mucous membranes	None Listed	None Listed	None Listed

(Continued)

Chemical	Effects	Target Organs	Immediately Dangerous To Life or Health (IDLH) (ppm)	National Institute of Occupational Safety and Health (NIOSH) Relative Exposure Limit (REL) (ppm)	Occupational Safety and Health Administration (OSHA) Relative Exposure Limit (PEL) (ppm)
Aniline*	Headache, lassitude (weakness, exhaustion), dizziness; cyanosis; ataxia; dyspnea (breathing difficulty) on effort; tachycardia; irritation eyes; methemoglobinemia; cirrhosis; (potential occupational carcinogen)	Blood, cardiovascular system, eyes, liver, kidneys, respiratory system	100	None listed	5
Benzaldehyde*	Causes eye irritation. Causes skin irritation. Harmful if swallowed. May cause gastrointestinal irritation with nausea, vomiting and diarrhea. May cause central nervous system depression, characterized by excitement, followed by headache, dizziness, drowsiness and nausea. Advanced stages may cause collapse, unconsciousness, coma and possible death due to respiratory failure. Inhalation of high concentrations may cause central nervous system effects characterized by nausea, headache, dizziness, unconsciousness and coma. May cause respiratory tract irritation. May cause narcotic effects in high concentration. Prolonged or repeated skin contact may cause dermatitis. May cause kidney injury.	Kidneys, central nervous system	None Listed	None Listed	None Listed
Benzene*	Irritation eyes, skin, nose, respiratory system; dizziness; headache, nausea, staggered gait; anorexia, lassitude (weakness, exhaustion); dermatitis; bone marrow depression; (potential occupational carcinogen).	Eyes, skin, respiratory system, blood, central nervous system, bone marrow Cancer Site (leukemia)	500	0.1	1

(Continued)

Chemical	Effects	Target Organs	Immediately Dangerous To Life or Health (IDLH) (ppm)	National Institute of Occupational Safety and Health (NIOSH) Relative Exposure Limit (REL) (ppm)	Occupational Safety and Health Administration (OSHA) Relative Exposure Limit (PEL) (ppm)
Benzyl Chloride*	Irritation eyes, skin, nose; lassitude (weakness, exhaustion); irritability; headache; skin eruption; pulmonary edema	Eyes, skin, respiratory system, central nervous system	10	1	1
Benzyl Cyanide*	Causes eye irritation. Causes skin irritation. Harmful if absorbed through the skin. May be metabolized to cyanide, which in turn acts by inhibiting cytochrome oxidase, impairing cellular respiration. Harmful if swallowed. May cause irritation of the digestive tract. Metabolism may release cyanide, which may result in headache, dizziness, weakness, collapse, unconsciousness and possible death. Ingestion may result in symptoms similar to cyanide poisoning, which is characterized by asphyxiation. May be fatal if inhaled. May cause effects similar to those described for ingestion.	Blood, kidneys, liver, spleen, brain	None Listed	None Listed	5 mg/m^3
Bromobenzene*	Causes eye irritation. Causes skin irritation. May be absorbed through the skin in harmful amounts. If absorbed, may cause liver injury. Causes gastrointestinal irritation with nausea, vomiting and diarrhea. May cause central nervous system depression. Inhalation of high concentrations may cause central nervous system effects characterized by nausea, headache, dizziness, unconsciousness and coma. Causes respiratory tract irritation. May cause narcotic effects in high concentration. May cause liver abnormalities. Vapors may cause dizziness or suffocation. May cause blood changes.	Blood, central nervous system, liver	None Listed	None Listed	None Listed

(Continued)

Chemical	Effects	Target Organs	Immediately Dangerous To Life or Health (IDLH) (ppm)	National Institute of Occupational Safety and Health (NIOSH) Relative Exposure Limit (REL) (ppm)	Occupational Safety and Health Administration (OSHA) Relative Exposure Limit (PEL) (ppm)
Bromoethane	The substance irritates the eyes, the skin and the respiratory tract. Inhalation of the substance may cause lung edema. Rapid evaporation of the liquid may cause frostbite. The substance may cause effects on the central nervous system, kidneys and lungs. Exposure to high concentrations may result in death. The effects may be delayed.	Central nervous system, kidneys and lungs	None Listed	None Listed	None Listed
Carbon Dioxide	Headache, dizziness, restlessness, paresthesia; dyspnea (breathing difficulty); sweating, malaise (vague feeling of discomfort); increased heart rate, cardiac output, blood pressure; coma; asphyxia; convulsions; frostbite (liquid, dry ice).	Respiratory system, cardiovascular system	40,000	5000	5000
Carbon Tetrachloride*	Irritation eyes, skin; central nervous system depression; nausea, vomiting; liver, kidney injury; drowsiness, dizziness, incoordination; (potential occupational carcinogen).	Central nervous system, eyes, lungs, liver, kidneys, skin	200	2	10
Cellulose	Causes irritation eyes, skin, mucous membrane.	Eyes, skin, respiratory system	None Listed	10	15

(Continued)

Chemical	Effects	Target Organs	Immediately Dangerous To Life or Health (IDLH) (ppm)	National Institute of Occupational Safety and Health (NIOSH) Relative Exposure Limit (REL) (ppm)	Occupational Safety and Health Administration (OSHA) Relative Exposure Limit (PEL) (ppm)
Chloro-2-propanone (Chloroacetone)	Causes eye burns. Lachrymator (substance which increases the flow of tears). May cause chemical conjunctivitis and corneal damage. May be fatal if absorbed through the skin. Causes skin burns. May cause cyanosis of the extremities. May cause skin rash (in milder cases), and cold and clammy skin with cyanosis or pale color. Harmful if swallowed. May cause severe and permanent damage to the digestive tract. May cause gastrointestinal irritation with nausea, vomiting and diarrhea. May cause liver and kidney damage. May cause perforation of the digestive tract. Ingestion of large amounts may cause central nervous system depression. May cause spleen damage. May cause systemic effects. May be fatal if inhaled. Causes chemical burns to the respiratory tract. Aspiration may lead to pulmonary edema. Vapors may cause dizziness or suffocation. May cause systemic effects. Vapors are extremely irritating to the respiratory tract. May cause burning sensation in the chest.	Kidneys, central nervous system, liver, spleen	None Listed	None Listed	None Listed
Citric Acid	Irritation, nausea, headaches, shortness of breath.		None Listed	None Listed	None Listed

(Continued)

Chemical	Effects	Target Organs	Immediately Dangerous To Life or Health (IDLH) (ppm)	National Institute of Occupational Safety and Health (NIOSH) Relative Exposure Limit (REL) (ppm)	Occupational Safety and Health Administration (OSHA) Relative Exposure Limit (PEL) (ppm)
Copper Oxide	Causes eye irritation. May result in corneal injury. May cause conjunctivitis. Causes skin irritation. May cause skin discoloration. May cause central nervous system depression, kidney damage and liver damage. May cause gastrointestinal irritation with nausea, vomiting and diarrhea. May cause circulatory system failure. May cause vascular collapse and damage. Causes respiratory tract irritation. May cause ulceration and perforation of the nasal septum if inhaled in excessive quantities. Inhalation of fumes may cause metal fume fever, which is characterized by flu-like symptoms with metallic taste, fever, chills, cough, weakness, chest pain, muscle pain and increased white blood cell count.	Kidneys, central nervous system, liver, red blood cells	100 mg / m^3	1 mg / m^3	1 mg / m^3

(Continued)

Chemical	Effects	Target Organs	Immediately Dangerous To Life or Health (IDLH) (ppm)	National Institute of Occupational Safety and Health (NIOSH) Relative Exposure Limit (REL) (ppm)	Occupational Safety and Health Administration (OSHA) Relative Exposure Limit (PEL) (ppm)
Copper Sulfate	Exposure to particulates or solution may cause conjunctivitis, ulceration and corneal abnormalities. Causes eye irritation and possible burns. May cause skin sensitization, an allergic reaction, which becomes evident upon re-exposure to this material. Causes skin irritation and possible burns. May cause itching eczema. Harmful if swallowed. May cause severe gastrointestinal tract irritation with nausea, vomiting and possible burns. Ingestion of large amounts of copper salts may cause bloody stools and vomit, low blood pressure, jaundice and coma. Ingestion of copper compounds may produce systemic toxic effects to the kidney and liver and central nervous excitation followed by depression. May cause ulceration and perforation of the nasal septum if inhaled in excessive quantities. Causes respiratory tract irritation with possible burns.	Blood, kidneys, liver.	100 mg/m^3	1 mg/m^3	1 mg/m^3

(Continued)

Chemical	Effects	Target Organs	Immediately Dangerous To Life or Health (IDLH) (ppm)	National Institute of Occupational Safety and Health (NIOSH) Relative Exposure Limit (REL) (ppm)	Occupational Safety and Health Administration (OSHA) Relative Exposure Limit (PEL) (ppm)
Cyclohexanone*	May result in corneal injury. Vapors may cause eye irritation. Contact produces irritation, tearing and burning pain. Causes skin irritation. Harmful if absorbed through the skin. Causes gastrointestinal irritation with nausea, vomiting and diarrhea. May cause liver and kidney damage. May cause central nervous system depression, characterized by excitement followed by headache, dizziness, drowsiness and nausea. Advanced stages may cause collapse, unconsciousness, coma and possible death due to respiratory failure. May be harmful if swallowed. Inhalation of high concentrations may cause central nervous system effects characterized by nausea, headache, dizziness, unconsciousness and coma. May cause liver and kidney damage. May cause narcotic effects in high concentration. Inhalation may be fatal as a result of spasm, inflammation, edema of the larynx and bronchi, chemical pneumonitis and pulmonary edema. May cause irritation of the mucous membranes.	Eyes, skin, respiratory system, central nervous system, liver, kidneys	700	25	50

(Continued)

Chemical	Effects	Target Organs	Immediately Dangerous To Life or Health (IDLH) (ppm)	National Institute of Occupational Safety and Health (NIOSH) Relative Exposure Limit (REL) (ppm)	Occupational Safety and Health Administration (OSHA) Relative Exposure Limit (PEL) (ppm)
Dichloroethane*	Contact with eyes may cause severe irritation and possible eye burns. May be absorbed through the skin. Causes irritation with burning pain, itching and redness. Prolonged exposure may result in skin burns. Causes gastrointestinal irritation with nausea, vomiting and diarrhea. May cause central nervous system depression, characterized by excitement followed by headache, dizziness, drowsiness and nausea. Advanced stages may cause collapse, unconsciousness, coma and possible death due to respiratory failure. May be harmful if swallowed. Inhalation of high concentrations may cause central nervous system effects characterized by nausea, headache, dizziness, unconsciousness and coma. Causes respiratory tract irritation. May cause narcotic effects in high concentration. Vapors may cause dizziness or suffocation. May cause blood changes. Overexposure may cause an increase in carboxyhemoglobin levels in the blood. Can produce delayed pulmonary edema.	Blood, heart, central nervous system, liver, pancreas	3000	100	100

(Continued)

Chemical	Effects	Target Organs	Immediately Dangerous To Life or Health (IDLH) (ppm)	National Institute of Occupational Safety and Health (NIOSH) Relative Exposure Limit (REL) (ppm)	Occupational Safety and Health Administration (OSHA) Relative Exposure Limit (PEL) (ppm)
Ephedrine	May cause eye irritation. May cause skin irritation. Contact with the skin may cause a local anesthetic effect. May cause irritation of the digestive tract. The toxicological properties of this substance have not been fully investigated. May cause respiratory tract irritation. The toxicological properties of this substance have not been fully investigated.	Heart, nerves	None Listed	None Listed	None Listed
Erythritol	None Listed	None Listed	None Listed	None Listed	None Listed
Ethyl Acetate	Causes eye irritation. Vapors may cause eye irritation. May cause skin irritation. Prolonged and/or repeated contact may cause irritation and/or dermatitis. May cause irritation of the digestive tract. May cause liver and kidney damage. Ingestion of large amounts may cause central nervous system depression. May cause headache, nausea, fatigue and dizziness. May cause respiratory tract irritation. May be harmful if inhaled. Inhalation of high concentrations may cause narcotic effects.	Kidneys, central nervous system, liver	2000	400	400
Formamide*	Causes eye irritation. Causes skin irritation. May cause irritation of the digestive tract. Inhalation of high concentrations may cause central nervous system effects characterized by nausea, headache, dizziness, unconsciousness and coma. May cause respiratory tract irritation.	Central nervous system	None Listed	10	None Listed

(Continued)

Chemical	Effects	Target Organs	Immediately Dangerous To Life or Health (IDLH) (ppm)	National Institute of Occupational Safety and Health (NIOSH) Relative Exposure Limit (REL) (ppm)	Occupational Safety and Health Administration (OSHA) Relative Exposure Limit (PEL) (ppm)
Formic Acid	Contact with liquid is corrosive to the eyes and causes severe burns. Lachrymator (substance which increases the flow of tears). May cause corneal edema, ulceration and scarring. May cause skin sensitization, an allergic reaction, which becomes evident upon re-exposure to this material. Contact with liquid is corrosive and causes severe burns and ulceration. Absorbed through the skin. May cause erythema (redness) and blistering. Causes severe digestive tract burns with abdominal pain, vomiting and possible death. May be harmful if swallowed. May cause central nervous system depression. Ingestion may produce corrosive ulceration and bleeding and necrosis of the gastrointestinal tract accompanied by shock and circulatory collapse. May cause asthmatic attacks due to allergic sensitization of the respiratory tract. Causes chemical burns to the respiratory tract. Aspiration may lead to pulmonary edema. Vapors may cause dizziness, nausea, itching, burning and swelling of the eyes.	Kidneys, central nervous system, liver, respiratory system, eyes, skin	30	5	5
Glycerol	Causes irritation eyes, skin, respiratory system; headache, nausea, vomiting; kidney injury.	Eyes, skin, respiratory system, kidneys	None Listed	None Listed	15 mg/m³
Hexamine	May cause allergic skin reaction. Symptoms of allergic reaction may include rash, itching, swelling, trouble breathing, tingling of the hands and feet, dizziness, lightheadedness, chest pain, muscle pain or flushing.	None Listed	None Listed	None Listed	None Listed

(Continued)

Chemical	Effects	Target Organs	Immediately Dangerous To Life or Health (IDLH) (ppm)	National Institute of Occupational Safety and Health (NIOSH) Relative Exposure Limit (REL) (ppm)	Occupational Safety and Health Administration (OSHA) Relative Exposure Limit (PEL) (ppm)
Hydrazine	irritation eyes, skin, nose, throat; temporary blindness; dizziness, nausea; dermatitis; eye, skin burns; in animals: bronchitis, pulmonary edema; liver, kidney damage; convulsions; (potential occupational carcinogen).	Eyes, skin, respiratory system, central nervous system, liver, kidneys	50	0.03	1
Hydriodic Acid (HI)*	Causes eye burns. Causes skin burns. May cause severe and permanent damage to the digestive tract. Causes gastrointestinal tract burns. May cause irritation of the respiratory tract with burning pain in the nose and throat, coughing, wheezing, shortness of breath and pulmonary edema. Causes chemical burns to the respiratory tract. Inhalation may be fatal as a result of spasm, inflammation, edema of the larynx and bronchi, chemical pneumonitis and pulmonary edema.	None known	None Listed	None Listed	None Listed
Hydrobromic Acid (HBr)*	May result in corneal injury. Causes severe eye irritation and burns. Causes severe skin irritation. May be absorbed through the skin. Contact with liquid is corrosive and causes severe burns and ulceration. Causes gastrointestinal tract burns. May cause respiratory failure. May cause circulatory system failure. May cause hemorrhaging of the digestive tract. May cause corrosion and permanent tissue destruction of the esophagus and digestive tract. Irritation may lead to chemical pneumonitis and pulmonary edema. Causes chemical burns to the respiratory tract. May cause effects similar to those described for ingestion.	None	30	3	3

(Continued)

Chemical	Effects	Target Organs	Immediately Dangerous To Life or Health (IDLH) (ppm)	National Institute of Occupational Safety and Health (NIOSH) Relative Exposure Limit (REL) (ppm)	Occupational Safety and Health Administration (OSHA) Relative Exposure Limit (PEL) (ppm)
Hydrochloric Acid (HCl)	May cause irreversible eye injury. Vapor or mist may cause irritation and severe burns. Contact with liquid is corrosive to the eyes and causes severe burns. May cause painful sensitization to light. May be absorbed through the skin in harmful amounts. May cause skin sensitization, an allergic reaction, which becomes evident upon re-exposure to this material. Contact with liquid is corrosive and causes severe burns and ulceration. May cause circulatory system failure. Causes severe digestive tract burns with abdominal pain, vomiting and possible death. May cause corrosion and permanent tissue destruction of the esophagus and digestive tract. May be harmful if swallowed. May cause severe irritation of the respiratory tract with sore throat, coughing, shortness of breath and delayed lung edema. Causes chemical burns to the respiratory tract. Exposure to the mist and vapor may erode exposed teeth. Causes corrosive action on the mucous membranes.	Toxic by inhalation. Causes severe burns. Corrosive. Mutagen	50	5	5
Hydrogen pdc	Defined as a simple asphyxiant. Inhalation of high concentrations of hydrogen may cause dizziness, headaches, deeper breathing due to air hunger, possible nausea and eventual unconsciousness.	Lungs and central nervous system			

(Continued)

Chemical	Effects	Target Organs	Immediately Dangerous To Life or Health (IDLH) (ppm)	National Institute of Occupational Safety and Health (NIOSH) Relative Exposure Limit (REL) (ppm)	Occupational Safety and Health Administration (OSHA) Relative Exposure Limit (PEL) (ppm)
Hydrogen Peroxide	Contact with liquid is corrosive to the eyes and causes severe burns. Contact with the eyes may cause corneal damage. Causes severe skin irritation and possible burns. May cause discoloration, erythema (redness), swelling, and the formation of papules and vesicles (blisters). Causes gastrointestinal irritation with nausea, vomiting and diarrhea. Causes gastrointestinal tract burns. May cause vascular collapse and damage. May cause damage to the red blood cells. May cause difficulty in swallowing, stomach distension, possible cerebral swelling and death. Ingestion may result in irritation of the esophagus, bleeding of the stomach and ulcer formation. Causes chemical burns to the respiratory tract. May cause ulceration of nasal tissue, insomnia, nervous tremors with numb extremities, chemical pneumonia, unconsciousness and death. At high concentrations, respiratory effects may include acute lung damage and delayed pulmonary edema.	Blood, central nervous system	75	1	1

(Continued)

Chemical	Effects	Target Organs	Immediately Dangerous To Life or Health (IDLH) (ppm)	National Institute of Occupational Safety and Health (NIOSH) Relative Exposure Limit (REL) (ppm)	Occupational Safety and Health Administration (OSHA) Relative Exposure Limit (PEL) (ppm)
Hydroxylamine HCl* pdc	Corrosive. Extremely destructive to the mucous membranes and upper respiratory tract. Symptoms may include burning sensation, coughing, wheezing, laryngitis, shortness of breath, headache, nausea and vomiting. Inhalation may be fatal as a result of spasm, inflammation and edema of the larynx and bronchi, chemical pneumonitis and pulmonary edema. May convert hemoglobin to methemoglobin, producing cyanosis. May also cause nausea, vomiting, fall in blood pressure, headache, vertigo, ringing in the ears, shortness of breath, severe blood oxygen deficiency and convulsions. High concentrations cause coma and death from circulatory collapse. Irritant and possible sensitizer. May cause burns. Corrosive to the eyes. May cause severe irritation and corneal damage.	Kidneys, central nervous system, eyes, blood, skin, liver and lungs	None Listed	None Listed	None Listed

(Continued)

Chemical	Effects	Target Organs	Immediately Dangerous To Life or Health (IDLH) (ppm)	National Institute of Occupational Safety and Health (NIOSH) Relative Exposure Limit (REL) (ppm)	Occupational Safety and Health Administration (OSHA) Relative Exposure Limit (PEL) (ppm)
Iodine	Causes severe eye irritation. May cause eye burns. Vapor or mist may cause irritation and severe burns. Causes skin burns. May cause skin sensitization, an allergic reaction, which becomes evident upon re-exposure to this material. May cause gastrointestinal irritation with nausea, vomiting and diarrhea. May cause kidney damage. May cause burns to the digestive tract. May cause thyroid abnormalities. May cause irritation of the respiratory tract with burning pain in the nose and throat, coughing, wheezing, shortness of breath and pulmonary edema. May cause epiphoria, which is an excessive flow of tears.	Kidneys, thyroid	None Listed	2	0.1
Isosafrole*	May cause eye irritation. May cause skin irritation. May cause irritation of the digestive tract. May cause liver damage. May cause cyanosis (bluish discoloration of skin due to deficient oxygenation of the blood), weakness, acidosis and shock. May be harmful if swallowed. May cause respiratory tract irritation. May cause effects similar to those described for ingestion.	Liver	None Listed	None Listed	None Listed
Lead Acetate	May damage fertility or the unborn child. May cause damage to organs (kidneys, liver, blood, brain) through prolonged or repeated exposure (oral).	Kidneys, liver, blood, brain	100 mg/m³	0.05 mg/m³	0.05 mg/m³

(Continued)

Chemical	Effects	Target Organs	Immediately Dangerous To Life or Health (IDLH) (ppm)	National Institute of Occupational Safety and Health (NIOSH) Relative Exposure Limit (REL) (ppm)	Occupational Safety and Health Administration (OSHA) Relative Exposure Limit (PEL) (ppm)
Lead Nitrate		Reproductive toxicant, central nervous system	100 mg/m^3	0.05 mg/m^3	0.05 mg/m^3
Lithium*	There is no known long-term hazard from lithium in its solid state. However, lithium metal is extremely reactive with body moisture and is corrosive to the skin, nose, throat and eyes.	None	None Listed	None Listed	None Listed
Lithium Aluminum Hydroxide*	Causes eye burns. When substance becomes wet or comes in contact with moisture of the mucous membranes, it will cause irritation. May cause chemical conjunctivitis and corneal damage. Causes skin burns. Contact with skin causes irritation and possible burns, especially if the skin is wet or moist. May cause skin rash (in milder cases) and cold and clammy skin with cyanosis or pale color. May cause severe and permanent damage to the digestive tract. Causes gastrointestinal tract burns. May cause perforation of the digestive tract. May cause systemic effects. Causes chemical burns to the respiratory tract. Aspiration may lead to pulmonary edema. May cause systemic effects.	None	None Listed	15 mg/m^3	10 mg/m^3

(*Continued*)

Chemical	Effects	Target Organs	Immediately Dangerous To Life or Health (IDLH) (ppm)	National Institute of Occupational Safety and Health (NIOSH) Relative Exposure Limit (REL) (ppm)	Occupational Safety and Health Administration (OSHA) Relative Exposure Limit (PEL) (ppm)
Magnesium Turnings*	Dust may cause mechanical irritation. May cause skin irritation. Particles embedded in the skin may cause "chemical gas gangrene" with symptoms of persistent lesions, inflammation and gas bubbles under the skin. May cause irritation of the digestive tract. May cause respiratory tract irritation. Inhalation of fumes may cause metal fume fever, which is characterized by flu-like symptoms with metallic taste, fever, chills, cough, weakness, chest pain, muscle pain and increased white blood cell count.	None	None Listed	None Listed	None Listed
Manganous Carbonate*	Acute poisoning can occur from excessive inhalation causing symptoms noted under Chronic Exposure. Extremely large oral dosages may produce gastrointestinal disturbances and acute poisoning as noted under Chronic Exposure. No adverse effects expected to dermal exposure. Eye Contact: No adverse effects expected but dust may cause mechanical irritation.	None	None Listed	None Listed	5 mg/m^3

(*Continued*)

Chemical	Effects	Target Organs	Immediately Dangerous To Life or Health (IDLH) (ppm)	National Institute of Occupational Safety and Health (NIOSH) Relative Exposure Limit (REL) (ppm)	Occupational Safety and Health Administration (OSHA) Relative Exposure Limit (PEL) (ppm)
Manganous Chloride*	Inhalation: Inhalation can cause a flu-like illness (metal fume fever). This 24- to 48-hour illness is characterized by chills, fever, aching muscles, dryness in the mouth and throat, and headache. May irritate the respiratory tract. May increase the incidence of upper respiratory infections (pneumonia). Absorption of inorganic manganese salts through the lungs is poor but may occur in chronic poisoning.Ingestion: May cause abdominal pain and nausea. Although they are poorly absorbed through the intestines, inorganic manganese salts may produce hypoglycemia and decreased calcium blood levels should absorption occur. May cause irritation with redness and pain.	Brain, kidney, blood			

(Continued)

Chemical	Effects	Target Organs	Immediately Dangerous To Life or Health (IDLH) (ppm)	National Institute of Occupational Safety and Health (NIOSH) Relative Exposure Limit (REL) (ppm)	Occupational Safety and Health Administration (OSHA) Relative Exposure Limit (PEL) (ppm)
Mercuric Chloride*pdc	Causes irritation to the respiratory tract. Symptoms include sore throat, coughing, pain, tightness in chest, breathing difficulties, shortness of breath and headache. Pneumonitis may develop. Can be absorbed through inhalation with symptoms to parallel ingestion. Vapor inhalation can burn the mucous membrane of the nose and throat. Highly Toxic! Average lethal dose for inorganic mercury salts is about 1 gram. May cause burning of the mouth and pharynx, abdominal pain, vomiting, corrosive ulceration, bloody diarrhea. May be followed by a rapid and weak pulse, shallow breathing, paleness, exhaustion, central nervous system problems, tremors and collapse. Delayed death may occur from renal failure. Causes irritation and burns to skin. Symptoms include redness and pain. May cause skin allergy and sensitization. Can be absorbed through the skin with symptoms to parallel ingestion. Causes irritation and burns to eyes. Symptoms include redness, pain, blurred vision; may cause serious and permanent eye damage.	Eyes, skin, respiratory system, central nervous system, kidneys	10 mg/m^3	0.05 mg/m^3	0.1 mg/m^3

(Continued)

Chemical	Effects	Target Organs	Immediately Dangerous To Life or Health (IDLH) (ppm)	National Institute of Occupational Safety and Health (NIOSH) Relative Exposure Limit (REL) (ppm)	Occupational Safety and Health Administration (OSHA) Relative Exposure Limit (PEL) (ppm)
Mercury*	Exposure to mercury or mercury compounds can cause discoloration on the front surface of the lens, which does not interfere with vision. Causes eye irritation and possible burns. Contact with mercury or mercury compounds can cause ulceration of the conjunctiva and cornea. May be absorbed through the skin in harmful amounts. May cause skin sensitization, an allergic reaction, which becomes evident upon re-exposure to this material. Causes skin irritation and possible burns. May cause skin rash (in milder cases), and cold and clammy skin with cyanosis or pale color. May cause severe and permanent damage to the digestive tract. May cause perforation of the digestive tract. May cause effects similar to those for inhalation exposure. May cause systemic effects. Causes chemical burns to the respiratory tract. Inhalation of fumes may cause metal fume fever, which is characterized by flu-like symptoms with metallic taste, fever, chills, cough, weakness, chest pain, muscle pain and increased white blood cell count. May cause central nervous system effects including vertigo, anxiety, depression, muscle incoordination and emotional instability. Aspiration may lead to pulmonary edema. May cause systemic effects. May cause respiratory sensitization.	Blood, kidneys, central nervous system, liver, brain	10 mg/m^3	0.05 mg/m^3	0.1 mg/m^3

(Continued)

Chemical	Effects	Target Organs	Immediately Dangerous To Life or Health (IDLH) (ppm)	National Institute of Occupational Safety and Health (NIOSH) Relative Exposure Limit (REL) (ppm)	Occupational Safety and Health Administration (OSHA) Relative Exposure Limit (PEL) (ppm)
Methanol	Causes moderate eye irritation. Vapors may cause eye irritation. May cause painful sensitization to light. May cause skin irritation. May be absorbed through the skin. May be fatal or cause blindness if swallowed. May cause irritation of the digestive tract. May cause kidney damage. May cause central nervous system depression, characterized by excitement followed by headache, dizziness, drowsiness and nausea. Advanced stages may cause collapse, unconsciousness, coma and possible death due to respiratory failure. May cause respiratory tract irritation. May cause adverse central nervous system effects including headache, convulsions and possible death. May cause visual impairment and possible permanent blindness. May cause effects similar to those described for ingestion. May cause kidney damage.	Kidneys, central nervous system, eyes	6000	200	200

(Continued)

Chemical	Effects	Target Organs	Immediately Dangerous To Life or Health (IDLH) (ppm)	National Institute of Occupational Safety and Health (NIOSH) Relative Exposure Limit (REL) (ppm)	Occupational Safety and Health Administration (OSHA) Relative Exposure Limit (PEL) (ppm)
Methylamine*	Causes eye burns. May result in corneal injury. May cause chemical conjunctivitis and corneal damage. May cause tearing, conjunctivitis and corneal edema when vapor is absorbed into the tissue of the eye. Causes skin burns. May be absorbed through the skin. May cause dermatitis. Methylamine is readily absorbed through the skin and may cause malaise, discomfort, injury and death unless treated promptly. Harmful if swallowed. Causes gastrointestinal tract burns. Causes chemical burns to the respiratory tract. May cause pulmonary edema and severe respiratory disturbances. May cause liver abnormalities. Inhalation of methylamine may cause coughing, nausea and pulmonary edema. Allergic or chemical bronchitis was reported in a worker exposed to methylamine in an unpublished report. It is unclear from this report what the actual exposure concentrations were.	Liver, respiratory system, eyes, skin	100	10	10
Methyl Ethyl Ketone	Irritation eyes, skin, nose; headache; dizziness; vomiting; dermatitis.	Eyes, skin, respiratory system, central nervous system	30,000	200	200
Methyl Ethyl Ketone Peroxide	Irritation eyes, skin, nose, throat; cough, dyspnea (breathing difficulty), pulmonary edema; blurred vision; blisters, scars skin; abdominal pain, vomiting, diarrhea; dermatitis; in animals: liver, kidney damage.	Eyes, skin, respiratory system, liver, kidneys	None Listed	0.2	None Listed

(Continued)

Chemical	Effects	Target Organs	Immediately Dangerous To Life or Health (IDLH) (ppm)	National Institute of Occupational Safety and Health (NIOSH) Relative Exposure Limit (REL) (ppm)	Occupational Safety and Health Administration (OSHA) Relative Exposure Limit (PEL) (ppm)
Methylformamide*	Causes eye irritation. May cause chemical conjunctivitis. Causes skin irritation. May cause gastrointestinal irritation with nausea, vomiting and diarrhea. Causes respiratory tract irritation. Can produce delayed pulmonary edema.	Reproductive system	None Listed	None Listed	None Listed
Nitric Acid	Irritation eyes, skin, mucous membrane; delayed pulmonary edema, pneumonitis, bronchitis; dental erosion	Eyes, skin, respiratory system, teeth	25	2	2
Nitroethane	Causes eye irritation. May cause chemical conjunctivitis and corneal damage. Causes skin irritation. May cause dermatitis. May cause cyanosis of the extremities. May cause gastrointestinal irritation with nausea, vomiting and diarrhea. Methemoglobinemia is characterized by dizziness, drowsiness, headache, breath shortness, cyanosis with bluish skin, rapid heart rate and chocolate-brown colored blood. Ingestion of large amounts may cause central nervous system depression. May be harmful if swallowed. May form methemoglobin, which in sufficient concentration causes cyanosis (bluish discoloration of skin due to deficient oxygenation of the blood). Causes respiratory tract irritation. Aspiration may lead to pulmonary edema. Vapors may cause dizziness or suffocation. May cause burning sensation in the chest.	Kidneys, central nervous system, liver, respiratory system, skin	1000	100	100

(Continued)

(Continued)

Chemical	Effects	Target Organs	Immediately Dangerous To Life or Health (IDLH) (ppm)	National Institute of Occupational Safety and Health (NIOSH) Relative Exposure Limit (REL) (ppm)	Occupational Safety and Health Administration (OSHA) Relative Exposure Limit (PEL) (ppm)
Norpseudoephedrine*	May cause eye irritation. May cause skin irritation. Contact with the skin may cause a local anesthetic effect. May cause irritation of the digestive tract. The toxicological properties of this substance have not been fully investigated. May cause respiratory tract irritation. The toxicological properties of this substance have not been fully investigated.	Heart, nerves	None Listed	None Listed	None Listed
Palladium Sulfate* pdc	May be harmful by inhalation, ingestion or skin absorption. May cause eye/skin irritation. To the best of the manufacturer's knowledge the toxicological properties have not been thoroughly investigated.	None Listed	None Listed	None Listed	None Listed
Perchloric Acid* pdc	Causes eye burns. May cause retinal damage. Causes skin burns. May cause skin sensitization, an allergic reaction, which becomes evident upon re-exposure to this material. May cause deep, penetrating ulcers of the skin. Causes gastrointestinal tract burns. May be harmful if swallowed. Ingestion may produce corrosive ulceration and bleeding and necrosis of the gastrointestinal tract accompanied by shock and circulatory collapse. May cause severe irritation of the respiratory tract with sore throat, coughing, shortness of breath and delayed lung edema. Inhalation may be fatal as a result of spasm, inflammation, edema of the larynx and bronchi, chemical pneumonitis and pulmonary edema.	Eyes, skin, mucous membranes	None Listed	None Listed	None Listed

Chemical	Effects	Target Organs	Immediately Dangerous To Life or Health (IDLH) (ppm)	National Institute of Occupational Safety and Health (NIOSH) Relative Exposure Limit (REL) (ppm)	Occupational Safety and Health Administration (OSHA) Relative Exposure Limit (PEL) (ppm)
Phenylacetic Acid*	Causes eye irritation and possible burns. May cause chemical conjunctivitis. Causes skin irritation and possible burns. May cause gastrointestinal irritation with nausea, vomiting and diarrhea. Causes respiratory tract irritation. Can produce delayed pulmonary edema.	None Listed	None Listed	None Listed	None Listed
Phenylacetone*	NA	NA	NA	NA	NA
Phenylmagnesium Bromide*	NA	NA	NA	NA	NA
Phosphorus*	Not considered highly toxic but acute exposure may cause coughing, bronchitis, possible liver or kidney impairment if contaminated with yellow phosphorus. Red phosphorus is not readily absorbed and, in pure form, is considered non-poisonous. However, possible contamination with the yellow form must be considered, and symptoms such as nausea, vomiting, abdominal pain or garlic odor on breath will indicate poisoning by the latter. The estimated lethal adult human dose for white phosphorus is 50–100 mg. Red phosphorus is not harmful to skin. If contaminated with white phosphorus, however, contact may cause deep, slow-healing burns. Red phosphorus causes eye irritation. If contaminated with yellow phosphorus, eye contact can cause severe irritation and burns.	Kidneys, liver	5 mg/m³	0.1 mg/m³	0.1 mg/m³

(Continued)

Chemical	Effects	Target Organs	Immediately Dangerous To Life or Health (IDLH) (ppm)	National Institute of Occupational Safety and Health (NIOSH) Relative Exposure Limit (REL) (ppm)	Occupational Safety and Health Administration (OSHA) Relative Exposure Limit (PEL) (ppm)
Phosphorus Pentachloride*	Causes severe eye burns. Causes skin burns. Causes digestive tract burns with immediate pain, swelling of the throat, convulsions and possible coma. May be harmful if swallowed. May cause severe irritation of the respiratory tract with sore throat, coughing, shortness of breath and delayed lung edema. Causes chemical burns to the respiratory tract.	None known	70 mg/m^3	1 mg/m^3	1 mg/m^3
Piperidine*	Contact with liquid or vapor causes severe burns and possible irreversible eye damage. Contact may cause ulceration of the conjunctiva and cornea. Eye damage may be delayed. May cause conjunctivitis. May cause blindness. Harmful if absorbed through the skin. May be absorbed through the skin. If absorbed, causes symptoms similar to those of ingestion. Penetration may continue for several days. Causes severe skin irritation and burns. Harmful if swallowed. May cause severe and permanent damage to the digestive tract. Causes gastrointestinal tract burns. Can cause nervous system damage. May cause tremors and convulsions. May cause severe irritation of the respiratory tract with sore throat, coughing, shortness of breath and delayed lung edema. Causes chemical burns to the respiratory tract. May cause effects similar to those described for ingestion. Damage may be delayed. May cause bronchial pneumonia.	Nervous system	None Listed	None Listed	None Listed

(Continued)

Chemical	Effects	Target Organs	Immediately Dangerous To Life or Health (IDLH) (ppm)	National Institute of Occupational Safety and Health (NIOSH) Relative Exposure Limit (REL) (ppm)	Occupational Safety and Health Administration (OSHA) Relative Exposure Limit (PEL) (ppm)
Piperonal*	Dust may cause mechanical irritation. Causes skin irritation. May cause irritation of the digestive tract. May cause central nervous system depression, characterized by excitement followed by headache, dizziness, drowsiness and nausea. Advanced stages may cause collapse, unconsciousness, coma and possible death due to respiratory failure. Inhalation of dust may cause respiratory tract irritation.	Central nervous system	None Listed	None Listed	None Listed
Platinum	Irritation skin, respiratory system; dermatitis	Eyes, skin, respiratory system	None Listed	1 mg/m^3	None Listed
Platinum Chloride*	Exposure can cause severe allergies affecting the nose, skin and lungs. Irritation and even ulcers can develop in the nose. May cause platinosis. Symptoms include wheezing, coughing, tightness of the chest, shortness of breath, cyanosis and pronounced asthmatic symptoms. May cause vomiting and bloody diarrhea. May cause platinosis; symptoms may include severe irritation, eczema, urticaria, itching and dermatitis. May cause skin allergy. May cause irritation, itching, and conjunctival vasodilation.	None	None Listed	None Listed	None Listed

(Continued)

Chemical	Effects	Target Organs	Immediately Dangerous To Life or Health (IDLH) (ppm)	National Institute of Occupational Safety and Health (NIOSH) Relative Exposure Limit (REL) (ppm)	Occupational Safety and Health Administration (OSHA) Relative Exposure Limit (PEL) (ppm)
Platinum Oxide*	Causes eye irritation. May cause chemical conjunctivitis. Causes skin irritation. May cause gastrointestinal irritation with nausea, vomiting and diarrhea. The toxicological properties of this substance have not been fully investigated. Causes respiratory tract irritation. The toxicological properties of this substance have not been fully investigated. Can produce delayed pulmonary edema.	None	None Listed	None Listed	None Listed
Potassium Carbonate	Causes irritation to the respiratory tract. Symptoms may include coughing, shortness of breath. Causes irritation to the gastrointestinal tract. Symptoms may include nausea, vomiting and diarrhea. May have moderate toxic effects if consumed in large enough quantities. Ingestion of large amounts may be corrosive to mouth, throat and gastrointestinal tract and produce abdominal pains, vomiting, diarrhea and circulatory collapse. Contact with dry material causes irritation. In aqueous solution it is a strong caustic and as such may have corrosive effects on the skin. Causes extreme irritation, redness, pain and possibly corneal damage.	Eyes, skin, mucous membranes	None Listed	None Listed	None Listed
Potassium Chlorate	Contact can cause eye and skin irritation and burns, irritate the nose, throat and lungs, causing sneezing, coughing and sore throat. High levels can interfere with the ability of the blood to carry oxygen, causing headache, weakness, dizziness and a blue color to the skin (methemoglobinemia). Higher levels can cause trouble breathing, collapse and even death	May affect the kidneys and nervous system	None Listed	None Listed	None Listed

(Continued)

Chemical	Effects	Target Organs	Immediately Dangerous To Life or Health (IDLH) (ppm)	National Institute of Occupational Safety and Health (NIOSH) Relative Exposure Limit (REL) (ppm)	Occupational Safety and Health Administration (OSHA) Relative Exposure Limit (PEL) (ppm)
Potassium Cyanide*	Contact with eyes may cause severe irritation and possible eye burns. May be absorbed through the skin in harmful amounts. Contact with skin causes irritation and possible burns, especially if the skin is wet or moist. If absorbed, causes symptoms similar to those of ingestion. Skin absorption may cause unconsciousness. Absorption into the body may cause cyanosis (bluish discoloration of skin due to deficient oxygenation of the blood). May be fatal if swallowed. Causes gastrointestinal tract burns. May cause tissue anoxia, characterized by weakness, headache, dizziness, confusion, cyanosis (bluish skin due to deficient oxygenation of the blood), weak and irregular heartbeat, collapse, unconsciousness, convulsions, coma and death. Contains cyanide. Human fatalities have been reported from acute poisoning. Large doses of cyanide may result in sudden loss of consciousness and prompt death; small doses will prolong the above symptoms 1 to 2 hours. Can cause central nervous system damage and death. Inhalation of high concentrations may cause central nervous system effects characterized by nausea, headache, dizziness, unconsciousness and coma. Causes respiratory tract irritation. May cause effects similar to those described for ingestion. Contains cyanide. Inhalation may result in symptoms similar to cyanide poisoning, which include tachypnea, hyperpnea (abnormally rapid or deep breathing) and dyspnea (labored breathing) followed rapidly by respiratory depression. Pulmonary edema may occur.	Central nervous system, respiratory system, cardiovascular system	25 mg/m^3	5 mg/m^3	5 mg/m^3

(Continued)

Chemical	Effects	Target Organs	Immediately Dangerous To Life or Health (IDLH) (ppm)	National Institute of Occupational Safety and Health (NIOSH) Relative Exposure Limit (REL) (ppm)	Occupational Safety and Health Administration (OSHA) Relative Exposure Limit (PEL) (ppm)
Potassium Hydroxide	Severe irritant. Effects from inhalation of dust or mist vary from mild irritation to serious damage of the upper respiratory tract, depending on the severity of exposure. Symptoms may include coughing, sneezing, damage to the nasal or respiratory tract. High concentrations can cause lung damage. Toxic! Swallowing may cause severe burns of mouth, throat and stomach. Other symptoms may include vomiting, diarrhea. Severe scarring of tissue and death may result. Estimated lethal dose: 5 grams. Corrosive! Contact with skin can cause irritation or severe burns and scarring with greater exposures. Highly Corrosive! Causes irritation of eyes with tearing, redness, swelling. Greater exposures cause severe burns with possible blindness resulting.	Eyes, skin, respiratory system	None Listed	2 mg/m^3	None Listed
Potassium Nitrate	Contact can cause eye and skin irritation. Breathing potassium nitrate can irritate the nose and throat, causing sneezing and coughing. High levels can interfere with the ability of the blood to carry oxygen, causing headache, fatigue, dizziness and a blue color to the skin and lips (methemoglobinemia). Higher levels can cause trouble breathing, collapse and even death.	May affect the kidneys and cause anemia. Reproductive system	None Listed	None Listed	None Listed
Potassium Perchlorate	Short-term exposure to high doses may cause eye, skin and respiratory tract irritation, coughing, nausea, vomiting and diarrhea.	May affect the thyroid, kidneys and cause anemia	None Listed	None Listed	None Listed

(Continued)

Chemical	Effects	Target Organs	Immediately Dangerous To Life or Health (IDLH) (ppm)	National Institute of Occupational Safety and Health (NIOSH) Relative Exposure Limit (REL) (ppm)	Occupational Safety and Health Administration (OSHA) Relative Exposure Limit (PEL) (ppm)
Pumice	NA	NA	NA	NA	NA
Pyridine*	Irritation eyes; headache, anxiety, dizziness, insomnia; nausea, anorexia; dermatitis; liver, kidney damage.	Eyes, skin, central nervous system, liver, kidneys, gastrointestinal tract	1000	5	5
Raney Nickel*	NA	NA	NA	NA	NA
Sodium*	Inhalation produces damaging effects on the mucous membranes and upper respiratory tract. Symptoms may include irritation of the nose and throat, and labored breathing. May cause lung edema, a medical emergency. Extremely dangerous, corrosive material. Will react immediately with saliva to cause serious burns and possible local combustion and even explosion of hydrogen in the mouth or esophagus. The metal's low melting point can cause further complications. Corrosive, can cause serious burns due to almost immediate reaction with water, especially on moist skin. If metal ignites, very deep burns and tissue destruction can occur. Corrosive. May cause redness, pain, blurred vision and damage from severe alkali burns.	None Listed	None Listed	None Listed	None Listed

(Continued)

Chemical	Effects	Target Organs	Immediately Dangerous To Life or Health (IDLH) (ppm)	National Institute of Occupational Safety and Health (NIOSH) Relative Exposure Limit (REL) (ppm)	Occupational Safety and Health Administration (OSHA) Relative Exposure Limit (PEL) (ppm)
Sodium Acetate*	May cause irritation to the respiratory tract. Symptoms may include coughing, sore throat, labored breathing and chest pain. Large doses may produce abdominal pain, nausea and vomiting. May cause irritation with redness and pain. Contact may cause irritation, redness and pain.	None	None Listed	None Listed	None Listed
Sodium Azide	Irritation eyes, skin; headache, dizziness, lassitude (weakness, exhaustion), blurred vision; low blood pressure, bradycardia; kidney changes.	Eyes, skin, central nervous system, cardiovascular system, kidneys	None Listed	0.1	None Listed
Sodium Bisulfate*	Causes eye burns. When substance becomes wet or comes in contact with moisture of the mucous membranes, it will cause irritation. May cause chemical conjunctivitis and corneal damage. Causes skin burns. Contact with skin causes irritation and possible burns, especially if the skin is wet or moist. May cause skin rash (in milder cases), and cold and clammy skin with cyanosis or pale color. May cause severe and permanent damage to the digestive tract. Causes gastrointestinal tract burns. May cause perforation of the digestive tract. May cause systemic effects. May cause severe irritation of the respiratory tract with sore throat, coughing, shortness of breath and delayed lung edema. Causes chemical burns to the respiratory tract. Aspiration may lead to pulmonary edema. May cause systemic effects.	None Listed	None Listed	None Listed	None Listed

(Continued)

Chemical	Effects	Target Organs	Immediately Dangerous To Life or Health (IDLH) (ppm)	National Institute of Occupational Safety and Health (NIOSH) Relative Exposure Limit (REL) (ppm)	Occupational Safety and Health Administration (OSHA) Relative Exposure Limit (PEL) (ppm)
Sodium Chlorate	Sodium chlorate may affect blood cells and damage the kidneys. Acute exposure to the compound may also damage the liver and have negative effects on the respiratory system. The kidney tubules may be severely damaged without producing detectable methemoglobinemia.	Liver, kidneys, circulatory system	None Listed	None Listed	None Listed
Sodium Hydroxide	Causes eye burns. May cause chemical conjunctivitis and corneal damage. Causes skin burns. May cause deep, penetrating ulcers of the skin. May cause skin rash (in milder cases), and cold and clammy skin with cyanosis or pale color. May cause severe and permanent damage to the digestive tract. Causes gastrointestinal tract burns. May cause perforation of the digestive tract. Causes severe pain, nausea, vomiting, diarrhea and shock. May cause systemic effects. Irritation may lead to chemical pneumonitis and pulmonary edema. Causes severe irritation of upper respiratory tract with coughing, burns, breathing difficulty and possible coma. Causes chemical burns to the respiratory tract. Aspiration may lead to pulmonary edema. May cause systemic effects.	Eyes, skin, mucous membranes	10 mg/m^3	2 mg/m^3	2 mg/m^3
Sodium Nitrate	Acute oral toxicity – Irritation of mucous membranes, nausea, vomiting, diarrhea. Acute inhalation toxicity – Possible damages, mucosal irritations.	None Listed	None Listed	None Listed	None Listed

(Continued)

Chemical	Effects	Target Organs	Immediately Dangerous To Life or Health (IDLH) (ppm)	National Institute of Occupational Safety and Health (NIOSH) Relative Exposure Limit (REL) (ppm)	Occupational Safety and Health Administration (OSHA) Relative Exposure Limit (PEL) (ppm)
Sodium Sulfate	May cause eye irritation. May cause skin irritation. Ingestion of large amounts may cause gastrointestinal irritation. Low hazard for usual industrial handling. May cause respiratory tract irritation. Low hazard for usual industrial handling.	None Listed	None Listed	None Listed	None Listed
Sucrose (Sugar)	Irritation eyes, skin, upper respiratory system; cough	Eyes, respiratory system	None Listed	10 mg/m^3	15 mg/m^3
Sulfuric Acid	Inhalation produces damaging effects on the mucous membranes and upper respiratory tract. Symptoms may include irritation of the nose and throat, and labored breathing. May cause lung edema, a medical emergency. Corrosive. Swallowing can cause severe burns of the mouth, throat and stomach, leading to death. Can cause sore throat, vomiting, diarrhea. Circulatory collapse with clammy skin, weak and rapid pulse, shallow respirations and scanty urine may follow ingestion or skin contact. Circulatory shock is often the immediate cause of death. Corrosive. Symptoms of redness, pain and severe burn can occur. Circulatory collapse with clammy skin, weak and rapid pulse, shallow respirations and scanty urine may follow skin contact or ingestion. Circulatory shock is often the immediate cause of death. Corrosive. Contact can cause blurred vision, redness, pain and severe tissue burns. Can cause blindness.	Eyes, skin, respiratory system, teeth	15	1	1

(Continued)

Chemical	Effects	Target Organs	Immediately Dangerous To Life or Health (IDLH) (ppm)	National Institute of Occupational Safety and Health (NIOSH) Relative Exposure Limit (REL) (ppm)	Occupational Safety and Health Administration (OSHA) Relative Exposure Limit (PEL) (ppm)
Thionyl Chloride*	Corrosive. Extremely destructive to tissues of the mucous membranes and upper respiratory tract. Symptoms may include burning sensation, coughing, wheezing, laryngitis, shortness of breath, headache, nausea and vomiting. Inhalation may be fatal as a result of spasm, inflammation and edema of the larynx and bronchi, chemical pneumonitis and pulmonary edema. Corrosive. May cause burning pain in throat, abdominal pain, nausea and vomiting. Corrosive. Liquid contact may cause blistering burns, irritation and pain. Vapors may be severely irritating to the skin. Corrosive! Vapors are irritating and may cause damage to the eyes. Contact may cause severe burns and permanent eye damage.	None Known	None Listed	1	None Listed
Thorium Nitrate*	Thorium nitrate is a radioactive material. EYES: Irritation and possible damage. SKIN: Irritation and dermatitis in sensitive persons. INGESTION/INHALATION: Not known.	None Listed	None Listed	None Listed	None Listed
Toluene	Extreme fatigue, mental confusion, exhilaration, nausea, headache and dizziness.	Eyes, skin, respiratory system, central nervous system, liver, kidneys	500	100	200
Urea	None Listed	None Listed	None Listed	None Listed	None Listed

Appendix G
Worksheets

Worksheet Guides

Date	**Administrative Information**	Case #

Location					
Address		City		State	
House	Apartment / Hotel	Out Building	Commercial Structure	Rural	Forest / Jungle
Lab Information					
Methamphetamine	Amphetamine	PCP	Cocaine	MDA/ MDMA	Explosive	Other
Operational		Non Operational		Stored	

Team Assignments			
Case Officer		Command Officer	
Site Control		Site Safety	
Chemist		EOD	
Photography		Spokes Person	
Scene Sketch		Evidence Log	
Entry Team	Assessment Team	Search Team	Security Team

Support Services Contact Information	
Fire Department	Waste Disposal Company
Ambulance	Hospital
Health Department	Environmental Quality
Child Protective Services	Explosives Disposal

Date	**Hazard Assessment**	Case #

Location

Address		City		State		

House	Apartment / Hotel	Out Building	Commercial Structure	Rural	Forest / Jungle

Lab Information

Methamphetamine	Amphetamine	PCP	Cocaine	MDA/ MDMA	Explosive	Other

Operational	Non Operational	Stored

Hazard Assessment

Hazard Type	Specific Hazard	Location

Physical Hazards		

PPE Requirements

Zone	LOCATION	A	B	C	D
Hot					
Warm					
Cold					

Date		Scene Sketch	Case #
Location			
Investigator/Chemist			

Sketch

Legend

Date		Evidence Log			Case #
Location					
Investigator/Chemist					
Ex. #	Sampled Y/N	Photo Y/N	Disposed Y/N		Description

Date			Non Chemical Inventory	Case #	
Location					
Investigator/Chemist					
Ex. #	Disposed Y/N	Photo Y/N	Item Description	Total Number

Date		Chemical Inventory		Case #
Location				
Investigator/Chemist				
Ex. #	Sampled Y/N	Description / Label Information (Field Test Results)		Estimated Total Amount

Date	Top of page	Case #
Location		
Investigator/Chemist		

PHOTO LOG

Roll #	Photo #	Time	Location	Description

EQUIPMENT/CHEMICAL INVENTORY

Equipment Type & Size	Total Number

Chemical	Estimated Total Amount

EVIDENCE LOG

Ex. #	Sampled Y/N	Disposed Y/N	Location	Description

Your Initials	Bottom of Each Page	Page x of y

Appendix H
Sphere Volume Estimates

Liquid Height (cm)	1000 ml (12.5 cm)	2000 ml (15.6 cm)	3000ml (17.9 cm)	5000 ml (27.2 cm)	12000 ml (28.4 cm)	22000 ml (34.8 cm)	50000 ml (45.7 cm)	72000 ml (51.6 cm)
0.5	5	6	7	9	11	14	18	21
1.0	19	24	28	33	45	55	73	83
1.5	42	53	62	74	100	123	163	185
2.0	79	93	107	129	176	217	288	326
2.5	109	141	164	198	271	336	447	506
3.0	152	199	232	280	385	478	638	724
3.5	200	264	309	348	518	644	862	979
4.0	252	336	394	481	667	832	1117	1270
4.5	308	414	488	597	833	1042	1402	1596
5.0	366	497	589	723	1015	1273	1717	1957
5.5	426	585	696	859	1212	1524	2061	2351
6.0	487	676	808	1002	1422	1794	2433	2778
6.5	549	770	925	153	1646	2082	2832	3238
7.0	609	866	1046	1310	1882	2388	3258	3728
7.5	669	963	1170	1473	2129	2710	3709	4249
8.0	726	1061	296	1641	2388	3048	4185	4799
8.5		1158	1423	1812	2556	3401	4685	5378
9.0		1253	1551	1987	2933	3769	5208	5984
9.5		1347	1678	2165	3218	4150	5753	6618
10.0			1804	2344	3510	4542	6319	7278
10.5			1929	2523	3810	4948	6909	7963
11.0			2050	2703	4115	5364	7513	8673
11.5			2168	2881	4424	5790	8139	9407
12.0			2281	3058	4738	6225	8783	10163
12.5				3232	5055	6668	9444	10942
13.0				3402	5357	7120	10121	10163
13.5				3569	5696	7577	10815	12562
14.0				3731	6017	8041	11523	13402
14.5					6340	8510	12245	14261
15.0					6661	8984	12980	15137
15.5					6980	9860	13728	16031
16.0					7297	9940	14487	16941
16.5					7610	10421	15257	17867

(*Continued*)

Liquid Height (cm)	1000 ml (12.5 cm)	2000 ml (15.6 cm)	3000ml (17.9 cm)	5000 ml (27.2 cm)	12000 ml (28.4 cm)	22000 ml (34.8 cm)	50000 ml (45.7 cm)	72000 ml (51.6 cm)
17.0					7920	0903	16036	18807
17.5					8224	11358	16825	19761
18.0					8522	11867	17622	20729
18.5						12347	18426	21708
19.0						12825	19237	22699
19.5						13299	20054	23700
20.0						13771	20875	24711
20.5						14236	21701	25731
21.0						14697	22530	26758
21.5							23361	27794
22.0							24194	28835
22.5							25028	29882
23.0							25862	30934
23.5							26695	31990
24.0							27526	33049
24.5							28354	34110
25.0							29190	35173
25.5							30001	36236
26.0							30817	37299
26.5							31627	38361
27.0							32431	39421
27.5								40479
28.0								42533
28.5								42583
29.0								43627
29.5								44666
30.0								45967

Microgram, 24, 7, 184 (July 1991)

Appendix I
Color Test
Reagents

Reagent	Composition/Preparation
Acetic Acid (20% Aqueous)	To 20 ml acetic acid, add enough water to make 100 ml solution.
Acetic Acid (67% Aqueous)	To 67 ml acetic acid, add enough water to make 100 ml solution.
Ammoniated Acetone (1.5% solution)	To 1.5 ml concentrated ammonium hydroxide, add enough acetone to make 100 ml solution.
Ammoniated Methanol (1.5% solution)	To 1.5 ml concentrated ammonium hydroxide, add enough methanol to make 100 ml solution.
Aniline Sulfate	Dissolve 0.1 g aniline sulfate in 100 ml concentrated sulfuric acid.
Aniline Sulfate (Aqueous)	Dissolve 5.0 g aniline sulfate in 100 ml deionized water.
Barium Chloride	Dissolve 5.0 g barium chloride in 100 ml deionized water (5% solution).
Benedict's	• Dissolve 1.73 g copper sulfate in 10 ml deionized water. • Dissolve 17.3 g sodium citrate and 10 g anhydrous sodium carbonate in 60 ml deionized water. • Filter the citrate solution. • Add the copper sulfate solution slowly to the citrate solution. • Dilute the mixture to 100 ml with deionized water.
Bismuth	• Prepare a solution of 12.5 ml deionized water and 2.5 ml concentrated sulfuric acid. • Mix the following ingredients together until dissolved: 1. 3.75 g potassium iodide 2. 1.24 g bismuth nitrate 3. 0.40 g sodium hypophosphite • Refrigerate. Over a period of time, the orange-red solution will darken as the iodide decomposes to iodine, at which point the reagent should be discarded.

(Continued)

Reagent	Composition/Preparation
Brucine Sulfate	Dissolve 5.0 g brucine sulfate in 100 ml concentrated sulfuric acid.
n-Butanol/Acetic Acid/Water	• Mix 2 parts n-butanol with 1 part acetic acid and 1 part deionized water. • Prepare fresh before use.
Chen's	Reagent 1: • Dissolve 1.0 g of copper sulfate in a solution of 1 ml acetic acid and 100 ml of deionized water. Reagent 2: • 2N solution of sodium hydroxide.
Cobalt Thiocyanate	Method 1: • Dissolve 2.0 g of cobalt thiocyanate in 100 ml of deionized water. Method 2: • Dissolve 2.0 g of cobalt thiocyanate in a solution of 100 ml glycerine and 100 ml of deionized water.
Davis	Add 15 ml ethylenediamine to 100 ml 10% aqueous silver nitrate solution.
Dilli-Koppanyi	Reagent 1: • Dissolve 0.1 g of cobaltous acetate tetrahydrate in a solution of 100 ml of methanol and 0.2 ml of glacial acetic acid. Reagent 2: • Mix 5.0 ml of isopropyl amine with 95 ml methanol.
m-Dinitrobenzene	2% solution of m-dinitrobenzene in reagent alcohol.
Diphenylamine	• Dissolve 0.68 g diphenylamine in 45 ml concentrated sulfuric acid. • Place in ice bath and cautiously add 22.5 ml glacial acetic acid.
Dragendorff Spray	• Mix 2.0 g bismuth subnitrate in a solution of 25 ml acetic acid and 100 ml deionized water. • Dissolve 40 g potassium iodide in 100 ml deionized water. • Mix 10 ml of each of solution with 20 ml acetic acid and 100 ml deionized water.
Duquenois	Dissolve 5 drops acetaldehyde and 0.4 g vanillin in 20 ml of 95% ethanol.
Ehrlich's	Dissolve 5.0 g of p-dimethylaminobenzaldehyde in a solution of 50 ml ethanol and 50 ml of concentrated hydrochloric acid.
Ethanol/Heptane (95:5)	Mix 95 parts absolute ethanol with 5 parts heptane.
Ethanol/Hexane (9:1)	Mix 9 parts absolute ethanol with 1 part hexane.

(Continued)

Reagent	Composition/Preparation
Ethylenediamine (15%)	Mix 15 ml of ethylenediamine in 85 ml of deionized water.
Fehling's	• Dissolve 7.5 g copper sulfate in 100 ml deionized water. • Dissolve 35 g sodium tartrate and 25 g potassium hydroxide in 100 ml deionized water. • Mix equal volumes of each solution.
Ferric Chloride	Mix 1 g ferric chloride in 10 ml deionized water.
Fiegel's (Sodium Nitroprusside)	• Dissolve 1.0 g sodium nitroprusside (sodium nitroferricyanide) in 60 ml deionized water. • Add 10 ml acetaldehyde. • Dilute solution to 100 ml with deionized water.
Froehde's	Dissolve 100 mg of sodium molybdate in 20 ml concentrated sulfuric acid.
Hydrochloric Acid (0.1 N)	1 ml concentrated hydrochloric acid diluted to 120 ml with deionized water.
Hydrochloric Acid (3.0 N)	125 ml concentrated hydrochloric acid diluted to 500 ml with deionized water.
Hydrochloric Acid (15%)	Mix 7.9 ml concentrated HCl into 12.1 ml deionized water.
Iodoplatinate Spray	• 10 ml of 10% aqueous solution of chloroplatinic acid. • 250 ml of 4% aqueous potassium iodide. • Combine solutions and add 500 ml deionized water. • Add 0.75 ml concentrated hydrochloric acid.
Le Rosen	Mix 75 ml of concentrated sulfuric acid in 1.5 ml 37% formaldehyde.
Liebermann's	Dissolve 10 g potassium nitrite in 100 ml of concentrated sulfuric acid.
Mandelins	Dissolve 1.0 g ammonium meta in 100 ml concentrated sulfuric acid.
Marquis' (Premixed)	Add 8–10 drops of 40% formaldehyde solution for each 10 ml concentrated sulfuric acid used.
Mecke's	Dissolve 0.25 g selenious acid (H_2SeO) in 25 ml concentrated sulfuric acid.
Mercuric Chloride	Dissolve 1.25 m mercuric chloride in 25 ml deionized water.
Mercuric Iodide	• Dilute 27 ml of concentrated hydrochloric acid to 100 ml with deionized water. • Add enough mercuric iodide to saturate the acid water solution.
Molish Solution (a-Naphthol)	• Dissolve 1.25 g of a-naphthol in reagent alcohol. • Refrigerate.

(*Continued*)

Reagent	Composition/Preparation
Nessler	• Dissolve 20 g potassium hydroxide in 50 ml deionized water. • Dissolve 10 g mercuric iodide and 5 g of potassium iodide into 50 ml deionized water. • Combine solutions.
Nitron	Dissolve 3.75 g nitron (diphenylenedianilhydrotriazole) in 75 ml of 88% formic acid.
p-DMBA	Dissolve 1.25 g p-dimethylaminobenzaldehyde in 25 ml concentrated acetic acid.
Silver Nitrate	Dissolve 3.75 g of silver nitrate in 75 ml of deionized water.
Sodium Hydroxide (0.5 N)	Dissolve 1.5 g NaOH in 75 ml deionized water.
Sodium Hydroxide (2.0 N)	Dissolve 6.0 g NaOH in 75 ml deionized water.
Stannous Chloride	• Dissolve 5.0 g stannous chloride in 10 ml concentrated hydrochloric acid. • Dilute to 100 ml with deionized water. • The stannous chloride must be completely dissolved in the HCl before diluting with water.
Starch	Saturated solution of hydrolyzed starch in deionized water.
Sulfuric Acid (75%)	Mix 56.25 ml concentrated sulfuric acid into 18.75 ml deionized water.
Sulfuric Acid (0.1 N)	Dissolve 1.0 ml concentrated sulfuric acid in 360 ml deionized water.
Tartaric Acid (2.5%)	Make 1.25 g of tartaric acid in 25 ml deionized water.
Thymol	Dissolve 0.25 g thymol in 25 ml methanol.
Toluene/Acetic Acid (9:1)	Mix 9 parts toluene with 1 part acetic acid.
Triphenyl Selenium Chloride	Saturated solution in 60 ml deionized water.
Triphenyltetrazolium Chloride	Dissolve 0.38 g triphenyltetrazolium chloride in 75 ml distilled water.
Van Urk's Spray	• Dissolve 1.0 g p-dimethylaminobenzaldehyde in 100 ml of ethanol. • Add 10 ml concentrated hydrochloric acid.
Wagenaar's	• Dissolve 1.25 gm copper sulfate in 25 ml deionized water. • Add sufficient ethylenediamine to turn the solution a dark violet color.
Wagner's	• Dissolve 1.27 g iodine and 2.75 g potassium iodide in 5.0 ml deionized water. • Dilute to 100 ml with deionized water.
Zwikker's	Reagent 1: • Dissolve 0.125 g of copper sulfate in 25 ml of deionized water. Reagent 2: • Mix 2.5 ml of chloroform with 22.5 ml pyridine.

Appendix J
Crystal Test Reagents

1 Isomer Crystal Reagents

Reagent	Composition/Preparation
Gold Bromide (Aqueous)	Dissolve 1 g of gold chloride ($HAuCl_4 * 3 H_2O$, i.e., chloroauric acid) and 1 g of sodium bromide in 20 ml of deionized water.
Gold Bromide (in H_2SO_4)	• Combine 1 g of gold chloride with 1.5 ml of 40% HBr. • Add 28.5 ml of a 2:3 solution of concentrated sulfuric acid and water.
Gold Chloride (Aqueous)	5% (w/v) solution of gold chloride in water.
Gold Chloride (in H_3PO_4)	5% (w/v) solution of gold chloride in a 1:3 solution of concentrated phosphoric acid and water.
Platinum Bromide (Aqueous)	Dissolve 1 g platinic chloride ($H_2PtCl_6 * 6 H_2O$, i.e., chloroplatinic acid) with 1 g of sodium bromide in 20 ml deionized water.
Platinum Bromide (in H_3PO_4)	• Combine 1 g of platinic chloride with 1.7 ml of 40% HBr. • Add 18.3 ml of a 1:3 solution of concentrated phosphoric acid and water.
Platinum Chloride (Aqueous)	5% (w/v) solution of platinic chloride in water.
Platinum Chloride (in H_3PO_4)	5% (w/v) solution of platinic chloride in a 1:3 solution of concentrated phosphoric acid and water.

2 Inorganic Crystal Reagents

Reagent	Composition/Preparation
Ammonium Molybdate	Saturated solution of ammonium molybdate [$(NH_4)_6Mo_7O_{24} * 4H_2O$] in concentrated nitric acid.
Cropen	Solution A • 5 g of zinc sulfate, 4 g of potassium nitrate dissolved in 20 ml of deionized water. Solution B • 0.015% methylene blue in deionized water.

(Continued)

367

Reagent	Composition/Preparation
Nitron	1 g of Nitron (1,2 dihydro 1,4 diphenyl 3,5 phenylimino 1,2,3 triazol) in 20 ml of formic acid.
Platinum Chloride	5% (w/v) aqueous solution of platinic chloride (chloroplatinic acid, $H_2PtCL_6 * 6 H_2O$).
Squaric Acid	Saturated aqueous solution of squaric acid (1,2 dihydroxycyclobutenedione).
Strychnine Sulfate	Saturated aqueous solution of strychnine sulfate.
Uranyl Acetate	Solution A (Best for Na^+ and K^+)
	• Saturated solution of glacial acetic acid containing 50/50 mixture of uranyl acetate $[UO_2(C_2H_3O_2) * 2 H_2O]$ and zinc acetate $[Zn(C_2H_3O_2) * 2 H_2O]$
	Or
	Solution B (Best for NH_4^+)
	• Saturated solution of glacial acetic acid containing uranyl acetate $[UO_2(C_2H_3O_2) * 2 H_2O]$.

Appendix K
Optical Properties

Compound	Crystal Form	n
Aluminum Powder	NA	NA
Aluminum Chloride (AlCl$_3$. 6H$_2$O)	col, rhomb, del	1.6
Aluminum Oxide (Al$_2$O$_3$)	col, hex	1.768, 1.760
Ammonium Chloride (NH$_4$Cl)	col, cube	1.642
Ammonium Nitrate (NH$_4$NO$_3$)	col, rho, (monocl > 32.1°)	
Antimony Sulfide (Sb$_2$S$_3$)	blk, rho	3.194, 4.064, 4.303
Barium Carbonate (BaCO$_3$)	wht, rho	1.529, 1.676, 1.677
Barium Chlorate (Ba(ClO$_3$)$_2$)	col, monocl	1.562, 1.577, 1.635
Barium Nitrate (Ba(NO$_3$)$_2$)	col, cub	1.572
Barium Perchlorate (Ba(ClO$_4$)$_2$)	col, hex	1.533
Barium Sulfate (BaSO$_4$)	wht, rho, (monocl)	1.637, 1.638, 1.649
Calcium Carbonate (CaCO$_3$ * 6H$_2$O)	col, monocl	1.460, 1.535, 1.545
Chromium Trioxide (CrO$_3$)	rd, rho, del	
Copper Sulfate (CuSO$_4$ * 5H$_2$O)	blu, tricl	1.514, 1.537, 1.543
Magnesium Powder	NA	NA
Magnesium Sulfate (MgSO$_4$) anhydrous	col, rho, cr	1.560
Magnesium Sulfate (MgSO$_4$ * 7H$_2$O) (Epsom salts)	col, rho or monocl	1.433, 1.455, 1.461
Manganous Chloride (MnCl$_2$ * 4H$_2$O)	rose, monocl, del	
Manganous Carbonate (MnCO$_3$)	lt brn, rho	
Mercuric Chloride (HgCl$_2$)	col, rho, or wht pdr	1.859
Palladium Chloride (PdCl$_2$)	rd, cu need, del	
Palladium Sulfate (PdSO$_4$ * 2H$_2$O)	rd-brn, cr, del	
Phosphorus Oxychloride (POCl$_3$) or (P$_2$O$_3$Cl$_4$)	col, fum liq	
Phosphorus Pentoxide (P$_2$O$_5$)	wht, monocl or pdr, del	
Potassium Bicarbonate (KHCO$_3$)	col, monocl	1.482
Potassium Carbonate (K$_2$CO$_3$)	col, monocl, hygro	1.531
Potassium Chlorate (KClO$_3$)	col, monocl	1.409, 1.517, 1.524
Potassium Chromate (K$_2$CrO$_4$)	hel, rho	1.74
Potassium Cyanide (KCN)	col, cub, wh gran	1.410
Potassium Dichromate (K$_2$Cr$_2$O$_7$)	red, monocl or tri	1.738

(*Continued*)

Compound	Crystal Form	n
Potassium Iodide (KI)	col or wht, cub	1.677
Potassium Nitrate (KNO_3)	col, rho or trig	1.335, 1.505, 1.506
Potassium Perchlorate ($KClO_4$)	col, rho	1.471, 1.472, 1.476
Sodium Acetate ($NaC_2H_3O_2$)	wht, monocl	1.464
Sodium Bicarbonate ($NaHCO_3$)	wht, monocl, pr	1.526
Sodium Borohydrate ($NaBH_4$)	wht, cub	1.524
Sodium Carbonate (Na_2CO_3)	wht, monocl	1.405, 1.425, 1.440
Sodium Chloride (NaCl)	wht, cub	1.544
Sodium Hydride (NaH)	silver, need	1.470
Sodium Nitrate ($NaNO_3$)	col, trig or rho, bdr	1.587, 1.336
Sodium Sulfate ($NaSO_4$)	monocl	1.480
Strontium Nitrate ($Sr(NO_3)_2$)	Col, cub	
Sulfur (S)	yel, rho	1.957
Thorium Nitrate ($Th(NO_3)_4$)	plates, del	
Zinc Powder	NA	NA
Zinc Chloride ($ZnCl_2$)	wht, hex, del	1.681, 1.713

Abbreviations

brn	brown	col	colorless	cub	cubic
del	deliquescent	hex	hexagonal	hygro	hygroscopic
monocl	monoclinic	need	needles	ppl	purple
pdr	powder	rd	red	rho	rhombic
tetr	tetragonal	trig	trigonal	wht	white
cr	crystals	pr	prisms	fum	fuming

CRC Handbook of Chemistry and Physics (68th ed.), CRC Press, Boca Raton, FL.

Appendix L
Extraction
Procedures

1 Extraction Solubility Guidelines

- Basic drugs are soluble in acidic (pH < 7) water solutions. Acidic drugs are not.
- Acidic drugs are soluble in basic (pH > 7) water solutions. Basic drugs are not.
- Neutral drugs can be soluble in both.
- Free base, free acid and neutral drugs are soluble in organic solvents.
- Free base and free acid drugs are insoluble in water.
- To determine whether a drug is acidic or basic, look at the salt form name. Basic drugs have an acid as part of their salt form (e.g., hydrochloride [HCl], sulfate [H_2SO_4], acetate [CH_3COOH], etc.). Acidic drugs have an alkali metal as part of their salt form name (e.g., sodium [Na] or potassium [K]). Neutral drugs do not have an associated salt form.

2 Particle Picking

- Remove particles that appear to be the substance in question with tweezers.
- Analyze by infrared (IR).

3 Dry Extraction

- Place powder sample in an organic solvent in which the component of interest is soluble, and the diluents and adulterants are not.
- Isolate and analyze the solvent
  - Evaporate the solvent and analyze the residue via IR or X-ray diffraction (XRD).
  - Solvent can be analyzed via gas chromatography mass spectrometry (GC/MS).

4 Dry Wash

- Place powder sample in an organic solvent in which the component of interest is NOT soluble, and the diluents and adulterants ARE soluble.
- Remove the solvent from the insoluble solid.
  - Multiple solvents may be required to remove all diluents and adulterants.
- Analyze the insoluble solids via IR or XRD.

5 General Liquid/Liquid Extraction Procedure

- Dissolve the sample in an acidic (pH < 7) water solution. (The basic drugs will dissolve into the acidic aqueous solution.)
- Add an organic solvent and agitate. (The acidic and neutral drugs will dissolve into the organic solvent.)
- Allow the liquids to separate.
- Remove and discard the organic solvent.
- Add a concentrated basic liquid to the solution until the pH is >7. (The basic drugs will come out of solution.)
- Add an organic solvent and agitate. (The basic drugs will dissolve into the organic solvent.)
- Allow the liquids to separate.
- Remove and retain the organic solvent.
- Analyze the solvent.
  - The organic solvent can be analyzed by GC/MS.
  - Or the solvent can be evaporated and the residue can be analyzed via IR or XRD.

6 General Ion Pairing Extraction Technique

- Dissolve the sample in an acidic (pH < 7) water solution with a high concentration of halide ions. (The use of HCl or the addition of a chloride salt will provide the needed environment. The basic drugs will dissolve into the acidic solution.)
- Add a chlorinated solvent (chloroform) and agitate. (The ion-pairing drug will extract into the organic solvent along with the acidic and neutral drugs.)

- Allow the liquids to separate.
- Remove and save the organic solvent.
- Add an acidic solvent, void of chloride ions, to the organic solution and agitate. (The ion-pairing drugs will dissolve into the acidic solution.)
- Allow the liquids to separate.
- Remove and discard the organic solvent.
- Add a concentrated basic liquid to the solution until the pH is >7. (The ion-pairing drugs will not be soluble in the basic aqueous solution.)
- Add an organic solvent and agitate. (The ion-pairing drugs will dissolve into the organic solvent.)
- Allow the liquids to separate.
- Remove and retain the organic solvent. (The organic solvent can be analyzed by GC/MS. The solvent can be evaporated, and IR can be used to analyze the residual free base drug.)

7 Non-Polar Extraction

- Place sample into a container with acetonitrile and agitate.
- Transfer the acetonitrile to a clean container.
- Add hexane or petroleum ether to the acetonitrile and agitate.
- Allow the liquids to separate.
- Remove and discard the hexane (top) layer.
  - High concentrations of non-polar compounds may require multiple washes.
  - Avoid additional washes with steroid esters with large, non-polar acids will lower the yield.
- Analyze the acetonitrile via GC/MS.

8 Amphoteric Drug (Morphine) Extraction

- Place sample into a container with methanol and 1–2 drops of concentrated acid (HCl or H_2SO_4), agitate.
- Transfer the liquid to a clean container and evaporate to dryness.
- Reconstitute the sample with deionized water and 1–2 drops of concentrated hydrochloric or sulfuric acid.
- Add dichloromethane and agitate.
  - Addition of a small amount of methanol will aid in morphine recovery.

- Allow liquids to separate.
- Transfer the water layer to a clean container.
  - Retain the dichloromethane for additional testing if required.
- Add a sodium bicarbonate solution to the water to pH of approximately 9.
  - pH is important for morphine recovery.
- Add dichloromethane and agitate.
  - Addition of 2–3 drops of methanol will aid in morphine recovery.
- Remove dichloromethane and place it into a clean container.
  - Morphine and amphoteric drugs will be in the organic layer.
- Analyze via GC/MS.

Additional alkaloids and basic drugs can be extracted from the water by:

- Adding drops of concentrated sodium hydroxide to the water layer to a pH > 12.
- Add dichloromethane and agitate.
- Remove dichloromethane and analyze via GC/MS.
  - This extract can be combined with the amphoteric extract for testing purposes.

Appendix M
Reaction Mixture Components

Chemical	PK1	PK2	PK3	PK4	PK5	MW	Drugs
E-1,5-DIPHENYL-2-METHYL-1-PENTENE-4-ONE	43	91	129	250	57	250	P1 P3
Z-1,5-DIPHENYL-2-METHYL-1-PENTENE-4-ONE	43	250	129	57	56	250	P1 P3
PYRROLIDINE	43	70	71	42	41	71	PC2
N-ACETYLAMPHETAMINE	44	86	119	91	65	177	A4
AMPHETAMINE	44	91	43	42	65	135	M5
METHYLENEDIOXYAMPHETAMINE (MDA)	44	136	135	77	51	179	MD3
N-METHYL 1,2-(METHYLENEDIOXY)-4-(3-AMINOPROPYL) BENZENE	44	162	65	77	135	193	MD5
CYCLOHEXANONE	55	42	98	69	70	98	ALL PC meth
1,2-(METHYLENEDIOXY)-4-(2-N-METHYLAMINOPROPYL)-BENZENE	56	191	135	77	57	191	MD3
MORPHOLINE	57	87	56	86	42	87	PC3
EPHEDRINE	58	69	79	41	59	165	M2 M3
METHYLENEDIOXYMETHAMPHETAMINE (MDMA)	58	136	135	59	77	193	MD3 MD5
N-METHYL-1-[1-(HYDROXY)-2-METHOXY)]-4-(2-AMINOPROPYL) BENZENE	58	51	77	137	94	195	MD5
N-METHYL-1-[1,2-(DIMETHOXY)-4-(2-AMINOPROPYL) BENZENE	58	152	51	151	59	209	MD5
1-(3,4-METHYLENEDIOXY)PHENYL-2-METHOXYPROPANE	59	194	135	136	77	194	MD5
N-FORMYLAMPHETAMINE	72	44	118	91	65	163	A1 M5
N,N DIMETHYLAMPHETAMINE	72	44	42	91	56	163	M1 M5
3,4-(METHYLENEDIOXY)-N,N-DIMETHYLAMPHETAMINE	72	56	44	73	58	207	MD3
BROMOBENZENE	77	71	50	156	158	156	PC1A, 2A, 3A
BENZENE	78	77	50	51	52	77	A5
a-METHYLBENZYL ALCOHOL	79	77	43	107	51	122	P1
PIPERIDINE	84	85	55	57	42	84	

(Continued)

Chemical	PK1	PK2	PK3	PK4	PK5	MW	Drugs
N-FORMYLMETHAMPHETAMINE	86	58	91	56	65	177	M5
3,4-(METHYLENEDIOXY)-N-N-METHYLETHYLAMPHETAMINE	86	58	87	56	44	221	MD3
PHENYL-2-PROPANONE	91	134	92	43	65	134	numerous A&M
DIBENZYL KETONE	91	65	39	119	92	210	num A, M & P
N,N-DI(b-PHENYLISOPROPYL)-AMINE	91	44	162	119	65	253	A1 M5
DI-(b-PHENYLISOPROPYL)-METHYLAMINE	91	58	176	119	42	267	A1 M5
N,N-DI-(b-PHENYLISOPROPYL)-FORMAMIDE	91	190	119	72	191	281	A1
N-(b-PHENYLISOPROPYL)-BENZYL METHYL KETIMINE	91	119	160	41	65	251	A4
1-OXY-1-PHENYL-2-(b-PHENYLISOPROPYLIMINO)-PROPANE	91	119	43	105	77	265	A4
2,4-DIHYDROXY-1,5-DIPHENYL-4-METHYL-1-PENTENE	91	159	131	65	115	268	A4
2-PHENYLMETHYLAZIRIDINE	91	104	132	78	51	133	A2 A3
2-METHYL-3-PHENYLAZIRIDINE	91	132	42	105	92	133	A2 A3
BENZYL METHYL KETOXIME	91	41	92	65	39	149	A3
BENZYL ACETATE	91	150	65	59	105	150	P1
Z-1,3-DIPHENYL-2-METHYL-1-PENTENE-4-ONE	91	130	115	65	159	250	P1 P3
E-1,3-DIPHENYL-2-METHYL-1-PENTENE-4-ONE	91	131	65	105	159	250	P1 P3
ISOSAFROLE GLYCOL	93	151	65	123	152	196	MD1
PHENOL	94	66	65	95	67	94	
CAMPHOR	95	81	41	108	152	152	MD5
1-[1-(2'-THIENYL)CYCLOHEXYL]-PIPERIDINE	97	165	165	206	84	249	PC1B

(Continued)

Chemical	PK1	PK2	PK3	PK4	PK5	MW	Drugs
1-[1-(2-THIENYL)CYCLOHEXYL]-MORPHOLINE	97	165	251	208	123	251	PC3B
1-[3,4-(METHYLENEDIOXY)-PHENYL]-2-NITRO-1-PROPENE	103	160	207	77	102	207	MD2
PHENYL NITROPROPENE	105	115	91	77	116	163	A2
1-PHENYL-2-PROPANOL	105	106	77	79	91	136	P6
METHYL BENZOATE	105	136	77	51	117	136	P1
BENZALDEHYDE	106	105	77	51	63	106	P1 P3
1-MORPHOLINOCYCLOHEXENE	108	81	167	109	152	167	PC3
Z-1-PHENYL-2-BENZYL-1-PROPENE	115	208	91	193	134	208	P1 P3
1-PHENYL-2-BENZYL-2-PROPENE	115	208	91	193	178	208	P1 P3
E-1-PHENYL-2-BENZYL-2-PROPENE	117	115	91	208	129	208	P1 P3
DIBENZYL METHYLAMINE	120	91	42	77	102	211	A1 M5
1-CYCLOHEXYLPIPERIDINE	125	41	167			167	PC1
1-PHENYLCYCLOPENTENE	129	44	43	128	115	129	PC1
1-PHENYLCYCLOHEXENE	129	158	115	130	142	158	PC1
E-1,3-DIPHENYL-2-METHYL-2-PENTENE-4-ONE	131	103	77	91	65	250	P1 P3
N-(b-PHENYLISOPROPYL)-BENZALDIMINE	132	105	91	77	65	223	A4
1-PHENYLCYCLOHEXANOL	133	105	176	55	120	176	PC1
a--BENZYL-N-METHYLPHENETHYLAMINE	134	91	42	119	65	225	M5
Z-PHENYL-2-PROPANONE ENOL ACETATE	134	43	91	119	105	176	P1
E-PHENYL-2-PROPANONE ENOL ACETATE	134	91	43	119	105	176	P1
4-METHYL-1,2-(METHYLENEDIOXY)-BENZENE	135	136	77	79	51	136	MD3
1,2-(METHYLENEDIOXY)-4-PROPYLBENZENE	135	77	164	51	79	164	MD1 MD3
3,4-(METHYLENEDIOXY)-BENZYL-N-METHYLAMINE	135	42	51	77	136	165	MD3
3,4-(METHYLENEDIOXY)-PHENYLPROPANONE	135	77	51	43	178	178	MD1 MD3

(Continued)

Chemical	PK1	PK2	PK3	PK4	PK5	MW	Drugs
1-[3,4-(METHYLENEDIOXY)]-PHENYL-2-PROPANOL	135	136	77	51	106	180	MD3 MD5
3,4-(METHYLENEDIOXY)-BENZYLMETHYLKETOXIME	135	178	176	77	136	283	MD2
N-{b-[3,4-(METHYLENEDIOXY)]-PHENYLMETHYL}-3,4-(METHYLENEDIOX Y)-BENZALDIMINE	135	178	176	77	136	283	MD2
N,N-DI-[3,4-(METHYLENEDIOXY)-PHENYLMETHYL]-AMINE	135	150	136	77	51	285	MD2
1-[3,4(METHYLENEDIOXY)]-4-(2-BROMOPROPYL)-BENZENE	135	77	242	244	51	242	MD5
2-THIENYLCYCLOHEXENE	135	164	165	136	122	164	PC1B, 2B, 3B
1-PYRROLIDINOCYCLOHEXANE CARBONITRILE	135	97	110	136	121	178	PC2
2-METHOXY-4-(2-BROMOPROPYL)-PHENOL	137	244	246	165	135	244	MD5
1,2-DIMETHYL-3-PHENYLAZIRIDINE	146	105	42	132	91	147	M2 M3
HYDROXYSKATOLE	147	146	62	63	89	147	MD2
PIPERONAL	149	150	121	63	65	150	MD2 MD3
1-[3,4(METHYLENEDIOXY)]-4-(3-BROMOPROPYL)-BENZENE	149	119	91	163	242	242	MD5
1-PIPERIDINICYCLOHEXYL CARBONITRILE (PCC)	149	150	191	164	124	192	PC1
1-PYRROLIDINOCYCLOHEXENE	150	151	122	136	95	151	PC2
1-PIPERIDINOCYCLOHEXENE	150	164	165	136	122	165	PC1
1,2-DIMETHOXY-4-(2-BROMOPROPYL)-BENZENE	151	179	107	258	260	258	MD5
1-MORPHOLINOCYCLOHEXANE CARBONITRILE	151	124	81	136	108	194	PC3
3,4-(METHYLENEDIOXY)-PHENYLMETHANOL	152	137	93	65	151	152	MD2
1,3-DIPHENYL-2-METHYL-2-PENTENE-4-ONE	159	91	144	160	141	250	P1 P3
SAFROLE	162	104	131	103	77	162	MD1 MD3 MD5
1,2-(DIMETHOXY)-4-PROPENYLBENZENE	162	163	178	147	135	178	MD3
N-FORMYL-METHYLENEDIOXYAMPHETAMINE	162	135	72	44	77	207	MD4

(Continued)

Chemical	PK1	PK2	PK3	PK4	PK5	MW	Drugs
DL-[3,4-(METHYLENEDIOXY)]-PHENYLPROPANONE	163	135	105	133	77	298	MD2
N-{b-[3,4-(METHYLENEDIOXY)]-PHENYLISOPROPYL}-3,4-(METHYLENEDIOXY)-BENZYLKETIMINE	163	204	135	105	77	339	MD2
DL-[1-3,4-(METHYLENEDIOXY)-PHENYL-2-PROPYL]-AMINE	163	135	206	105	133	341	MD1
DL-[3,4-(METHYLENEDIOXY)-PHENYL-2-PROPYL]-METHYLAMINE	163	220	135	105	58	355	MD1
EUGENOL	164	77	55	103	149	164	MD5
1,2-DIMETHOXY-4-(3-BROMOPROPYL)-BENZENE	165	162	258	260	119	258	MD5
DIPHENYLMETHANE	167	168	152	153	91	168	P3
4-BENZYLPYRIMIDINE	169	170	91	115	142	170	A1
4-METHYL-5-PHENYLPYRIMIDE	170	169	102	115	116	170	A1
N-{b-[3,4-(METHYLENEDIOXY)]-PHENYLISOPROPYL}-3,4-(METHYLENEDIOXY)-BENZALDIMINE	176	149	177	77	135	311	MD2
4-ALLYL-1,2-(DIMETHOXY)-BENZENE	178	91	107	103	147	178	MD5
1,2-(DIMETHOXY)-4-PROPENYLBENZENE	178	107	163	91	103	178	MD5
BIBENZYL	179	180	91	165	89	180	P3
1,3,5-TRIPHENYL-2,4,6-TRIMETHYL BENZENE	179	91	257	348	178	348	P1 P3
cis OR trans STILBENE	180	179	178	165	89	180	P1 P3
TRIMETHOXY-4-(2-BROMOPROPYL) BENZENE	181	209	288	290	148	277	MD5
2-METHYL-2-PHENYLMETHYL-5-PHENYL-2,3-DIHYDROPYRID-4-ONE	186	91	158	143	187	277	A1
1-(1-PHENYLCYCLOHEXYL)-PYRROLIDINE (PCPy)	186	91	70	152	229	229	PC2A
1-[1-(2-THIENYL)-CYCLOHEXYL]-PYRROLIDINE	192	97	165	70	235	235	PC2B
1-(1-PHENYLCYCLOHEXYL)-PIPERIDINE (PCP)	200	91	242	243	186	243	PC1A
1-(1-PHENYLCYCLOHEXYL)-MORPHOLINE	202	91	245	244	117	245	PC3A
4-ALLYL-TRIMETHOXYBENZENE	208	193	161	133	105	208	MD5

(Continued)

Chemical	PK1	PK2	PK3	PK4	PK5	MW	Drugs
1-BENZYL-3-METHYLNAPHTHALENE	232	217	108	215	202	232	A1
1,3-DIMETHYL-2-PHENYLNAPHTHALENE	232	215	217	108	202	232	A1
4-METHYL-5-PHENYL-(2-PHENYLMETHYL)-PYRIDINE	258	259	243	244	260	259	A1
2-METHYL-3-PHENYL-6-PHENYLMETHYL-PYRIDINE	258	259	180	244	260	259	A1
2,4-DIMETHYL-3,5-DIPHENYLPYRIDINE	259	260	244	115	215	259	A1
2,6-DIMETHYL-3,5-DIPHENYLPYRIDINE	259	260	115	244	101	259	A1
2,4-DIMETHYL-3-PHENYL-6-PHENYLMETHYL-PYRIDINE	272	273	258	55	57	273	A1
2,4-DIPHENYL-3,5-DIMETHYLPHENOL	274	259	101	165	152	274	

DRUG KEY

A1	Amphetamine via Leuckart reaction
A2	Amphetamine via benzaldehyde/nitroethane
A3	Amphetamine via phenylacetone/hydroxylamine
A4	Amphetamine via phenylacetone/ammonia
M2 M3	Methamphetamine via ephedrine reduction
M5	Methamphetamine via Leuckart reaction
P1	Phenylacetone via phenylacetic acid/acetic anhydride
P3	Phenylacetone via phenylacetic acid/lead acetate
PC1	Piperidine/cyclohexane intermediate
PC1A	Phenyl addition
PC1B	Thionyl addition
PC2	Pyrrolidine/cyclohexane intermediate
PC2A	Phenyl addition
PC2B	Thionyl addition
PC3	Morpholine/cyclohexane intermediate
PC3A	Phenyl addition
PC3B	Thionyl addition
MD2	MDA via piperonal/nitroethane
MD3	MDMA via amination of MD-P-2-P
MD5	MDMA via safrole/HBr reaction

Appendix N
Calculation Equations

---

1 Geometric Shape Volumes

1.1 Cylinder
$= \pi$ * radius * radius * height
$= 3.1415$ * (diameter/2) * (diameter/2) * height
$= 0.78$ * diameter * diameter * height

1.2 Cone
$= 0.33$ * π * radius * radius * height
$= 0.33$ * 3.1415 * (diameter/2) * (diameter/2) * height
$= 0.26$ * diameter * diameter * height

1.3 Sphere
$= 1.33$ * π * radius * radius * radius
$= 1.33$ * 3.1415 * (diameter/2) * (diameter/2) * (diameter/2)
$= 0.522$ * diameter * diameter * diameter

2 Quantitation Equations

2.1 Gravimetric Quantitation

2.1.1 Concentration$_{Sample}$
$=$ Weight$_{Extracted\ Methamphetamine}$/Volume$_{Sample}$

2.1.2 Weight$_{Original\ Container}$
$=$ Original Volume*Concentration$_{Sample}$

2.2 Serial Dilution Quantitation

2.2.1 Line Equation
$Y = (m$*$x) + b$

2.2.2 Line Slope
$=$ (Concentration$_{Max\ -\ Min}$)/(GC Response$_{Max}$ $-$ GC Response$_{Min}$)

2.2.3 Concentration (y)
$=$ [Line Slope (m) * peak area (x)] $+$ Y intercept (b)
$=$ [(Concentration^{Max} $-$ Concentration^{Min})/(GC Response^{Max} $-$ GC Response^{Min})] * Peak Area^{unknown} $+$ 0(assumed)

2.2.4 Percentage
$=$ (Concentration$_{Calculated}$/Concentration$_{Original\ sample}$)*100

2.3 Single Standard Solution

2.3.1 Concentration$_{unknown}$
$=$ (Area$_{unknown}$ * Concentration$_{Standard}$)/Area$_{Standard}$

2.3.2 Concentration$_{unknown}$
$=$ (Area$_{unknown}$ * Area$_{Internal\ Standard\ of\ Standard}$ * Concentration$_{Standard}$)/ (Area$_{Standard}$ * Area$_{Internal\ Standard\ of\ Unknown}$)

2.4 Production Estimates

2.4.1 Conversion Factor (n)
$=$ Molecular weight$_{final\ product}$/Molecular weight$_{precursor\ chemical}$

2.4.2 Weight$_{Theoretical}$
$=$ n * Weight$_{precursor\ chemical}$

2.4.3 Weight$_{Actual}$
$=$ n * Weight$_{precursor\ chemical}$ * Percentage Yield$_{estimated}$

Appendix O
Drug
Reference List

General Information

"Safety Alert re. Red Phosphorus to Yellow Phosphorus", *Microgram*, 25, 12, 305 (Dec 1992).

"Safety Alert re. Hydriodic Acid Synthesis Using Hydrogen Sulfide", *Microgram*, 25, 12, 305 (Dec 1992).

"The Rise of Crack and Ice: Experience in Three Locales", National Institute of Justice Research Brief (March 1993).

"Controlling Chemicals Used in to Make Illegal Drugs: The Chemical Action Task Force and the Domestic Chemical Action Group", National Institute of Justice Research Brief (January 1993).

"Glossary of Common MSDS Terms", Chemical Hygiene Plan, Arizona Department of Public Safety.

Frank, R. S., "The Clandestine Drug Laboratory Situation in the United States", *Journal of Forensic Science*, 28, 1, 18 (Jan 1983).

Some, W. H., "Contamination of Clandestinely Prepared Drugs with Synthetic By-Products", *NIDA research monograph*, 95, 44 (1989).

Courtney, M., Ekis, T., "A Protocol for the Forensic Chemist in the Field Investigation of Clandestine Amphetamine Labs", *Southwestern Association of Forensic Scientists Journal*, 10, 1, 20 (March 1988).

Johnson, S. B., et.al., "Evidentiary Aspects of Clandestine Laboratories", 1, 13.

Manchester, R., Pearce, P., "Safe Meth Lab Raids: Police and Security News", *The Narc Officer*, 65 (Oct 1989).

Gregory, P., Lazarous, B., "Safety and Seizure Aspects of Clandestine Drug Laboratories", 105.

Ruple, T. M., Hoffman, C., "Clandestine Laboratory Production Estimates", Presentation at the Clandestine Laboratory Investigating Chemists Association Training Seminar (Sept. 1991).

Lewin, R., "Trail of Ironies to Parkinson~ s Disease", *Science*, 224, 1083 (8 June 1984).

Ekis, T. R., "The Efficacy of Latent Print Examinations in Clandestine Lab Seizures", *Southwestern Association of Forensic Scientist Journal*, 13, 1, 34 (April 1991).

Connors, E., "Hazardous Chemicals from Clandestine Drug Labs Pose a Threat to Law Enforcement", *Narcotics Control Technical Assistance Newsletter*, 3, 1, 1 (July 1989).

James, R. D., "Hazards of Clandestine Drug Laboratories", *FBI Law Enforcement Bulletin*, Vol. 58, Issue 4 (April 1989) Pages: 16-21.

Wilkin, G., "The New Midnight Dumpers", *US News and World Report*, 57, (9 January 1989).

Evens, H. K., Kelley, P. M., "Clandestine Laboratory Trends in Southern California", San Bernadino County Sheriffs Office.

Couteur, G. R., "Investigation of a Clandestine Laboratory", *The Australian Police Journal*, 29, 4, 267 (October 1975).

Roberton, R. J., "Designer Drugs - The Analog *Game*", *The Narc Officer*, 65 (Winter 1985).

Dal Cason, T. A., et.al., "Investigations of Clandestine Drug Manufacturing Laboratories", *Analytical Chemistry*, 52, 804A (1980).

Garmon, L., "Sluething Clandestine Chemistry", *Science News*, 118, 3, 44 (July 1980) 1980 Page 44.

Roberton, R. J., "The Analog Game: Designer Drugs Killing Crippling Users", *Narcotics Control Digest*, 15, 7, 2 (April 1985).

Howard, H. A., "Clandestine Drug Labs", *Fire Engineering*, 16 (August 1986).

Largent, D. A., "'ICE' Crystal Methamphetamine", California Department of Justice, Bureau of Narcotics Enforcement (September 1989).

Simpson, N. L., "Recent - Federal Decisions: Lab Prosecutions and a Few Others", Clandestine Laboratory Investigating ChemistsAssociation Training Conference (September 1991).

Clandestine Laboratory Safety Guide. U.S. Drug Enforcement Administration, Washington, D.C. (June 1987).

Clandestine Laboratory Investiative Guide. U.S. Drug Enforcement Administration, Washington, D.C. (1989).

"Drug Synthesis: A Supplement for the Clandestine Laboratory Investigator", U.S. Drug Enforcement Administration, Washington, D.C. (1989)

Clandestine Lab Recertification Manual. U.S. Drug Enforcement Administration, Washington, D.C. (1993).

Clandestine Laboratoratory Enforcement Tramning and Assistance Proaram. National Sheriffs' Association, Washington, D.C. (1988).

Gundersen, M., "A Glossary of Clandestine Lab Terms", Western States Information Network.

Some, W. H., "Clandestine Drug Synthesis", *Medical Research Reviews*, 6, 1, 41 (Jan 1986).

Amphetamine / Methamphetamine

Huizer, H. H., "Di-(b-Phenylisopropyl)amine in illicit Amphetamine", *Journal of Forensic Science*, 30, 4, 1022 (April 1985).

Nguyen, M., Forjohn, H., "Separation of Methamphetamine and Phenylacetone from Clandestine Laboratory Samples by HPLC", *Southwestern Association of Forensic Scientists Journal*, 8, 1, 26 (March 1986).

Simpson, B. J., et. al., "Microcrystalloscopic Differentiation of 3,4-Methylene-dioxyamphetamine and Related Amphetamine Derivatives", *Journal of Forensic Science*, 36, 3, 908 (May 1991).

Ely, R. A., "An Investigation of the Extraction of Methamphetamine from Chicken Feed, and Other Myths", *Journal of Forensic Science*, 30, 6, 363 (1990).

Keil, R. D., Summerhays, L. R., "The' Ephedrine/HI Reaction: Mechanism and Variations", Clandestine Laboratory Investigating Chemists Association Training Seminar (September 1991).

Abercrombie, J. T., "Analytical Data' from Modifications of the Ephedrine I HI Synthetic Route for Methamphetamine: 1. Substitutes for Hydriodic Acid", Clandestine Laboratory Investigating Chemists Association Training Seminar (September 1991).

Abercrombie, J. T., "Empirical Study of the Effects of Initial Precursor Amount in Regard to Final Yield, Ratio of By-Products and Other Information in the Ephedrine *I* HI I Red Phosphorus Synthetic Route", Clandestine Laboratory Investigating Chemists Association Training Seminar (September 1991).

Ely, R. A., et.al., "Lithium-Ammonia Reduction of Ephedrine to Methamphetamine: An Unusual Clandestine Synthesis", *Journal of Forensic Science*, 35, 3, 720 (May 1990).

Courtney, M., et.al., "The Leuckart Synthesis in the Clandestine Manufacture of Amphetamines", Clandestine Laboratory Investigating Chemists Association Training Seminar (September 1992).

Ely, R. A., "Serial Dry Extraction of Illicit Methamphetamine Powders for the Identification of Adulterants and Diluents by Infrared Spectroscopy", *Journal of the Clandestine Laboratory Investigating Chemists Association*, 3, 1, 22 (January 1993).

Skinner, H. F., "Methamphetamine Synthesis via Reductive Alkylation Hydrogenolysis of Phenyl-2-Propanone with n-Benzylmethylamine", *Forensic Science International*, 60, 155 (1993).

Skinner, H. F., "Methamphetamine Synthesis via Hydriodic / Red Phosphorus Reduction of Ephedrine", *Forensic Science International*, 48, 123 (1990).

Verweij, A. M. A., "Impurities in Illicit Drug Preparations: Amphetamine and Methamphetamine", *Forensic Science Review*, 1, 1, 2 (June 1989).

Timmons, J. E., "Five Ion Table of Leuckart Reaction Related Compounds", Arizona Department of Public Safety in House Data (1989).

Allen, A. C., et.al., "Methamphetamine from Ephedrine: I. Chloroephedrines and Aziradines", *Journal of Forensic Science*, 32, 4, 953 (July 1987).

Cantell, T. S., et.al., "A Study of Impurities Found in Methamphetamine Synthesized from Ephedrine", *Forensic Science International*, 39, 39 (1988).

"A Review of the Synthesis and Analysis of Phenyl-2-Propanone, Amphetamine andMethamphetamine, Volume 1: Origin of Reactions and Production Estimates from the Literature", Clandestine Laboratory Investigating Chemists Association TrainingSeminar (September 1993).

"A Review of the Synthesis and Analysis of Phenyl-2-Propanone, Amphetamine and Methamphetamine, Volume 2: Reaction Impurities, Profiling and History", Clandestine Laboratory Investigating Chemists Association Training Seminar (September 1993).

Allen, A. C., Cantrell, T. S., "Synthetic Reductions in Clandestine Amphetamine and Methamphetamine Laboratories: A Review", *Forensic Science International*, 42, 183 (1989).

Eaton, D. K., Harbison, G. C. "Isolation and Identification of Major Products and Reactants of Clandestine Amphetamine Laboratories", *Southwestern Association of Forensic Scientists Journal*, 8, 1, 18 (March 1986).

Noggle, F. T., et.al., "Methods for Differentiation of Methamphetamine from Regioisomeric Phenethylamines", *Journal of Chromatographic Science*, 29, 1, 31 (January 1991).

Noggle, F. T., "Methods of the Identification of the 1-phenyl-3-Butanines: Homologs of the Amphetamines", *Microgram*, 24, 8, 197 (August 1991).

Noggle, F. T., "Comparative Analytical Profiles for Regioisomeric Phenethylamines Related to Methamphetamine", *Microgram*, 24, 4, 76 (April 1991).

Kalchik, M. F., "Oxazolidine Impurities in Methamphetamine", Clandestine Laboratory Investigating Chemists Association Training Seminar (September 1991).

By, A. W., et.al., "The Synthesis and Spectra of 4-ethoxyamphetamine and Its Isomers", *Forensic Science International*, 49, 159 (1991).

Christian, D. R., Schneider, R. S., "Metbamphetamine via the Pressure Cooker", *Southwestern Association of Forensic Scientists Journal*, 13, 1, 42 (April 1991).

Christian, D. R., Schneider, R. S., "Methamphetatnine via the Pressure Cooker", *Journal of the Clandestine Laboratory Investigating Chemists Association*, 1, 3, 10 (July 1992).

Phenylacetone

Allen, A. C., et.al., "Differentiation of Illicit Phenyl-2-Propanone Synthesized from Phenylacetic Acid with Acetic Anhydride vs. Lead (II) Acetate", *Journal of Forensic Science*, 37, 1, 301 (January 1992).

"A Review of the Synthesis and Analysis of Phenyl-2-Propanone, Amphetamine and Methamphetamine, Volume 1: Origin of Reactions and Production Estimates from the Literature", Clandestine Laboratory Investigating Chemists Association Training Seminar (September 1993).

"A Review of the Synthesis and Analysis of Phenyl-2-Propanone, Amphetamine and Methamphetamine, Volume 2: Reaction Impurities, Profiling and History", Clandestine Laboratory Investigating Chemists Association Training Seminar (September 1993).

Kiser, W. O., "A Field Test for Phenyl-2-Propanone", *Microgram*, 15, 8, 127 (August 1982).

Ekis, T. R., Courtney, M., "Who Needs Regulated Chemical? Phenylacetone Synthesis Through Friedel-Crafts Alkylation", Clandestine Laboratory Investigating Chemists Association Training Seminar (September 1992).

Ekis, T. R., et.al., "Phenylacetone Synthesis and Clandestine Laboratories", *Southwestern Association of Forensic Scientist Journal*, 12, 1, 19 (April 1990).

Dal Cason, T. A., et.al., "A Clandestine Approach to the Synthesis of Phenyl-2-Propanone from Phenylpropenes", *Journal of Forensic Science*, 29, 4, 1187 (October 1984).

Netwal, T., Battles, J., "Production of Phenyl-2-Propanone via the 1-(Phenyl)-2-Nitropropene Intermediate as Encountered by the Colorado Bureau of Investigation", *Southwestern Association of Forensic Scientist Journal*, 12, 2, 22 (October 1990).

Christian, D. R., "A Case Sudy of Precursor Manufacture: Mandelic Acid to Phenylacedic Acid", *Southwestern Association of Forensic Scientists Journal*, 16, 1, 26 (April 1994).

Schnieder, R. S., Johnson, R. A., "Synthesis of Phenylacetic Acid via Mandelic Acid", *Journal of the Clandestine Laboratory Investigating Chemists Association*, 3, 1, 15 (January 1993).

MDA/MDMA

Verweij, A. M. A., "Impurities in Illicit Drug Preparation: 3,4-(Methylenedioxy)amphetamine and 3,4-(Methylenedioxy)methamphetamine", *Forensic Science Review*, 4, 2, 138 (December 1992).

Renten, R. J., Cowie, J. S., "A Study of the Precursors, Intermediates and Reaction By-products in the Synthesis of 3,4-(Methylenedioxy)methamphetamine and Its Application to Forensic Science", *Forensic Science International*, 60, 189 (1993).

Antoine, M. A., et.al., "A Note About Some Impurities in Commercially Available Piperonymethylketone", *Microgram*, 26, 9, 209 (September 1993).

Verweij, A. M. A., "Clandestine Manufacture of 3,4-(Methylenedioxy)methamphetamine (MDMA) by Low Pressure Reductive Amination: A Mass Spectral Study of the Reaction Mixtures", *Forensic Science International*, 45, 91 (1990).

Dal Cason, T. A., "An Evaluation of the Potential for the Clandestine Manufacture of 3,4-Methylenedioxyamphetamine (MDA) Analogs and Homologs", *Journal of Forensic Science*, 35, 3, 675 (May 1990).

Noggle, F. T., et.al., "Gas Chromatographic and Mass Spectrometric Analysis of Samples from Clandestine Laboratories involved in the Synthesis of Ecstacy form Sassafras Oil", *Journal of Chromatographic Science*, 29, 4, 76 (April 1991).

Simpson, B. J., et.al., "Microcrystalloscopic Differentiation of 3 ,4-Methylenedioxyamphetamine and Related Amphetamine Derivatives", *Journal of Forensic Science*, 36, 3, 908 (May 1991).

Cathinone/Methcathinone

Semkin, E. P., et.al., "Examination of Ephedrone", *Microgram*, 26, 1, 11 (January 1993).

Killips, R., "Methcathinone: A Law Enforcement Challenge", Michigan State Police (June 1993).

Dal Cason, T. A., "The Identification of Cathinone and Methcathinone", *Microgram*, 25, 12, 313 (December 1992).

Zhingle, K. Y., et.al., "Ephedrone: 2-Methylamino- 1 -Phenylpropan- 1-one (Jeff)", *Forensic Science*, 36, 3, 915 (May 1991).

Phencyclidine

Timmons, J. E., "Five Ion Table of PCP Related Compounds", Arizona Department of Public Safety in House Data (1989).

Angelos, S. A., et.al., "The Identification of Unreacted Precursors Impurities and By-products of Clandestinely Produced Phencyclidine Preparations", *Journal of Forensic Science*, 35, 6, 1297 (November 1992).

Lodge, B. A., et.al., "New Street Analogs of Phencyclidine", *Forensic Science International*, 55, 13 (1992).

Timmons, J. E., "The Synthesis and Mass Spectral Characterization of PCC, PyCC and MCC", *Southwestern Association of Forensic Scientists Journal*, 9, 1, 27 (March 1987).

Aniline, O., et.al., "Incidental Intoxication with Phencyclidine", *Journal of Clinical Psychiatry*, 41, 11, 393 (November 1980).

PiUs, F. N., Occupational Intoxication and Long-term Persistence of Phencyclidine (PCP) in Law Enforcement Personnel", *Clinical Toxicology*, 18, 9, 1015 (September 1981).

Robinson, B., Yates, A., "Angel Dust: Medical and Psychiatric Aspects of Phencyclidine Intoxication", *Arizona Medicine*, 41, 12, 808 (December 1984).

Fentanyl

Stanley, T. H., "The History and Development of the Fentanyl Series", *Journal of Pain and Symptom Management*, 7, 3 Supplement, S3–S7 (1992).

Stanley, T. H., "The Fentanyl Story", *The Journal of Pain*, 15, 12, 1215–1226 (2014).

Cooper, D., Jacob, M., Allen A., "Identification of Fentanyl Derivatives", *Journal of Forensic Science*, 31, 2, 511–528 (1986).

Moore, J. M., Allen, A. C., Cooper, D. A., Carr, S. M., "Determination of Fentanyl and Related Compounds by Capillary Gas Chromatography with Electron Capture Detection", *Analytical Chemistry*, 58, 1656–1660 (1986).

Gupta, P. K., Ganesan, K., Pande, A., Malhotra, R. C., "A Convenient One Pot Synthesis of Fentanyl", *Journal of Chemical Research*, 36, 49, 452–453 (2005).

Lurie, I. S., Berrier, A. L., Casale, J. F., Iio, R., Bozenko, J. S., "Profiling of Illicit Fentanyl Using UHPLC-MS/MS", *Forensic Science International*, 220, 191–196 (2012).

Girón, R., Abalo, R., Goicoechea, C., et.al., "Synthesis and Opioid Activity of New Fentanyl Analogs", *Life Sciences*, 71, 9, 1023–1034 (2002). doi:10.1016/s0024-3205(02)01798-8

Suh, Y. G., Cho, K. H., Shin, D. Y., "Total Synthesis of Fentanyl", *Archives of Pharmacal Research*, 21, 1, 70–72 (1998). doi:10.1007/BF03216756103.

Valdez, C. A., Leif, R. N., Mayer, B. P., "An Efficient, Optimized Synthesis of Fentanyl and Related Analogs", *PLoS ONE*, 9, 9, e108250 (2014). doi:10.1371/journal. pone.0108250

Cocaine

Casale, J. F., Klein, R. F. X., "Illicit Production of Cocaine", *Forensic Science Review*, 5, 95–107 (1993).

Casale, J. F., "A Practical Total Synthesis of Cocaine's Enantiomers", *Forensic Science International*, 33, 275–298 (1987).

Coca Cultivation and Cocaine Processing: An Overview, U. S. Drug Enforcement Administration, Washington, D.C. (1993).

General Analysis

Courtney, M., Ekis, T. R., "Laboratory Analysis of Clandestine Lab Chemicals, Reaction Mixtures and Raw Products", *Southwestern Association of Forensic Scientists Journal*, 11, 1, 16 (April 1989).

"Appendix 2, Quantitative Methodology", *DEA Basic Training for Forensic Chemists*, Second Edition. U.S. Drug Enforcement Administration, Washington, D.C. (1989)

Ely, R. A., "A Spreadsheet Program for the Determination of Volumes of One and Two Phase Liquids in Round Bottom Reaction Flasks", *Microgram*, 24, 7, 182 (July 1991).

Courtney, M., "Procedure for Volume Estimates in Clandestine Laboratory Reaction Vessels: Part 2", *Southwestern Association of Forensic Scientist Journal*, 12, 1, (1990).

Lomonte, J. N., "Determination of Volumes in Laboratory Vessels", *Journal of Forensic Science*, 37, 5, 1380 (April 1992).

Churchill, K. T., "Theoretical Yields from Precursors in Clandestine Laboratory Investigations", *Microgram*, 25, 4, 95 (April 1992).

Appendix P
Explosive
References

The following is a condensed version of the explosive references compiled by the Fire Debris and Explosives Subcommittee of the Organization of Scientific Area Committees for Forensic Science. For brevity, most of the sections only address publications from 2000 forward. The complete reference list can be found at: https://www.nist.gov/system/files/documents/2020/03/27/OSAC %20-%20Explosives%20References%20List%20Feb%2019%202020.pdf

Books

Naouum, P. P., and Montgomery, E., *Nitroglycerine and Nitroglycerine Explosives.* Williams and Wilkins Co, Baltimore, 1928.

Marshall, A., *Explosives*, Volumes I–III, 2nd Edition, Gordon Press Reprint, Great Briton, 1981.

Davis, T. L., *Chemistry of Powder and Explosives*, Volumes I and II, John Wiley and Sons Inc., New York.

Weingart, G. W., *Pyrotechnics.* 2nd Edition, Chemical Publishing Company Inc., New York.

Fedoroff, B. T. and Sheffield, O. E., Encyclopedial of Explosives and Related Items, *PATR 2700.* Picatinny Arsenal, Dover, NJ (vols. 1–10, different authors and years of publication), 1960.

Ellern, E., *Modern Pyrotechnics – Fundamentals of Applied Physical Pyrochemistry*, 1st Edition. Chemical Publishing Company, 1961.

The B.D.H Spot-Test Outfit Handbook. British Drug Houses Limited, 2nd Edition, 1962.

DuPont Blaster's Handbook. 14th Edition. Wilmington, DE: E. I. Du Pont de Nemours & Company (Inc.), Prepared by the Technical Service Section of the Explosives Department, 1963.

Urbanski, T., *Chemistry and Technology of Explosives.* Pergamon Press, 1964.

Army Field Manual FM 5–25, *Explosives and Demolitions.* Department of the United States of America War Office, May 1967.

Urbanski, T., *Chemistry and Technology of Explosives: Vol III.* Pergamon Press, Oxford, England, 1967.

Ellern, H., *Military and Civilian Pyrotechnics.* Chemical Publishing Company Inc., New York, 1968.

Lange, N. A., Spot Tests for Explosives and Explosive Residues. *Handbook of Chemistry.* 10th edition, McGraw Hill, 1971.

Powell, W., *The Anarchist Cookbook.* Lyle Stuart, 1971.

Feigl, F., and Anger, V., *Spot Tests in Inorganic Analysis*. 6th Edition, Elsevier Science, 1972.

Brauer, K. O., *Handbook of Pyrotechnics*. Chemical Pub Co, 1974; Reprint edition Dec 1 1995.

Malone, H. E., *The Analysis of Rocket Propellants*, (Analysis of Organic Materials; no. 12). Academic Press, 1976.

McCrone, W. C., *The Particle Atlas*, Vol 1, Ann Arbor Science, Ann Arbor, MI, 1976.

Encyclopedia of Explosives and Related Items. Picatinny Arsenal, Dover.

Explosives and Homemade Bombs, Second Edition. Charles C. Thomas Publisher, 1977.

Improvised Munitions Black Book. Frankford Arsenal Philadelphia. Desert Publications.

Humphrey, D. A., *The Alchemist's Cookbook: 80 Demonstrations*, 3rd Edition, 1979.

Dupont. *Blaster's Handbook*. 175th Anniversary Edition.

Yallop, H. J., *Explosion Investigation*, 1st Edition. Harrogate, UK: Forensic Science Society and Edinburgh UK: Scottish Academic Press, 1980.

Fordham, S., *High Explosives and Propellants*. Pergamon Press, UK, 1980.

Yinon, J., and Zitron, S., *The Analysis of Explosives*. 1st Edition. Pergamon Press, Oxford.

Shimizu, T., *Fireworks: The Art, Science and Technique*. Pyrotechnica Publications, 1981.

Encyclopedia of Explosives and Related Items, Volumes 1–10. Picatinny Arsenal. Picatinny Arsenal (US AARADCOM), Dover, NJ, 1960–1982.

Kaye, S. M., *Encyclopedia of Explosives and Related Items*. Picatinny Arsenal, Vol 10, Dover 1983.

Dobratz, B. M., and Crawford, P. C., *LLNL Explosives Handbook*. Lawrence Livermore National Laboratory, CA, 1985.

Bailey, A., and Murray, S. G., *Explosives, Propellants and Pyrotechnics*. Brassey's, UK.

Olofsson, S. O., *Applied Explosives Technology for Construction and Mining*. Applex, Arla, Sweden, 1991.

Saxon, K., *The Poor Man's James Bond*. Vol 1. Desert Publications, 1991.

Yinon, J., *Advances in Analysis and Detection of Explosives*. Kluwer Academic Publishers, Norwell MA, 1992.

Lancaster, R., *Fireworks Principles and Practice*. 2nd Edition, 1992. Chemical Publishing Company Inc., New York.

McLean, D., *The Do-It-Yourself Gunpowder Cookbook*. Paladin Press, 1992.

Gregory, C. E., *Explosives for Engineers: Fourth Edition*. Clausthal - Zellerfeld, Zurich, 1993.

Yinon, J., and Zitrin, S., *Modern Methods and Applications in Analysis of Explosives*. John Wiley & Sons Ltd, West Sussex, England, 1993.

Wang, X., *Emulsion Explosives*. Metallurgical Industry Press, Cleveland, OH, 1994.

Kosanke, K. L., *The Illustrated Dictionary of Pyrotechnics*. Journal of Pyrotechnics Inc, 1995.

Cooper, P. W., *Explosives Engineering*. Wiley-VCH, New York, 1996.

Cooper, P. W., and Kurowski, S. R., *Introduction to the Technology of Explosives*. Wiley-VCH, 1996.

Jungreis, E., *Spot Test Analysis: Clinical, Environmental, Forensic and Geochemical Applications*, 2nd Edition. John Wiley and Sons, Inc, 1996.

Dolan, J. E., and Langer, S. S., *Explosives in the Service of Man: The Nobel Heritage*. The Royal Society of Chemistry, 1997.

Donner, J., *A Professionals Guide to Pyrotechnics*. Paladin Press, 1997.

Jungreis, E., *Spot Test Analysis - Clinical, Environmental, Forensic and Geochemical Applications*. 2nd Edition, Ed. J. D. Winefordner, New York: John Wiley and Sons, Inc. Chapter 4.3, 1997.

Black and Smokeless Powders: Technologies for Finding Bombs and the Bomb Makers. National Research Council, National Academies Press, 1998.

Hopler, R. B., *Blasters Handbook*, 17th Edition. International Society of Explosives Engineers, Cleveland, 1998.

Saferstein, R., *Criminalistics: An Introduction to Forensic Science*. 6th ed. Prentice-Hall, Englewood Cliffs, NJ, 1998.

Beveridge, A. (Ed)., *Forensic Investigation of Explosions*. Taylor and Francis, London, 1998.

Naoum, P., *Nitroglycerine and Nitroglycerine Explosives*. 5th Edition. Angriff Press.

Partington, J. R., *A History of Greek Fire and Gunpowder*. Johns Hopkins University Press, Baltimore, MD, 1999.

Yinon, J., *Forensic and Environmental Detection of Explosives*. John Wiley & Sons, Ltd, New York, NY, 1999.

Siegel, J., Knupfer, G., and Saukko, P. (Eds.), *Encyclopedia of Forensic Sciences*. Academic Press London, England, 2000.

Bartick, E. G., *Handbook of Vibrational Spectroscopy*. Wiley, Hoboken, NJ, 2002.

Kushner, H. W., *Encyclopedia of Terrorism*. SAGE Publications Inc, 2002.

Meyer, R., Kohler, J., and Homburg, A., *Explosives*, 5th Edition. John Wiley-VCH, Weinheim, 2002.

Saferstein, R., *Forensic Science Handbook*. Volume 1. 2nd edition. Prentice Hall, Upper Saddle River, NJ, 2002.

Lough, E., *The Firecracker Cookbook*. Delta Group Press, 2002.

Yinon, J., *Advances in Forensic Applications of Mass Spectroscopy*. CRC Press, Boca Raton, FL, 2003.

Gallagher, P. K. and Brown, M. E. (Eds.), *Handbook of Thermal Analysis and Calorimetry: Applications to Inorganic and Miscellaneous Materials*. Elsevier, Amsterdam, 2003.

Crowl, D. A., *Understanding Explosions*. Wiley, Hoboken, NJ, 2003.

Kirk-Othmer, *Encyclopedia of Chemical Technology*, 5th Edition. John Wiley & Sons, Hoboken, NJ, 2004.

Pickett, M., *Explosives Identification Guide*, 2nd Edition. Cengage Learning, Delmar, 2004.

Kelly, J., *Gunpowder – Alchemy, Bombards, & Pyrotechnics: The History of the Explosive that Changed the World*. Basic Books, New York, NY, 2004.

Kosanke, B. J., Sturman, B., Shimizu, T., Wilson, M. A., et al., *Pyrotechnic Chemistry*. Journal of Pyrotechnics Inc., 2004.

Ledgard, J. B., *The Preparatory Manual of Explosives*. 2nd Edition. The Paranoid Publications Group, 2004.

Teipel, U., *Energetic Materials: Particle Processing and Characterization*. Wiley-VCH, 2005.

Lancaster, R., *Fireworks Principles and Practice*, 4th Edition. Chemical Publishing Company Inc., New York.

Crippin, J. B., *Explosives and Chemical Weapons Identification*. CRC Press Taylor & Francis.

Woodfin, R. L. (Ed.), *Trace Chemical Sensing of Explosives*. John Wiley & Sons Inc., New Jersey, 2006.

Counter Terrorism for Emergency Responders, 2nd Edition. Taylor & Francis, Boca Raton, 2007.

Yinon, J. (Ed.), *Counterterrorist Detection Techniques of Explosives*. 1st Edition. Elsevier, Amsterdam, 2007.

Meyer, R., Köhler, J., and Homburg, A., *Explosives*. 6th Edition. Wiley-VCH, 2007.

Kubota, N., *Propellants and Explosives: Thermochemical Aspects of Combustion*, 2nd Edition. Wiley-VCH, 2007.

Agrawal, J. P., and Hodgson, R. D., *Organic Chemistry of Explosives*. J. Wiley & Sons, 2007.

Houghton, L. R., *Emergency Characterization of Unknown Materials*. CRC Press, Taylor and Francis Group, 2008.

Meyer, R., Kohler, J., and Homburg, A., *Explosives*. 6th Edition. Wiley-VCH.

Russell, M. S., *The Chemistry of Fireworks*. 2nd Edition. RSC Publishing, 2008.

Marshall, M., and Oxley, J. C. (Eds.), *Aspects of Explosives Detection*. Elsevier, Oxford, 2009.

Houghton, R., *Field Confirmation Testing for Suspicious Substances*. 1st Edition. CRC Press, Boca Raton, 2009.

Russell, M. S., *The Chemistry of Fireworks*. 2nd Edition, Royal Society of Chemistry Publishing.

Jamieson, A. and Moenssens, A. (Eds.), *Wiley Encyclopedia of Forensic Science*. Wiley, Chichester, UK, 2009.

Conkling, J. A., *Chemistry of Pyrotechnics Basic Principles and Theory*. CRC Press, Taylor and Francis Group LLC.

Saferstein, R., *Forensic Science Handbook*. Prentice Hall, Vol. 3, 2010.

Agrawal, J. P., *High Energy Materials: Propellants, Explosives and Pyrotechnics*, 2010.

Akhavan, J., *The Chemistry of Explosives: Edition 3*. Royal Society of Chemistry, 2011.

Thurman, J. T., *Practical Bomb Scene Investigation*. 2nd Edition. CRC Press.

Beveridge, A., *Forensic Investigation of Explosions*. 2nd Edition. CRC Press, Boca Raton, FL, 2012.

Chalmers, J. M., Edwards, H. G. M., and Hargreaves, M. D. (Eds.), *Infrared and Raman Spectroscopy in Forensic Science*. Wiley, 2012.

Oxley, J. C., Marshall, M., and Lancaster S. L., Principles and Issues in Forensic Analysis of Explosives. *Forensic Chemistry Handbook*, Ed. L. Kobilinsky, Wiley, Chapter 2; 23–39, 2012.

Matyas, R., and Pachman, J., *Primary Explosives*. Springer, Heidelberg, 2013.

Saferstein, R., *Criminalistics: An Introduction to Forensic Science*, 11th Edition, Pearson Education.

Ledgard, J. B., *The Preparatory Manual of Explosives*. 4th Edition. UVKCHEM, 2014.

Smith, J. (Author), *Brodie's Bombs and Bombings - A Handbook to Protection, Security, Detection, Disposal and Investigation for Industry, Police and Fire Departments*. 4th Edition. Charles C Thomas Pub Ltd, 2015.

Akhavan, J., *The Chemistry of Explosives*, 3rd Edition. The Royal Society of Chemistry, UK.

Houck, M. M. (Ed.), *Forensic Chemistry*. 1st Edition. Section 5 Explosives. Academic Press, 2015.

White, P. (Ed.), *Crime Scene to Court: The Essentials of Forensic Science*, 4th Edition. Royal Society of Chemistry, Cambridge, 2016.

Meyer, R. Kohler, J., and Homburg, A., *Explosives*. 7th Edition. Wiley-VCH.

Siegel, J. A., *Forensic Chemistry Fundamentals and Applications*. Chapter. 5 Explosives, Goodpaster, J., Ed. Wiley Blackwell, 2016.

Houck, M. (Ed.), *Materials Analysis in Forensic Science*, 1st Edition. Academic Press, 2016.

NFPA 921: Guide for Fire and Explosion Investigations. National Fire Protection Agency, Quincy, MA.

Thurman, J. T., *Practical Bomb Scene Investigation*. 3rd Edition, CRC Press, 2017.

Lucas, D. M. L., *A Life of Crime: My Career in Forensic Science*. Chapters 8–11. CRC Press, 2018.

Houghton, R., *Field Confirmation Testing for Suspicious Substances*. Hardback 2009, Paperback First Edition 2018. CRC Press, Taylor & Francis Group, Boca Raton.

Evans-Nguyen, K., and Hutches, K. (Eds.), *Forensic Analysis of Fire Debris and Explosives*. Springer Nature, Switzerland AG, 2019.

Reviews/Guides/Studies/On-line Resources

Technical Manual. 1967 No. 9-1300-214 Technical Order No. 11A-1-34.

Lenz, R. R., *Explosives and Bomb Disposal Guide*. Charles C. Thomas, 1970.

Newhouser, C. R., *Introduction to Explosives*. National Bomb Data Center, Gaithersburg, MD, 1972.

Brucker, E. W., *Blasting Cap Recognition and Identification Manual*. International Assn. Chiefs of Police, Washington, D.C., 1973.

Friend, R. C., *Explosives Training Manual: A Complete Illustrated Course dealing with the Safe Handling and Effective Use of Explosives*. ABA Publishing Co, Wilmington, DE.

Higgs, D. G., Jones, P. N., Markham, J. A., and Newton, E., "A review of explosives sabotage and its investigation in civil aircraft." *Journal of the Forensic Science Society* 1978; 18: 137.

Atlas Powder Co, *Explosives and Rock Blasting*. Field Technical Operations. Atlas Powder Co, Dallas, 1987.

Karpas, Z., "Forensic science applications of Ion Mobility Spectrometry." *Forensic Sci. Rev.* 1989; 1(2): 103–119, December.

Meyers, S., and Shanley, E. S., "Industrial explosives - A brief history of their development and use." *J. Hazard. Mater.* 1990; 23: 183–201.

Feraday, A. W., "The Semtex-H story." *Adv. Anal. Detect. Explos. Proc. 4th Inter. Symp. Anal. Detect. Explos., September 7–10, 1992 Jerusalem*, Israel Yinon, J. Ed. Kluwer Academic Publishers Dordrecht, Holland 1992 pp. 67–72.

Hubball, J., "The use of chromatography in forensic science." *Adv. Chromatogr.* Vol. 32 Giddings, J. C., Grushka, E., and Brown, P. R. (Eds.), Marcel Dekker, New York, pp. 131–172, 1992.

Hiley, R. W., "Investigations of thin layer chromatographic techniques used for forensic explosives analysis in the early 1970s." *J. Forensic Sci.* 1993; 38(4): 864–873, July.

Kolla, P. and Hohenstatt, P., "Stability of explosives traces on different supports." *Forensic Sci. Int.* June 1993; 60(1, 2): 127–137.

McCord, B. R., Hargadon, K. A., Hall, K. E., and Burmeister, S. G., "Forensic Analysis of Explosives Using Ion Chromatographic Methods." *Anal. Chim. Acta* 1994; 288(1–2): 43–56, March 30.

Donaldson, T. P., "Overview of United Kingdom Research." Compendium, Inter. Explos. Symp. Fairfax, VA September 18–22, 1995 Treasury Dept., BATF April 1996 pp. 111 –117.

Foulger, B., and Hubbard, P. J., "A review of techniques examined by U.K. authorities to prevent or inhibit the illegal use of fertiliser in terrorist devices." Compendium, Inter. Explos. Symp. Fairfax, VA September 18–22, 1995

Crowson, A., Cullum, H. E., Hiley, R. W., and Lowe, A. M., "A survey of high explosives traces in public places." *J. Forensic Sci.* 1996; 41(6): 980–989.

Fink, C. L., Micklich, B. J., Sagalovsky, L., Smith, D. L., and Yule, T. J., "Explosives detection studies using fast-neutron transmission spectroscopy." *Proc. 2nd Explos. Detect. Technol. Symp. Aviation Secur. Technol. Conf.* Makky, W. -Chair November 12–15, 1996 FAA Atlantic City, NJ pp. 142–147.

Fox, F., Sisk, S., DiBartolo, R., Miller, J. F., and Gandy, J., "Immersion studies of Aircraft parts exposed to plastic explosives." *Proc. 2nd Explos. Detect. Technol. Symp. Aviation Secur. Technol. Conf.* Makky, W. -Chair November 12-15, 1996 FAA Atlantic City, NJ, pp. 31–37.

Jones, M. L. and Lee, E., "Impact sensitivity of Nitroglycerin." *J. Energ. Mat.* 1997; 15: 193–204.

Black and Smokeless Powders: Technologies for Finding Bombs and the Bomb Makers. National Research Council. National Academy Press, 1998.

McCord, B. R. and Bender, E. C., "Chromatography of explosives." In: *Forensic Invest. Explos.* Beveridge, A.D., Ed. Taylor & Francis, London, pp. 231–265, 1998.

Hopler, R. B., "The history, development and characteristics of explosives and propellants." In: *Forensic Invest. Explos.* Beveridge, A.D. - Ed. London, Taylor & Francis, London, pp. 1–13, 1998.

Leach, C., Flower, P., Hollands, R., Flynn, S., Marshall, E., and Kendrick, J., "Plasticisers in energetic materials formulations. A UK overview." *Int. Annu. Conf. ICT* 1998 29th (Energetic Materials) 2.1-2.14.

McAvoy, Y., Backstrom, B., Janhunen, K., Stewart, A., and Cole, M. D. "Supercritical fluid chromatography in forensic science: A critical appraisal." *Forensic Sci. Int.* 1999; 99(2): 107–122.

Midkiff, C. R., *Analysis and Detection of Explosives, Published Papers, Results and Presentations: 1988–1998*. August 1999. Available at https://www.swgfex.com/publications.

Picket, M., *Explosives Identification Guide*. Delmar Publishers, Albany, NY, 1999.

National Institute of Justice, *A Guide for Explosion and Bombing Scene Investigation*. 2000. Available at https://www.swgfex.com/publications.

TWGFEX/NCFS, *A Guide for Explosion and Bombing Scene Investigation*. DOJ, Washington, DC, 2000.

A Guide for Explosion and Bombing Scene Investigation. National Institute of Justice, written and approved by the Technical Working Group for Bombing Scene Investigation. U.S. Department of Justice, 2000.

Walker, C., Cullum, H. E., Hiley, R. W., "An environmental survey relating to improvised and emulsion gel explosives." *J. Forensic Sci*, 2001: 46(2): 254–267.

Oxley, J. C., "The Thermal Stability of Explosives." In *Handbook of Thermal Analysis and Calorimetry: Applications to Inorganic and Miscellaneous Materials*, P. K., Gallagher and M. E., Brown, Eds, Elsevier, Chapter 8, Amsterdam, 2003 Vol. 2, pp. 349–369.

Oxley, J. C., Smith, J. L., Resende, E., Pearce, E., and Chamberlain, T. "Trends in explosive contamination." *J. Forensic Sci*. 2003; 48(2): 1–9.

Cullum, H. E., McGavigan, C., Uttley, C. Z., Stroud, M. A., and Warren, D. C., "A second survey of background levels of explosives and related compounds in the environment" *J. Forensic Sci* 2004; 49(4): 684–690.

Hopen, T. J., "Dr. Walter C. McCrone's contribution to the characterization and identification of explosives." *J Forensic Sci*. 2004 Mar; 49(2): 275–276.

Nambayah, M., and Quickenden, T., "A quantitative assessment of chemical techniques for detecting traces of explosives at counter-terrorist portals." *Talanta* 2004 May 28; 63(2): 461–467.

The Technical and Scientific Working Group on Fire and Explosion Analysis (T/SWGFEX). "Recommended Guidelines for Forensic Identification of Intact Explosives." July 2004. Available at https://www.swgfex.com/publications.

University of Rhode Island, "Explosives Database." Available at http://expdb.chm.uri.edu/

Technical/Scientific Working Group for Fire and Explosions/National Center for Forensic Science (T/SWGFEX/NCFS), "Smokeless Powder Database." Available at http://www.ilrc.ucf.edu/powders/

Borusiewicz, R., "A review of methods of preparing samples for chromotographic analysis for the presence of organic explosive substances." *Problems of Forensic Sciences* 2007; 69: 5–29.

Gaurav, D., Malik, A. K., and Rai, P. K., "High performance liquid chromatographic methods for the analysis of explosives." *Critical Reviews in Analytical Chemistry* 2007; 37: 227–268.

The Technical and Scientific Working Group on Fire and Explosion Analysis (T/SWGFEX). "Recommended guidelines for forensic identification of post blast explosive residues." 2007. Available at https://www.swgfex.com/publications.

Technical Support Working Group. "Indicators and Warnings for Homemade Explosives." First Edition, December 2007, *For Official Use Only*.

Lahoda, K. G., Collin, O. L., Mathis, H. E., LeClair, W. S. H., and McCord, B. R., "A survey of background levels of explosives and related compounds in the environment." *J. Forensic Sci.* 2008; 53(4): 802–806.

Tagliaro, F., and Bortolotti, F., "Recent advances in the Applications of CE to Forensic Sciences (2005 – 2007)." *Electrophoresis.* 2008 Jan; 29(1):260–8.

Burks, R. M., and Hage, D. S., "Current trends in the detection of peroxide-based explosives." *Anal Bioanal Chem.* 2009 Sep; 395(2):301–13. doi: 10.1007/s00216-009-2968-5. Epub 2009 Jul 31.

Gottfried, J. L., De Lucia, F. C., Munson, C. A., and Mizolek, A. W., "Laser-induced breakdown spectroscopy for detection of explosives residues: a review of recent advances, challenges, and future prospects." *Anal Bioanal Chem,* 2009; 395: 283–300. doi:10.1007/s00216-009-2802-0.

Naval Explosive Ordnance Disposal Technology Division, Explosives: Military, Commercial, Homemade, and Precursors Identification Guide, Version 2.0; 2009. For Official Use Only.

Oxley, J. C., Smith, J. L., Bernier, E., Moran, J. S., and Luongo, J., "Hair as forensic evidence of explosive handling." *Propellants, Explosives, Pyrotechnics* 2009; 34(4): 307–14.

Wallin, S., Pettersson, A., Ostmark, H., and Hobro, A., "Laser-based standoff detection of explosives: A critical review." *Anal Bioanal Chem.* 2009 Sep; 395(2): 259–74.

Nair, U.R., Asthana, S. N., Subhananda Rao, A., and Gandhe, B.R., "Advances in high energy materials." *Defence Science Journal* March 2010; 60(2): 137–51.

Fernandez de la Ossa, A., Lopez-Lopez, M., Torre, M., and Garcia-Ruiz, C., "Analytical techniques in the study of highly-nitrated nitrocellulose." *Trends in Analytical Chemistry* 2011; 30(11): 1740–1755.

Makinen, M., Nousiainen, M., and Sillanpaa, M., "Ion spectrometric detection technologies for ultra-traces of explosives: A review." *Mass Spectrometry Reviews* 2011; 30: 940–973.

Sorensen, A., and McGill, W. L., "What to look for in the aftermath of an explosion? A review of blast scene damage observables." *Engineering Failure Analysis* 2011; 18(3): 836–845.

Caygill, J. S., Davis, and Higson, S. P. J., "Current trends in explosive detection techniques." *Talanta,* 15 January 2012; 88: 14–29.

Ostmark, H., Wallin, A., and Ang, H., "Vapor pressure of explosives: a critical review." *Propellants Explos, Pyrotech* 2012; 37: 12–13.

Oxley, J. C., Smith, J. L., Kirschenbaum, L. J., Marimiganti, S., Efremenko, I., Zach, R., and Zeiri, Y., "Accumulation of explosives in hair – part 3: Binding site study." *Journal of Forensic Sciences* 2012; 57: 623–635.

Bureau of Alcohol, Tobacco, Firearms, and Explosives (ATF) Department of Justice "List of Explosives Materials." Available at https://www.atf.gov/file/97716/download

Morelato, M, Beavis, A., Kirkbride, P., and Roux, C., "Forensic applications of desorption electrospray ionisation mass spectrometry (DESI-MS)." *Forensic Science International* 2013; 226(1–3): 10–21.

Barron, L., and Gilchrist, E., "Ion chromatography-mass spectrometry: A review of recent technologies and applications in forensic and environmental explosives analysis." *Analytica Chimica Acta* 2014; 806: 27–54.

Fountain, A.I., Christesen, S., Moon, R., and Guicheteau, J., "Recent advances and remaining challenges for spectroscopic detection of explosives threats." *Appl. Spectrosc.* OA 2014; 68: 795–811.

Oxley, J.C., "Explosive detection: How we got here and where are we going?" *International Journal of Energetic Materials and Chemical Propulsion* 2014; 13(4): 373–381.

NFPA 921: *"Guide for Fire and Explosion Investigations,"* National Fire Protection Agency, Quincy, MA.

Smyth, A., and Sims, M. R., "Detection of fingermarks from post-blast debris: A review." *Journal of Forensic Identification* 2018; 63(8): 369–378.

Gillen, G., Verkouteren, J., Najarro, M., Staymates, M., Verkouteren, M., Fletcher, R., Muramoto, S., Staymates, J., Lawrence, J., Robinson, L., and Sisco, E. "Review of the National Institute of Standards and Technology Research Program in Trace Contraband Detection." *Homeland Security and Public Safety: Research, Applications and Standards* 2019 Oct. ASTM International. https://www.astm.org/DIGITAL_LIBRARY/STP/PAGES/STP161420180050.htm

Greibl, W., "End user commentary on advances in the analysis of explosives. emerging technologies for the analysis of Forensic traces." 2019; 241–243. https://link.springer.com/chapter/10.1007/978-3-030-20542-3_16

Romolo, F. S., and Palucci, A., "Advances in the analysis of explosives. Emerging Technologies for the analysis of forensic traces." 2019; 207–240. https://link.springer.com/chapter/10.1007/978-3-030-20542-3_15

Explosive Devices

Oxley, J. C., Smith, J. L., Resende E., Rogers, E., Strobel, R. A., and Bender, E. C., "Improvised explosive devices: Pipe bombs." *J Forensic Sci.* 2001; 46(3), 510–534.

Oxley, J. C., Smith, J. L., and Resende, E., "Determining explosivity part II: Comparison of small scale cartridge tests to actual pipe bombs." *J Forensic Sci.* 2001 Sep; 46(5): 1070–1075.

Walsh, G. A., Inal, O. T., and Romero, V. D., "A potential metallographic technique for the investigation of pipe bombings." *Journal of Forensic Sciences* 2003; 48(5): 484–500.

Chumbley, L. S., and Laabs, F. C., "Analysis of explosive damage in metals using orientation imaging microscopy." *J Forensic Sci* 2005; Jan 50(1): 104–111.

Quirk, A. T., Bellerby, J. M., Carter, J. F., Thomas, F. A., and Hill, J. C., "An initial evaluation of stable isotopic characterisation of post-blast plastic debris from improvised explosive devices." *Science and Justice* 2009; 49: 87–93.

Gregory, O., Oxley, J. C., Smith, J. L., Platek, M., Ghonem, H., Bernier, E., Downey, M., and Cumminskey, C., "Microstructural characterization of pipe bomb fragments." *Materials Characterization* 2010; 61(3): 347–354.

Kuzmin, V., Mikheev, D., and Kozak, G., "Detonability of Ammonium Nitrate and mixtures on its Base." *Central European Journal of Energetic Materials,* 2010; 7(4): 335–343.

Lichorobiec, S., "Development of alternative projectile to deactivate an improvised explosive device - Pipe bomb." *Komunikacie* 2011; 13(2): 20–25.

Ramasamy, A., Hill, A. M., Masouros, S., Gibb, I., Bull, A. M. J., and Clasper, J. C., "Blast-related fracture patterns: A forensic biomechanical approach." *Journal of the Royal Society Interface* 2011; 8(58):689–698.

Chakrobortty, A., Bagchi, S., and Lahiri, S. C., "Investigation on improvised explosive devices and the physicochemical nature of the fuel binders." *Journal of the Indian Chemical Society* 2012; 89(11): 1515–1524.

da Silva, L. A., Johnson, S., Critchley, R., Clements, J., Norris, K., and Stennett, C. "Experimental fragmentation of pipe bombs with varying case thickness." *Forensic Science International* January 2020; 306: 110034. https://doi.org/10.1016/j.forsciint.2019.110034

Low Explosives

Dujay, R. C., "Manufacturing and processing techniques affecting morphology of Pyrotechnic Oxidizer Particles." *Microscopy Society of America* 2001; 9(4): 8–13.

Phillips, S., "Pyrotechnic residue analysis – Detection and analysis of characteristic particles by scanning electron microscopy/energy dispersive spectroscopy." *Science & Justice* 2001; 41(2): 73–80.

Kosanke, K., Dujay, R., and Kosanke, B., "Characterization of pyrotechnic reaction residue particles by SEM/EDS." *Journal of Forensic Sciences* 2003; 48(2): 531–537.

Babu, E. S., and Kaur, S., "A DSC analysis of inverse salt-pair explosive composition." *Propellants, Explosives and Pyrotechnics* 2004; 29(1): 50–55.

Goodpaster, J. V., and Keto, R. O., "Identification of Ascorbic Acid and its degradation products in black powder substitutes." *J Forensic Sci* 2004; 49(3): 523–528.

Bradley, K. S., "Determination of elemental sulfur in explosives and explosive residues by gas chromatography-mass spectrometry." *J Forensic Sci.* 2005; 50(1): 96–103.

Kasamatsu, M., Suzuki, Y., Sugita, R., and Suzuki, S., "Forensic discrimination of match heads by elemental analysis with inductively coupled Plasma-Atomic emission spectrometry." *J Forensic Sci.* 2005; 50(4): 883–886.

Kosanke, K. L., Dujay, R. C., and Kosanke, B. J., "Pyrotechnic reaction residue particle analysis." *J Forensic Sci.* 2006 Mar; 51(2): 296–302.

Mahoney, C. M., Gillen, G., and Fahey, A. J., "Characterization of gunpowder samples using Time-of-Flight Secondary Ion Mass Spectrometry (TOF-SIMS)." *Forensic Sci Int.* 2006 Apr 20; 158(1): 39–51.

Laza, D., Nys, B., Kinder, J. D., Kirsch-De Mesmaeker, A., and Moucheron, C., "Development of a Quantitative LC-MS/MS Method for the Analysis of Common Propellant Powder Stabilizers in Gunshot Residue." *J Forensic Sci.* 2007 Jul; 52(4): 842–850.

Koch, E. C., "Special Materials in Pyrotechnics: V. Military Applications of Phosphorus and its compounds." *Propellants, Explosives, Pyrotechnics* June 2008; 33(3): 165–176.

Gentile, N., Siegwolf, R. T. W., and Delemont, O., "Study of isotopic variations in black powder: reflections on the use of stable isotopes in forensic science for source inference." *Rapid Commun. Mass Spectrom.* 2009; 23: 2559–2567.

Lang, G. H., and Boyle, K. M., "The analysis of black powder substitutes containing ascorbic acid by ion chromatography/mass spectrometry." *J Forensic Sci.* 2009 Nov; 54(6):1315–22. doi: 10.1111/j.1556-4029.2009.01144.x. Epub 2009 Aug 28.

Oxley, J. C., Smith, J. L., Higgins, C., Bowden, P., Moran, J. S., Brady, J., Aziz, C. E. and Cox, E., "Efficiency of perchlorate consumption in road flares, propellants and explosives." *Journal of Environmental Management* 2009; 90(11): 3629–3634.

Vermeij, E., Duvalois, W., Webb, R., and Koeberg, M., "Morphology and composition of Pyrotechnic residues formed at different levels of confinement." *Forensic Sci Int.* 2009 Apr 15; 186(1–3):68–74.

Bottegal, M., Lang, L., Miller, M., and McCord, B. "Analysis of ascorbic acid based black powder substitutes by high-performance liquid chromatography\electrospray ionization quadrupole time-of-flight mass spectrometry." *Rapid Commun. Mass Spectrom.* 2010; 24: 1377–1386.

Crawford, C. L., Boudries, H., Reda, R. J., Roscioli, K. M., Kaplan, K. A., Siems, W. F., and Hill Jr, H. H., "Analysis of black powder by Ion Mobility-Time-of-Flight mass spectrometry." *Anal. Chem.* 2010; 82: 387–393.

Dalby, O., and Birkett, J. W., "The evaluation of solid phase micro-extraction fibre types for the analysis of organic components in unburned propellant powders." *Journal of Chromatography A.* 2010; 1217(46): 7183–7188.

Damour, P. L., Freedman, A., and Wormhoudt, J., "Knudsen effusion measurement of organic peroxide vapour pressures." *Propellants, Explosives, Pyrotechnics.* 2010; 35(6), 514–520.

Castro, K., Fdez-Ortiz de Vallejuelo, S., Astondoa, I., and Madariaga, J. M., "Are these liquids explosive? Forensic analysis of confiscated indoor fireworks." *Analytical and Bioanalytical Chemistry.* July 2011; 400: 3065–3071.

Castro, K., Fdez-Ortiz de Vallejuelo, S., Astondoa, I., Goñi, F. M., and Madariaga, J. M., "Analysis of confiscated fireworks using Raman spectroscopy assisted with SEM-EDS and FTIR." *Journal of Raman Spectroscopy,* 2011; 42(11): 2000–2005.

Routon, B. J., Kocher, B. B., and Goodpaster, J. V. "Discriminating Hodgdon Pyrodex® and Triple Seven® using gas chromatography-mass spectrometry." *Journal of Forensic Sciences* 2011 Jan; 56(1): 194–199.

Martin-Alberca, C., de la Ossa, M. A. F., Saiz, J., Ferrando, J. L., and Garcia-Ruiz, C., "Anions in pre- and post-blast consumer fireworks by capillary electrophoresis." *Electrophoresis* 2014; 35(21–22): 3273–3280.

Martin-Alberca, C., and Garcia-Ruiz, C., "Analytical techniques for the analysis of consumer fireworks." *TrAC* 2014; 56: 27–36.

E2998-16 "Standard Practice for Characterization and Classification of Smokeless Powder." (2016). American Society for Testing and Materials International (ASTM).

E2999–17 "Standard Test Method for Analysis of Organic Compounds in Smokeless Powder by Gas Chromatography-Mass Spectrometry and Fourier Transform Infrared Spectroscopy." American Society for Testing and Materials International (ASTM), 2017.

Chabaud, K. R., Thomas, J. L., Torres, M. N., Oliveira, S., and McCord, B. R., "Simultaneous colorimetric detection of metallic salts contained in low explosives residue using a microfluidic paper-based analytical device (mPAD)." *Forensic Chemistry* 2018; 9: 35–41.

Bezemer, K. D. B., van Duin, L. V. A., Martin-Alberca, C., Somsen, G. W., Schoenmakers, P. J., Haselberg, R., and van Asten, A. C., "Rapid forensic chemical classification of confiscated flash banger fireworks using capillary electrophoresis." *Forensic Chemistry* 2019; 16: 100187. https://www.sciencedirect.com/science/article/abs/pii/S2468170919300840

Sampling and Sample Prep

Thompson, R., Fetterolf, D., Miller, M., and Mothershead, R., "Aqueous recovery from cotton swabs of organic explosives residue followed by solid phase extraction." *Journal of Forensic Sciences* 1999; 44(4): 795–804, Available at: https://doi.org/10.1520/JFS14555J. ISSN 0022-1198.

Warren, D., Hiley, R. W., Phillips, S. S., and Ritchie, K. "Novel technique for the combined recovery, extraction and clean-up of forensic organic and inorganic trace explosives samples." *Science & Justice* 1999; 39(1): 11–18, January-March.

Borusiewicz, R., "A review of methods of preparing samples for chromotographic analysis for the presence of organic explosive substances." *Problems of Forensic Sciences* 2007; 69: 5–29.

Duff, M. C., Crump, S. L., Ray, R. J., Cotham, W. E., LaMont, S., Beals, D., Mount, K., Koons, R. D., and Leggitt, J., "Solid phase microextraction sampling of high explosive residues in the presence of radionuclide surrogate metals." *Journal of Radioanalytical and Nuclear Chemistry* 2008; 275(3): 579–593.

Perret, D., Marchese, S., Gentili, A., Curini, R., Terracciano, A., Bafile, E., and Romolo, F., "LC-MS-MS determination of stabilizers and explosives residues in hand-swabs." *Chromatographia* 2008; 68(7/8): 517–524.

Tachon, R., Pichon, V., Le Borgne, M. B., and Minet, J-J., "Comparison of solid-phase extraction sorbents for sample clean-up in the analysis of organic explosives." *Journal of Chromatography A* 2008; 1185(1): 1–8.

Dalby, O., and Birkett, J. W., "The evaluation of solid phase micro-extraction fibre types for the analysis of organic components in unburned propellant powders." *Journal of Chromatography A*. 2010; 1217(46): 7183–7188.

Song-im N., Benson, S., and Lennard, C., "Evaluation of different sampling media for their potential use as a combined swab for the collection of both organic and inorganic explosive residues." *Forensic Science International* 2012; 222(1–3): 102–110.

Song-im, N., Benson, S., and Lennard C., "Establishing a universal swabbing and clean-up protocol for the combined recovery of organic and inorganic explosive residues." *Forensic Science International* 2012; 223(1–3): 136–147.

DeTata, D. A., Collins, P. A., and McKinley, A. J., "A comparison of common swabbing materials for the recovery of organic and inorganic explosive residues." *Journal of Forensic Sciences* 2013; 58(3): 757–763.

DeTata, D. A., Collins, P. A., and McKinley, A. J., "A comparison of Solvent extract cleanup procedures in the analysis of organic explosives." *Journal of Forensic Sciences* 2013; 58(2): 500–507.

Fan, W., and Almirall, J., "High-efficiency headspace sampling of volatile organic compounds in explosives using capillary microextraction of volatiles (CMV) coupled to gas chromatography-mass spectrometry (GC-MS)." *Anal Bioanal Chem* 2013, doi:10.1007/s00216-013-7410-3.

Szomborg, K., Jongekrijg, F., Gilchrist, E., Webb, T., Wood, D., and Barron, L., "Residues from low-order energetic materials: The comparative performance of a range of sampling approaches prior to analysis by ion chromatography." *Forensic Science International* 2013; 233(1–3): 55–62.

Romolo, F. S., Cassioli, L., Grossi, S., Cinelli, G., and Russo, M. V., "Surface-sampling and analysis of TATP by swabbing and gas chromatography/mass spectrometry." *Forensic Science International* 2013; 224(1–3): 96–100.

Bianchi, F., Gregori, A., Braun, G., Crescenzi, C., and Careri, M., "Micro-solid-phase extraction coupled to desorption electrospray ionization-high-resolution mass spectrometry for the analysis of explosives in soil." *Analytical and Bioanalytical Chemistry* 2015; 407(3): 931–938.

DeGreeff, L., Rogers, D. A., Katilie, C., Johnson, K., and Rose-Pehrsson, S., "Technical note: Headspace analysis of explosive compounds using a novel sampling chamber." *Forensic Science International* 2015; 248: 55–60.

Howa, J. D., Lott, M. J., Chesson, L. A., and Ehleringer, J. R., "Isolation of components of plastic explosives for isotope ratio mass spectrometry." *Forensic Chemistry* 2016; 1: 6–12.

Yu, H. A., Lewis, S. W., Beardah, M. S., and Nic Daeid, N., "Assessing a novel contact heater as a new method of recovering explosives traces from porous surfaces." *Talanta* 2016; 148: 721–728.

Daeid, N. N., Yu, H. A., and Beardah, M. S., "Investigating TNT loss between sample collection and analysis." *Science and Justice* 2017; 57(2): 95–100.

Yu, H. A., Becker, T., Daeid, N. N., and Lewis, S. W., "Fundamental studies of the adhesion of explosives to textile and non-textile surfaces." *Forensic Science International* 2017; 273: 88–95.

Lees, H., Zapata, F., Vaher, M., and Garcia-Ruiz, C., "Study of the adhesion of explosive residues to the finger and transfer to clothing and luggage." *Science and Justice* 2018; 58(6): 415–424.

Thomas, J. L., Donnelly, C. C., Lloyd, E. W., Mothershead, R. F., Miller, J. V., McCollam, D. A., and Miller, M. L., "Application of a co-polymeric solid phase extraction cartridge to residues containing nitro-organic explosives." *Forensic Chemistry* 2018; 11: 38–46.

Thomas, J. L., Donnelly, C. C., Lloyd, E. W., Mothershead, R. F., and Miller, M. L., "Development and validation of a solid phase extraction sample cleanup procedure for the recovery of trace levels of nitro-organic explosives in soil." *Forensic Science International* 2018; 284: 65–77.

Pagliano, E., "Versatile derivatization for GC-MS and LC-MS: alkylation with trialkyloxonium tetrafluoroborates for inorganic anions, chemical warfare agent degradation products, organic acids, and proteomic analysis." *Analytical and Bioanalytical Chemistry* 2020 Jan 9: 1–9. https://link.springer.com/article/10.1007/s00216-019-02299-8

Intact Explosives Analysis

Tamiri, T., "Explosives analysis." In *Encyclopedia of Forensic Sciences*. Siegel, J., Knupfer, G., and Saukko, P., (Eds). Academic Press London, England, 2000.

Sigman, M. E., and Ma, C. Y., "Detection limits for GC/MS analysis of organic explosives." *Journal of Forensic Sciences* 2001; 46(1): 6–11.

Casamento, S., Kwok, B., Roux, C., Dawson, M., and Dobie, P., "Optimization of the separation of organic explosives by capillary electrophoresis with artificial neural networks." *J Forensic Sci.* 2003 Sep; 48(5): 1075–1083.

Brown, H., Kirkbride, K. P., Pigou, P. E., and Walker, G. S., "New developments in SPME part 2: Analysis of Ammonium Nitrate-based explosives." *J Forensic Sci.* 2004 Mar; 49(2): 215–221.

Calderera, S., Gardebas, D., Martinez, F., and Khong, S. P., "Organic explosives analysis using on Column-Ion Trap EI/NICI GC-MS with and external source." *J Forensic Sci.* 2004 Sep; 49(5): 1005–1008.

Paull, B., Roux, C., Dawson, M., and Doble, P., "Rapid screening of selected organic explosives by High Performance Liquid Chromatography using Reversed-Phase Monolithic Columns." *J Forensic Sci.* 2004 Nov; 49(6): 1181–1186.

The Technical and Scientific Working Group on Fire and Explosion Analysis (T/SWGFEX). "Recommended Guidelines for Forensic Identification of Intact Explosives." July 2004. Available at https://www.swgfex.com/publications.

Agüí, L., Vega-Montenegro, D., Yáñez-Sedeño, P., and Pingarrón, J. M., "Rapid voltammetric determination of nitroaromatic explosives at electrochemically activated carbon-fibre electrodes." *Anal Bioanal Chem.* 2005 May; 382(2): 381–387. Epub 2005 Apr 14.

Bradley, K. S., "Determination of Elemental Sulfur in explosives and explosive residues by gas chromatography-mass spectrometry." *J Forensic Sci.* 2005 Jan; 50(1): 96–103.

Hodyss, R., and Beauchamp, J. L., "Multidimensional detection of nitroorganic explosives by Gas Chromatography-Pyrolysis-Ultraviolet Detection." *Anal. Chem.* 2005; 77(11): 3607–3610.

Perr, J. M., Furton, K. G., and Almirall, J. R., "Gas Chromatography Positive Chemical Ionization and Tandem Mass Spectrometry for the Analysis of organic high Explosives." *Talanta.* 2005 Aug 15; 67(2): 430–436.

Reardon M. R., and Bender, E. C., "Differentiation of composition C-4 based on the analysis of the process oil." *Journal of Forensic Sciences* 2005; 50: 564–570.

Sharma, S. P., and Lahiri, S. C., "Characterization and Identification of Explosives and Explosive Residues Using GC-MS, an FTIR Microscope and HPTLC." *Journal of Energetic Materials*, 2005; 23(4): 239–264.

Sharma, S. P., and Lahiri, S. C., "GC-MS and HPTLC analysis of constituents of an unexploded bomb." *Journal of the Indian Chemical Society* February 2005; 82(2): 131–133.

Alzate, L. F., et al., "The vibrational spectroscopic signature of TNT in clay minerals." *Vibrat, Spectrosc.* 42: 357–368.

Collin, O. L., Niegel, C., DeRhodes, K. E., McCord, B. R., and Jackson, G. P., "Fast gas chromatography of explosive compounds using a pulsed-discharge electron capture detector." *Journal of Forensic Sciences* July 2006; 51(4): 815–818.

Cotte - Rodriguez, I., and Cooks, R.G., "Non-proximate detection of explosives and chemical warfare agent simulants by desorption electrospray ionization mass spectrometry." *Chemical Communications* 2006; 28: 2968–2970.

De Tata, D., Collins, P., and Campbell, N., "The identification of the emulsifier component of emulsion explosives by liquid chromatography-mass spectrometry." *Journal of Forensic Sciences* 2006; 51: 303–307.

Oehrle, S. A., "Analysis of CL-20 and TNAZ in the Presence of other Nitroaromatic and Nitramine Explosives using HPLC with Diode Array (PDA) Detection." *Journal of Energetic Materials* 1994; 12(4): 211–222.

Reynolds, J., Nunes, P., Whipple, R., and Alcaraz, A., *On-Site Analysis of Explosives in Various Matrices*. NATO Security through Science Series. Springer, The Netherlands, 2006.

Schmidt, A. C., Herzschuh, R., Matysik, F. M., and Engewald, W., "Investigation of the Ionisation and fragmentation behaviour of different Nitroaromatic compounds occurring as polar metabolites of explosives using electrospray ionisation Tandem mass spectrometry." *Rapid Commun Mass Spectrom.* 2006; 20(15): 2293–2302.

Gaurav, D., Malik, A. K., and Rai, P. K., "High performance liquid chromatographic methods for the analysis of explosives." *Critical Reviews in Analytical Chemistry* 2007; 37: 227–268.

Pierrini, G., Doyle, S., Champod, C., Taroni, F., Wakelin, D., and Lock, C., Evaluation of Preliminary Isotopic Analysis (13C and 15N) of Explosives: A Likelihood Ratio Approach to Assess the Links between Semtex Samples." *Forensic Science International* 22 March 2007; 167(1): 43–48.

Reardon, M. R., and Proudfoot, J. E., "Oils and waxes in Composition C-4 and emulsions: A comparison of intact samples to post-blast residues." In Proceedings of the 9th International Symposium on the Analysis and Detection of Explosives, Paris, July 5, 2007.

Tachon, R., Pichon, V., Le Borgne, M. B., and Minet, J-J., "Use of Porous Graphitic Carbon for the Analysis of Nitrate Ester, Nitramine, and Nitroaromatic explosives and By-products by liquid chromatography-atmospheric pressure chemical Ionisation-Mass Spectrometry." *J Chromatogr A* 2007 Jun 22; 1154(1–2): 174–181.

Comanescu, G., Manka, C. K., Grun, J., Nikitin, S., and Zabetakis, D., "Identification of explosives with two-dimensional resonance Raman Spectroscopy." *Applied Spectroscopy* 2008; 62(8): 833–839.

Husakova, L., Sramkova, J., Stankova, J, Nemec, P., Vecera, M., Krejcova, A., Stancl, M., and Akstein, Z., "Characterization of industrial explosives based on the determination of metal oxides in the identification particles by microwave digestion and atomic absorption spectrometry method." *Forensic Sci Int.* 2008 Jul 4; 178(2–3): 146–152.

Johns, C., Shellie, R. A., Potter, O. G., et al. "Identification of homemade inorganic explosives by ion chromatographic analysis of post-blast residues." *J Chrom, A* 2008; 1182: 205–214.

Lai, H., Guerra, P, Joshi, M., and Almirall, J. R., "Analysis of volatile components of drugs and explosives by solid phase Microextraction-Ion mobility spectrometry." *J Sep Sci.* 2008 Feb; 31(2): 402–412.

Oehrle, S. A., "Analysis of explosives using Ultra Performance Liquid Chromatography (UPLC®) with UV and/or mass spectrometry detection." *Journal of Energetic Materials* 2008; 26: 197–206.

Benson, S. J., Lennard, C. J., Maynard, P., Hill, D. M., Andrew, A. S., and Roux, C., "Forensic analysis of explosives using Isotope Ratio Mass Spectrometry (IRMS) - Discrimination of Ammonium Nitrate sources." *Sci Justice* 2009 Jun; 49(2): 73–80. doi:10.1016/j.scijus.2009.04.005.

Douglas, T. A., Walsh, M. E., McGrath, C. J., and Weiss, C. A., "Investigating the fate of Nitroaromatic (TNT) and Nitramine (RDX and HMX) explosives in fractured and pristine soils." *J Environ Qual.* 2009 Oct 29; 38(6): 2285–2294.

Benson, S. J., Lennard, C. J., Hill, D. M., Maynard, P., and Roux, C., "Forensic analysis of explosives using isotope ratios mass spectrometry (IRMS) - Part1: Instrument validation of the Delta (plus) XP IRMS for bulk nitrogen isotope ratio measurements." *J Forensic Sci.* 2010 Jan; 55(1): 193–204. doi: 10.1111/j.1556-4029.2009.01241.x. Epub 2009 Dec 10.

Benson, S. J., Lennard, C. J., Maynard, P., Hill, D. Andrew, A. S., Neal, K., Stuart-Williams, H., Hope, J., Walker, G. S., and Roux, C., "Forensic analysis of explosives using isotope ratios mass spectrometry (IRMS) - Part2: Forensic inter-laboratory trial: Bulk carbon and nitrogen stable isotopes in a range of chemical compounds (Australia and New Zealand)." *J Forensic Sci.* 2010 Jan; 55(1): 205–12. doi: 10.1111/j.1556-4029.2009.01242.x. Epub 2009 Dec 10.

Brady, J., Judge, E., and Levis, R. J., "Identification of explosives and explosive formulations using laser electrospray mass spectrometry." *Rapid Communications in Mass Spectrometry* 2010; 24(11): 1659–1664.

Cook, G. W., La Puma, P. T., Hook, G. L., and Eckenrode, B. A., "Using gas chromatography with ion mobility spectrometry to resolve explosive compounds in the presence of interferents." *J Forensic Sci,* November 2010; 55(6): 1582–1591.

Lai, H., Leung, A., Magee, M., and Almirall, J. R., "Identification of volatile chemical signatures from plastic explosives by SPME-GC/MS and detection by Ion Mobility Spectrometry." *Anal Bioanal Chem.* 2010 Apr; 396(8): 2997–3007.

Mahoney, C. M., Fahey, A. J., Steffens, K. L., Benner, B. A., and Lareau, R. T., "Characterization of composition C4 explosives using time-of-flight secondary ion mass spectrometry and x-ray photoelectron spectroscopy." *Analytical Chemistry* 2010; 82(17): 7237–7248.

Nilles, J. M., Connell, T. R., Stokes, S. T., and Dupont Durst, H., "Explosives detection using Direct Analysis in Real Time (DART) mass spectrometry." *Propellants, Explosives, Pyrotechnics* 2010; 35(5): 446–451.

Verkouteren, J. R., Coleman, J. L., and Cho, I., "Automated mapping of explosives particles in composition C-4 Fingerprints." *J Forensic Sci* March 2010; 55(2): 344–340.

Cummins, J., Hull, J., Kits, K., and Goodpaster, J. V., "Separation and identification of anions using porous graphite carbon and electrospray ionization mass spectrometry: Application to inorganic explosives and their post blast residues. *Anal. Methods* 2011; 3: 1682–1687.

Dudek, K., Matyas, R., and Dorazil, T., "Detection and identification of explosive precursors and explosives." *New Trends Res. Energ. Mater, Proc. Semin.*, 14th Pardubice, April 13–15, 2011, 595–601.

Fernandez de la Ossa, M. T., Lopez-Lopez, M., Torre, M., and Garcia-Ruiz, C., "Analytical techniques in the study of highly-nitrated nitrocellulose." *TrAC* 2011; 30(11): 1740–1755.

Lucena, P., Dona, A., Tobaria, L. M., and Laserna, J. J., "New challenges and insights in the detection and spectral identification of organic explosives by laser induced breakdown spectroscopy." *Spectrochimica Acta Part B* 2011; 66: 12–20.

Moore, S., MacCrehan, W., and Schantz, M., "Evaluation of vapor profiles of explosives over time using ATASS (Automated Training Aid Simulation using SPME)." *Forensic Science International* 2011; 212(1–3): 90–95.

Routon, B. J., Kocher, B. B., and Goodpaster, J. V., "Discriminating Hodgdon Pyrodex® and Triple Seven® using gas chromatography-mass spectrometry," *Journal of Forensic Sciences* 2011 Jan; 56(1): 194–199.

Saiz, J., Bravo, J. C., Velasco, A. E., Torre, M., and Garcia-Ruiz, C., "Determination of Ethylene Glycol Dinitrate in Dynamites using HPLC: Application to the Plastic Explosive Goma-2 ECO." *J Sep Sci.* 2011 Dec; 34(23): 3353–3358.

Tian, F.-F., Yu, J., Hu, J.-H., Zhang, Y., Xie, M.-X., Liu, Y., Wang, X.-F., Liu, H.-L., and Han, J., "Determination of emulsion explosives with Span-80 as emulsifier by gas chromatography-mass spectrometry." *Journal of Chromatography A* 2011; 1218: 3521–3528.

Tyrrell, E., Dicinoski, G. W., Hilder, E. F., Shellie, R. A., Breadmore, M. C., Pohl, C. A., and Haddad, P. R., "Coupled reversed-phase and ion chromatographic system for the simultaneous identification of inorganic and organic explosives." *Journal of Chromatography A* 2011; 1218(20): 3007–3012.

Bansal, P., Gaurav, G., Nidhi, N., Malik, A. K., and Matysik, F. M., "Liquid chromatographic determination of 1, 3, 5-trinitroperhydro-1, 3, 5- triazine and 2, 4, 6-trinitrotoluene in human plasma and groundwater samples utilizing microextraction in packed syringe." *Chromatographia* 2012; 75(13–14): 739–745.

Brady, J. E., Smith, J. L., Hart, C. E., and Oxley, J. C., "Estimating ambient vapor pressures of low volatility explosives by rising-temperature thermogravimetry." *Propellants, Explosives, Pyrotechnics* 2012; 37(2): 215–222.

Nesvold, S., Pacholke, K., "Detecting and confirming the presence of road flare residue in fire investigations." In: Proceedings of the 5th international symposium on fire investigation science and technology, National Association of Fire Investigators, International, Sarasota, 2012.

Oxley, J. C., Smith, J. L., Brady, J. E., and Brown, A. C., "Characterization and analysis of tetranitrate esters." *Propellants, Explos, Pyrotech* 2012; 37: 24–39.

Tamini, T., and Zitrin, S., "Analysis of explosives by mass spectrometry." In *Forensic Investigation of Explosives*, 2nd Edition, ed. A. Beveridge. Boca Raton, FL. CRC Press (Taylor and Francis Group), 2012.

Tellez, H., Vadillo, J. M., and Laserna, J. J., "Secondary ion mass spectrometry of powdered explosive compounds for forensic evidence analysis." *Rapid Communications in Mass Spectrometry* 2012; 26: 1203–1207.

Carriere, J. T. A., Havermeyer, F., and Heyler, R. A., "THz-Raman spectroscopy for explosives, chemical and biological detection." *Proceedings of SPIE* 2013; 8710: 87100M.

Delgado, T., Alcantara, J. F., Vadillo, J. M., and Laserna, J. J., "Condensed-phase laser ionization time-of-flight mass spectrometry of highly energetic nitro-aromatic compounds." *Rapid Communications in Mass Spectrometry* 2013; 27(15): 1807–1813.

DeTata, D., Collins, P., and McKinley, A., "A fast liquid chromatography quadru-pole time-of-flight mass spectrometry (LC-QToF-MS) method for the identi-fication of organic explosives and propellants." *Forensic Science International* 2013; 233(1–3): 63–74.

DeTata, D. A., Collins, P. A., and McKinley, A. J., "A comparison of solvent extract cleanup procedures in the analysis of organic explosives." *Journal of Forensic Sciences* 2013; 58(2): 500–507.

Heyler, R. A., Carriere, J. T. A., and Havermeyer, F., "THz-Raman - Accessing molecular structure with Raman spectroscopy for enhanced chemical iden-tification, analysis and monitoring." *Proceedings of SPIE* 2013; 8726: 87260J.

Lopez-Lopez, M., Ferrando, J. L., and Garcia-Ruiz, C., "Dynamite analysis by Raman Spectroscopy as a unique analytical tool." *Anal. Chem.* 2013; 85(5): 2595–2600.

Lv, J., Feng, J., Zhang, W., Shi, R., Liu, Y., Wang, Z., and Zhao, M., "Identification of Carbonates as additives in pressure-sensitive adhesive tape substrate with Fourier Transform Infrared Spectroscopy (FTIR) and its application in three explosive cases." *Journal of Forensic Sciences* 2013; 58(1): 134–137.

Kaplan-Sandquist, K., LeBeau, M., and Miller, M., "Chemical analysis of pharma-ceuticals and explosives in fingermarks using matrix-assisted laser desorption ionization/time-of flight mass spectrometry." *Forensic Science International* 2014; 235: 68–77.

Xu, X., Koeberg, M., Kuijpers, C. J., and Kok, E., "Development and validation of highly selective screening and confirmatory methods for the qualitative foren-sic analysis of organic explosive compounds with high performance liquid chromatography coupled with (photodiode array and) LTQ ion trap/Orbitrap mass spectrometric detections (HPLC-(PDA)-LTQOrbitrap)." *Science & Justice* 2014; 54(1): 3–21.

Brust, H., van Asten, A., Koeberg, M., Dalmolen, J., van der Heijden, A., and Schoenmakers, P., "Accurate quantitation of pentaerythritol tetranitrate and its degradation products using liquid chromatography-atmospheric pressure chemical ionization-mass spectrometry." *Journal of Chromatography A* 2014; 1338: 111–116.

Brust, H., Willemse, S., Zeng, T., van Asten, A., Koeberg, M., van der Heijden, Bolck A., and Schoenmakers, P., "Impurity profiling of trinitrotoluene using vacuum-outlet gas chromatography-mass spectrometry." *Journal of Chromatography A* 2014; 1374: 224–230.

Howa, J. D., Lott, M. J., and Ehleringer, J. R., "Isolation and stable nitrogen iso-tope analysis of ammonium ions in ammonium nitrate prills using sodium tetraphenylborate." *Rapid Communications in Mass Spectrometry* 2014; 28(13): 1530–1534.

Howa, J. D., Lott, M. J., and Ehleringer, J. R., "Observations and sources of carbon and nitrogen isotope ratio variation of pentaerythritol tetranitrate (PETN)." *Forensic Science International* 2014; 244: 152–157.

Lopez-Lopez, M., and Garcia-Ruiz, C., "Infrared and Raman spectroscopy techniques applied to identification of explosives." *TrAC* 2014; 54: 36–44.

Schwarzenberg, A., Dossmann, H., Cole, R. B., Machuron-Mandard, X., and Tablet, J. C., "Differentiation of isomeric dinitrotoluenes and aminodinitrotoluenes using electrospray high resolution mass spectrometry." *Journal of Mass Spectrometry* 2014; 49(12): 1330–1337.

Xu, X., Koeberg, M., Kuijpers, C. J., and Kok, E., "Development and validation of highly selective screening and confirmatory methods for the qualitative forensic analysis of organic explosive compounds with high performance liquid chromatography coupled with (photodiode array and) LTQ ion trap/Orbitrap mass spectrometric detections (HPLC-(PDA)-LTQOrbitrap)." *Science and Justice* 2014; 54(1): 3–21.

Chakraborty, A., Bagchi, S., and Lahiri, S. C., "Studies of fire debris from bomb blasts using ion chromatography, gas chromatography-mass spectrometry and fluorescence measurements – evidence of ammonium nitrate, wax-based explosives and identification of a biomarker." *Australian Journal of Forensic Science*, 47: 83–94.

DeGreeff, L., Rogers, D. A., Katilie, C., Johnson, K., and Rose-Pehrsson, S., "Technical note: Headspace analysis of explosive compounds using a novel sampling chamber." *Forensic Science International* 2015; 248: 55–60.

Hernandes, V. V., Franco, M. F., Santos, J. M., Melendez-Perez, J. J., Morais, D. R., Rocha, W. F. D. C., Borges, R., de Souza, W., Zacca, J. J., Logrado, L. P. L., Eberline, M. N., and Correa, D. N. "Characterization of ANFO explosive by high accuracy ESI (±)-FTMS with forensic identification on real samples by EASI (-)-MS." *Forensic Science International* 2015; 249: 156–164.

Bezemer, K. D. B., Koeberg, M., van der Heijden, A. E. D. M., van Driel, C. A., Blaga, C., Bruinsma, J., and van Asten, A. C., "The Potential of Isotope Ratio Mass Spectrometry (IRMS) and Gas Chromatography-IRMS Analysis of Triacetone Triperoxide in Forensic Explosives Investigations." *Journal of Forensic Sciences* 2016; 5: 1198–1207.

Brensinger, K., Rollman, C., Copper, C., Genzman, A., Rine, J., Lurie, I., and Moini, M., "Novel CE-MS technique for detection of high explosives using perfluorooctanoic acid as a MEKC and mass spectrometric complexation reagent." *Forensic Science International* 2016; 258: 74–79.

Bridoux, M. C., Schwarzenberg, A., Schramm, S., and Cole, R. B. "Combined use of direct analysis in real-time/Orbitrap mass spectrometry and micro-Raman spectroscopy for the comprehensive characterization of real explosive samples." *Analytical and Bioanalytical Chemistry* 2016; 408(21): 5677–5687.

Chajistamatiou, A. S., and Bakeas, E. B., "A rapid method for the identification of nitrocellulose in high explosives and smokeless powders using GC-EI-MS." *Talanta* 2016; 151: 192–201.

Chesson, L. A., Howa, J. D., Lott, M. J., and Ehleringer, J. R., "Development of a methodological framework for applying isotope ratio mass spectrometry to explosive components." *Forensic Chemistry* 2016; 2: 9–14.

Leigh, B. S., Monson, K. L., and Kim, J. E., "Visible and UV resonance Raman spectroscopy of the peroxide-based explosive HMTD and its photoproducts." *Forensic Chemistry* 2016; 2: 22–28.

Almeida, M. R., Logrado, L. P. L, Zacca, J. J., Correa, D. N., and Poppi, R. J., "Raman hyperspectral imaging in conjunction with independent component analysis as a forensic tool for explosive analysis: The case of an ATM explosion." *Talanta* 2017; 174: 628–632.

Andrasko, J., Lagesson-Andrasko, L., Dahlen, J., and Jonsson, B.H., "Analysis of explosives by GC-UV." *Journal of Forensic Sciences* 2017; 62(4): 1022–1027.

DeGreeff, L. E., Malito, M., Katilie, C. J., Brandon, A., Conroy, M. W., Peranich, K., Ananth, R., and Rose-Pehrsson, S. L., "Passive delivery of mixed explosives vapor from separated components." *Forensic Chemistry* 2017; 4: 19–31.

Ostrinskaya, A., Kelley, J. A., and Kunz, R. R., "Characterization of nitrated sugar alcohols by atmospheric-pressure chemical-ionization mass spectrometry." *Rapid Communications in Mass Spectrometry* 2017; 31(4):333–343.

Tsai, C. W., Milam, S. J., and Tipple, C. A., "Exploring the analysis and differentiation of plastic explosives by comprehensive multidimensional gas chromatography-mass spectrometry (GC x GC–MS) with a statistical approach." *Forensic Chemistry* 2017; 6: 10–18.

Banas, K., Banas, A. M., Heussler, S. P., and Breese, M. B. H., "Influence of spectral resolution, spectral range and signal-to-noise ratio of Fourier transform infrared spectra on identification of high explosive substances." *Spectrochimica Acta - Part A: Molecular and Biomolecular Spectroscopy* 2018; 188: 106–112.

Boggess, A, Crump, S., Gregory, C., Young, J., and Kessinger, G., "Analytical method for nitroaromatic explosives in radiologically contaminated soil for ISO/IEC 17025 accreditation." *Forensic Chemistry* 2018; 7: 26–32.

Dunn, L., Obaidly, H. S. A. A., and Khalil, S. E., "Development and validation of fast liquid chromatography high-resolution mass spectrometric (LC-APCI-QToF-MS) methods for the analysis of hexamethylene triperoxide diamine (HMTD) and triacetone triperoxide (TATP)." *Forensic Chemistry* 2018; 10: 5–14.

Elbasuney, S., and El-Sharkawy, Y. H., "Instant identification of explosive material: Laser induced photoacoustic spectroscopy versus Fourier transform infrared." *TrAC* 2018; 108: 269–277.

Grimm, B.L., Stern, L. A., and Lowe, A. J., "Forensic utility of a nitrogen and oxygen isotope ratio time series of ammonium nitrate and its isolated ions." *Talanta* 2018; 178: 94–101.

Ostrinskaya, A., Kunz, R. R., Clark, M., Kingsborough, R. P., Ong, T. H., and Deneault, S., "Rapid quantitative analysis of multiple explosive compound classes on a single instrument via flow-injection analysis Tandem Mass Spectrometry." *Journal of Forensic Sciences* 2019; 64(1): 223–230. Available at: https://onlinelibrary.wiley.com/doi/full/10.1111/1556-4029.13827

Reiss, R., Ehlert, S., Heide, J., Putz, M., Forster, T., and Zimmermann, R., "Ambient pressure laser desorption-chemical ionization mass spectrometry for fast and reliable detection of explosives, drugs, and their precursors." *Applied Sciences* 2018; 8(6): 933.

Zapata, F., and Garcia-Ruiz, C., "The discrimination of 72 Nitrate, Chlorate and Perchlorate salts using IR and Raman spectroscopy." *Spectrochimica Acta - Part A: Molecular and Biomolecular Spectroscopy* 2018; 189: 535–542.

He, N., Ni, Y., Teng, J., Li, H., Yao, L., and Zhao, P. "Identification of inorganic oxidizing salts in homemade explosives using Fourier transform infrared spectroscopy." *Spectrochimica Acta Part A: Molecular and Biomolecular Spectroscopy* 2019; 221: 117164. https://www.sciencedirect.com/science/article/pii/S1386142519305542

Suppajariyawat, P., Elie, M., Baron, M., and Gonzalez-Rodriguez, J. "Classification of ANFO samples based on their fuel composition by GC——MS and FTIR combined with Chemometrics." *Forensic Science International*, 2019; 301(1): 415–425. https://www.sciencedirect.com/science/article/pii/S0379073818306583

Taranto, V., Ueland, M., Forbes, S. L., and Blanes, L. "The analysis of nitrate explosive vapour samples using Lab-on-a-chip instrumentation." *Journal of Chromatography A*, 2019; 1602: 467–473. https://www.sciencedirect.com/science/article/pii/S0021967319306004

Joubert, V., Silvestre, V., Ladroue, V., Besacier, F., Blondel, P., Akoka, S., Baguet, E., and Remaud, G. S., "Forensic application of position-specific isotopic analysis of trinitrotoluene (TNT) by NMR to determine 13C and 15N intramolecular isotopic profiles." *Talanta* 2020 Feb 13: 120819. https://www.sciencedirect.com/science/article/abs/pii/S0039914020301107

Schachel, T. D., Stork, A., Schulte-Ladbeck, R., Vielhaber, T., and Karst, U. "Identification and differentiation of commercial and Military Explosives via High Performance Liquid Chromatography–High Resolution Mass Spectrometry (HPLC-HRMS), X-Ray Diffractometry (XRD) and X-Ray Fluorescence spectroscopy (XRF): Towards a Forensic Substance Database on Explosives." *Forensic Science International*. 2020 Feb 4: 110180. https://www.sciencedirect.com/science/article/pii/S0379073820300426

Homemade/Improvised Explosives

Zhang, J., Oxley, J. C., Smith, J., and Cioffi, E., "Mass spectra of unlabeled and isotopically labeled hexamethylene triperoxide diamine (HMTD)." *Propellants, Explosives, Pyrotechnics* 2000; 25: 1–4.

Bartick, E. G., and Mount, K. H., "Analysis of a suspect explosive component: Hydrogen peroxide in hair coloring developer." *Forensic Sci. Commun.* 3(4).

McKay, G. L., "Forensic characterisation of organic peroxide explosives (TATP, HMTD, DADP)." *Kayaku Gakkaishi* 2002; 63(6): 323–329.

Oxley, J. C., Smith, J. L., and Chen, H., "Decomposition of multi-peroxidic compound: triacetone triperoxides (TATP)." *Propellants, Explosives, Pyrotechnics* 2002; 27: 209–216.

Oxley, J. C., Smith, J. L., Chen, H., and Cioffi, E., "Decomposition of multi-peroxidic compounds: Part II. Hexamethylene triperoxide diamine (HMTD)." *Thermochimica Acta* 2002; 388: 215–225.

Oxley, J. C., Smith, J. L., Rogers, E., and Yu, M., "Kinetic studies on Ammonium Nitrate formulations: the search for explosivity modifiers." *Thermochemica Acta* 2002; 384(1–2): 23–45.

Widmer, L., Watson, S., Schlater, K., and Crowson, A., "Development of an LC/MS method for the trace analysis of triacetone triperoxide." *Analyst* 2002; 127: 1627–1632.

Matyas, R., "Chemical Destruction of Triacetone Triperoxide and Hexamethylenetriperoxidediamine." *New Trends Res. Energ. Mater. Proc. Semin., 6th*, Pardubice, April 2003, 164–173.

Schulte-Ladbeck, R., and Karst, U., "Determination of Triacetonetriperoxide in ambient air." *Anal. Chim. Acta* 2003; 482: 183–188.

Schulte-Ladbeck, R., Kolla, P., and Karst, U., "Trace analysis of peroxide-based explosives." *Analytical Chemistry* 2003; 75(4): 731–735.

Muller, D., Levy, A., Shelef, R., Abramovich-Bar, S., Sonenfeld, D., and Tamiri T. "Improved method for the detection of TATP after explosion." *J Forensic Sci.* 2004 Sep; 49(5): 935–938.

Schreck, A., Knorr, A., Wehrstedt, K. D., Wandrey, P. A., Gmeinwieser, T., and Steinbach, J., "Investigation of the Explosive Hazard of mixtures containing Hydrogen Peroxide and Different Alcohols." *J Hazard Mater.* 2004 Apr 30; 108(1–2): 1–7.

Stambouli, A., El Bouri, A., Bouayoun, T., and Belliman, M. A., "Headspace GC-MS detection of TATP traces in post-explosion debris." *Forensic Sci. Int.* 2004; 146(Supplement): S191–S194.

Xu, X., van de Craats, A. M., Kok, E. M., and de Bruyn, P. C., "Trace analysis of peroxide explosives by high performance liquid chromatography-atmospheric pressure chemical ionization-tandem mass spectrometry (HPLC-APCI-MS/MS) for forensic applications." *J Forensic Sci.* 2004 Nov; 49(6): 1230–1236.

Almog, J., Klein, A., Tamiri, T., Shloosh, Y., and Abramovich-Bar, S., "A field diagnostic test for the improvised explosive urea nitrate." *Journal of Forensic Sciences* 2005; 50(3): 582–586.

Danekamp, C., Gottlieb, L., Tamiri, T., Tsoglin, A., Shilav, R., and Kapon, M., "Two separable conformers of TATP and analogues exist at room temperature." *Organic Letters* 2015; 7(12): 2461–2464.

Dubnikova, F., Kosloff, R., Almog, J., et al., "Decomposition of triacetone triperoxide is an entropic explosion." *Journal of the American Chemical Society* 2005; 127: 1146–1159.

Oxley, J. C., Smith, J. L., Shinde, K., and Moran, J., "Determination of the vapor density of triacetone triperoxide (TATP) using a gas chromatography headspace technique." *Propellants, Explosives, Pyrotechnics* 2005; 30: 127–130.

Pena-Quevedo, A. J., Figueroa, J., Pacheco-Londono, L. C., and Rivera-Montalvo, L. A., "Characterization and differentiation of High energy Cyclic Organic Peroxides by GC/FTIR, GC-MS, FTIR and Raman Microscopy." *Proc SPIE* June 2005; 5778: 347–358.

Tamiri, T., "Characterization of the improvised explosive urea nitrate using electrospray ionization and atmospheric pressure chemical ionization." *Rapid Communications in Mass Spectroscopy* 2005; 19: 2094–2098.

Greenway, G. M., Leelasattarathkul, T., Liawruangrath, S., Wheatley, R. A., and Youngvises, N., "Ultrasound-Enhanced Flow Injection Chemiluminescence for Determination of Hydrogen peroxide." *Analyst* 131, 4, 501-8 (2006)).

Lu, D., Cagan, A., Munoz, R., Tangkuaram, T., and Wang, J., "Highly Sensitive Electrochemical Detection of Trace Liquid Peroxide explosives at a Prussian-blue 'Artificial-Peroxidase' modified Electrode." *The Analyst* 2006; (12).

Menzel, E. R., "Fluorescence Detection of the Explosive Urea Nitrate with p-DMAC." *Journal of Forensic Identification*, May 2006; 56(3): 325–332.

Schulte-Ladbeck, R., Vogel, M., and Karst, U., "Recent methods for the determination of peroxide-based explosives." *Anal. Bioanal. Chem.* 2006; 386: 559–565.

Schulte-Ladbeck, R., Edelmann, A., Quintas, G., Lendl, B., and Karst, U., "Determination of peroxide based explosives using liquid chromatography with on-line infrared detection." *Analytical Chemistry* 2006; 78(23): 8150–8155.

Sigman, M. E., Clark, C. D., Fidler, R., Geiger, C. L., and Clausen, C. A., "Analysis of triacetone triperoxide by gas chromatography/mass spectrometry and gas chromatography/tandem mass spectrometry by electron and chemical ionization." *Rapid Commun. Mass Spectrom.* 2006; 20: 2851–2857.

Almog, J., Burda, G., Shloosh, Y., Abramovich-Bar, S., Wolf, E., and Tamiri, T., "Recovery and detection of urea nitrate in traces." *Journal of Forensic Sciences* 2007; 52(6): 1284–1290.

Hargather, M. J., and Settles, G. S., "Optical measurement and scaling of blasts from gram-range explosive charges." *Shock Waves* Dec 2007; 17(4): 215–223.

Hiyoshi, R. I., Nakamura, J., and Brill, T. B., "Thermal decomposition of Organic Peroxides TATP and HMTD by T-Jump/FTIR Spectroscopy." *Propellants, Explosives, Pyrotechnics* 2007; 32(2): 127–134. Available at: https://doi.org/10.1002/prep.200700002

Lemberger, N., and Almog, J., "Structure Elucidation of dyes that are formed in the Colorimetric detection of the improvised explosive Urea Nitrate." *J Forensic Sci* September 2007; 52(5): 1107–1110.

Pena-Quevedo, A. J., Mina-Camilde, N., Cody, R., and Ramos, M., "Characterization and differentiation of High energy Amine Peroxides by Direct Analysis in Real Time TOF/MS." *Proc SPIE* May 2007; 6538.

Technical Support Working Group. "Indicators and Warnings for Homemade Explosives." First Edition, December 2007, For Official Use Only.

Zeman, S., Trzcinski, W. A., and Matyas, R., "Properties of explosive mixtures containing peroxides Part1. Relative performance and detonation of mixtures with Triacetone Triperoxide." *Journal of Hazardous Materials* 154(2008): 192–198.

Zeman, S., and Bartei, C., "Some properties of explosive mixtures containing peroxides Part II. Relationships between detonation parameters and thermal reactivity of the mixtures with Triacetone Triperoxide." *Journal of Hazardous Materials* 154(2008): 199–203.

American Institute of Chemical Engineers. "Is sugar an explosive hazard?" *J Fail. Anal. and Preven.* 2008; 8: 311–312. doi:10.1007/s11668-008-9157-2

Cotte-Rodriguez, I., Hernandez-Soto, H., Chen, H., and Cooks, R. G., "In situ trace detection of peroxide explosives by desorption electrospray ionization and desorption atmospheric pressure chemical ionization." *Anal. Chem.* 2008; 80: 1512–1519.

Germain, M. E., and Knapp, M. J., "Turn-on Fluorescence Detection of H2O2 and TATP." *Inorganic Chemistry* 2008; 47(21): 4422–4429.

Johns, C., Shellie, R. A., Potter, O. G., O'Reilly, J. W., Hutchinson, J. P., Guijt, R.
    M., Breadmore, M. C., Hilder, E. F., Dicinoski, G. W., and Haddad, P. R.,
    "Identification of homemade inorganic explosives by ion chromatographic
    analysis of post-blast residues." *Journal of Chromatography A* 2008; 1182:
    205–214.
Kuzmin, V., Solovev, M., Tuzkov, Y., "Forensic investigation of some peroxides
    explosives." *Central European Journal of Energetic Materials* 2008; 5: 77–85.
Laine, D. F., and Cheng, I. F., "Electrochemical Detection of the Explosive,
    Hexamethylene Triperoxide Diamine (HMTD)." *Microchemical Journal*
    January 2009; 91(1): 125–128.
Matyas, R., Pachman, J., and Ang, H. G., "Study of TATP: spontaneous transforma-
    tion of TATP to DADP." *Prop. Explos. Pyrotech.* 2008; 33: 89–91.
Oxley, J. C., Smith, J. L., Brady, J., Dubnikova, F., Kosloff, R., Zeiri, L., and Zeiri, Y.,
    "Raman and infrared fingerprint spectroscopy of peroxide-based explosives."
    Applied Spectroscopy 2008; 62(8): 906–915.
Oxley, J. C., Smith, J. L., Naik, S., and Moran, J. S., "Decompositions of urea and
    guanidine nitrates." *Journal of Energetic Materials* 2008; 27(1): 17–39.
Sigman, M. E., Clark, C. D., Caiano, T., and Mullen, R., "Analysis of triacetone trip-
    eroxide (TATP) and TATP synthetic intermediates by electrospray ionization
    mass spectrometry." *Rapid Commun. Mass Spectrom.* 2008; 22: 84–90.
Anaya, D. R., et al. "Analysis of explosives' precursors by means of a portable Raman
    spectrometer." *Proc of SPIE* 2009; 7499: 749902.
Burks, R. M., and Hage, D. S., "Current trends in the detection of peroxide-based
    explosives." *Anal Bioanal Chem.* 2009 Sep; 395(2): 301–313. doi: 10.1007/
    s00216-009-2968-5. Epub 2009 Jul 31.
Banerjee, S., Mohapatra, S. K., Misra, M., and Mishra, I., "The detection of impro-
    vised non-military peroxide based explosives using a titania nanotube array
    sensor." *Nanotechnology* 2009; 20(7).
Egorshev, V., Sinditskii, V., Smirnov, S., Glinkovsky, E., and Kuzmin, V., "A
    Comparative Study on Cyclic Acetone Peroxides." *New Trends Re. Energ.
    Mater. Proc. Semin., 12th*, Pardubice, April 1–3, 2009, 113–123.
Ellenbogen, M., and Bijjani, R., "Liquids and Homemade Explosive Detection."
    Proc. of SPIE, Vol. 7306 73060Y–1.
Itozaki, H., and Yamauchi, Y., "Liquid explosive detection from outside of the bottle
    by NIR." Optically Based Biological and Chemical Detection for Defence V,
    Proc. of SPIE 2009; 7484: 748405. doi: 10.1117/12.830333
Matyas, R., and Selesovsky, J., "Power of TATP based explosives." *Journal of
    Hazardous Materials* 2009; 165: 95–99.
Matyas, R., and Pachman, J., "Study of TATP: spontaneous transformation of TATP
    to DADP." *Prop. Explos. Pyrotech.* 2009; 34: 484–488.
Naval Explosive Ordnance Disposal Technology Division, Explosives: Military,
    Commercial, Homemade, and Precursors Identification Guide, Version 2.0
    2009. *For Official Use Only.*
Oxley, J. C., Smith, J. L., Luo, W., and Brady, J., "Determining the vapor pressures
    of diacetone diperoxide (DADP) and hexamethylene triperoxide diamine
    (HMTD). *Propellants Explos. Pyrotech.* 2009; 34: 539–543.

Reany, O., Kapon, M., Botoshansky, M., and Keinan, E., "Rich Polymorphism in Triacetone Triperoxide." *Cryst. Growth Des.* 2009; 9(8): 3661–367. Available at: doi: 10.1021/cg900390y

Sigman, M. E., Clark, C. D., Painter, K., Milton, C., Simatos, E., and Frisch, J., "Analysis of oligomeric peroxides in synthetic triacetone triperoxide samples by tandem mass spectrometry." *Rapid Commun. Mass Spectrom.* 2009; 23: 349–356.

Tamiri, T., Rozin, R., Lemberger, N., and Almog, J., "Urea Nitrate, an exceptionally easy-to-make improvised explosive: Studies towards trace characterization." *Analytical and Bioanalytical Chemistry* 2009; 395: 421–428.

Fidler Albo, R. L., Legron, T., Elie, M. R., Saitta, E. H., Sigman, M. E., Geiger, C. L., and Clausen, C., "Degradation of Triacetone Triperoxide (TATP) Using Mechanically Alloyed Mg/Pd." *Propellants, Explosives and Pyrotechnics* 2010. Available at: https://doi.org/10.1002/prep.200900011

Matyas, R., and Pachman, J., "Study of TATP: Influence of reaction conditions on product composition." *Propellants. Explos. Pyrotech.* 2010; 35(1): 31–37.

Oxley, J. C., Smith, J. L., and Naik, S., "Determination of urea Nitrate and Guanidine Nitrate vapor pressures by isothermal thermogravimetry." *Propellants, Explosives, Pyrotechnics* 2010; 35(3): 278–283.

Partridge, A., Walker, S., and Armitt, D., "Detection of impurities in organic peroxide explosives from precursor chemicals." *Australian Journal of Chemistry* 2010; 63(1): 30–37.

Sarazin, C., Delaunay, N., Varenne, A., Costanza, C., Eudes, V., and Gareil, P., "Simultaneous capillary electrophoretic analysis of inorganic ions and cations in post-blast extracts of acid-aluminum mixtures." *J Sep Sci.* 2010 Oct; 33(20): 3177–3183.

Spidell, M. T., Gordon, J. M., Pitz, J., Gross, K. C., Perram, G. P., "High speed radiometric measurements of IED detonation fireballs." *Proceedings of SPIE* 2010; 7668: 76680C.

Aranda, R., Stern, L. A., Dietz, M. E., McCormick, M. C., Barrow, J. A., and Mothershead, M. F., "Forensic utility of isotope ratio analysis of the explosive Urea Nitrate and its precursors." *Forensic Sci Int.* 2011 Mar 20; 206(1–3):143–9.

Barnette, J. E., Lott, M. J., Howa, J. D., Podlesak, D. W., and Ehieringer, J. R., "Hydrogen and oxygen isotope values in hydrogen peroxide." *Rapid Commun Mass Spectrom.* 2011 May 30; 25(10): 1422–1488. doi: 10.1002/rcm.5004

de Perre, C., Prado, A., and McCord, B., "Rapid and specific detection of urea nitrate and ammonium nitrate by electrospray ionization time-of-flight mass spectrometry using infusion with crown ethers." *Rapid Commun. Mass Spectrom.* 2012; 26: 154–162.

Dudek, K., Matyas, R., and Dorazil, T., "Detection and identification of explosive precursors and explosives." *New Trends Res. Energ. Mater. Proc. Semin., 14th* Pardubice, April 13–15, 2011, 595–601.

Felix-Rivera, H., Ramirez - Cedeno, M. L., Sanchez - Cuprill, R. A., Hernadez - Rivera, S.P., "Triacetone triperoxide thermogravimetric study of vapor pressure and enthalpy of sublimation in 303–338 K temperature range." *Thermochimica Acta* 2011; 514: 37–43.

Fitzgerald, M., and Bilusich, D., "Sulfuric, Hydrochloric, and Nitric Acid-Catalyzed Triacetone Triperoxide (TATP) Reaction Mixtures: An Aging Study." *J Forensic Sci* September 2011; 56(5): 1143–1149.

Flanigan, P. M., Brady, J. J., Judge, E. J., and Levis, R. J., "Determination of inorganic improvised explosive device signatures using laser electrospray mass spectrometry detection with offline classification." *Anal. Chem.* 2011; 83: 7115–7122.

Girotti, S., Ferri, E., Maiolini, E., Bolelli, L., D'Elia, M., Coppe, D., and Romolo, F. S., "A quantitative chemiluminescent assay for analysis of peroxide-based explosives." *Anal. Bioanal. Chem.* 2011; 400: 313–320.

Haroune, N., Crowson, A., and Campbell, B., "Characterisation of triacetone triperoxide (TATP) conformers using LC NMR." *Sci. Justice* 2011; 51: 50–56.

Lichorobiec, S., "Development of alternative projectile to deactivate an improvised explosive device - Pipe bomb." *Komunikacie* 2011; 13(2): 20–25.

Pachman, J., and Matyas, R., "Study of TATP: Stability of TATP solutions." *Forensic Science International* 2011; 207: 212–214.

Pena-Quevedo, A. J., Laramee, J. A., Durst, H. D., and Hernandez-Rivera, S. P., "Cyclic organic peroxides characterization by mass spectrometry and Raman microscopy." *IEEE Sensors Journal* 2011; 11(4): 1053–1060.

Rozin, R., and Almog, J., "Colorimetric detection of urea nitrate: The missing link." *Forensic Science International* 2011; 208(1–3): 25–28.

Tarvin, M., McCord, B., Mount, K., and Miller, M. L., "Analysis of hydrogen peroxide field samples by HPLC/FD and HPLC/ED in DC mode." *Forensic Science International* 2011; 209(1–3): 166–172.

Amani, M., Chu, Y., Waterman, K. L., Hurley, C. M., Platek, M. J., and Gergory, O. J., "Detection of triacetone triperoxide (TATP) using a thermodynamic based gas sensor." *Sensors and Actuators B: Chemical* 2012; 162(1): 7–13.

Fan, W., Young, M., Canino, J., Smith, J., Oxley, J., and Almirati, J. R., "Fast detection of triacetone triperoxide (TATP) from headspace using planar solid-phase microextraction (PSPME) coupled to and IMS detector." *Analytical and Bioanalytical Chemistry* 2012; 403(2): 401–408.

Fitzgerald, M., and Bilusich, D., "The Identification of Chlorinated Acetones in Analyses of Aged Triacetone Triperoxide (TATP)." *Journal of Forensic Sciences* 2012; 57(5): 1299–1302.

Lock, C. M., Brust, H., Van Breukelen, M., Dalmolen, J., Koeberg, M., and Stoker, D. A., "Investigation of isotopic linkages between precursor materials and the improvised high explosive product hexamethylene triperoxide diamine." *Analytical Chemistry* 2012; 84(11): 4984–4992.

Matyas, R., Selesovsky, J., and Musil, T., "Sensitivity to friction for primary explosives." *J. Hazard. Mater.* 2012; 213–214: 236–241.

Oxley, J. C., Smith, J. L., Vadlamannati, S., Brown, A. C., Zhang, G., Swanson, D. S., and Canino, J., "Synthesis and characterization of urea nitrate and nitrourea." *Propellants, Explosives, Pyrotechnicsd* 2013; 38(3): 335–344.

Kunzel, M., Nemec, O., and Matyas, R., "Erythritol tetranitrate in ammonium nitrate based explosives." *Cent Eur J Energ Mater* 2013; 10: 351–358.

Matyas, R., and Chylkova, J., "Study of TATP: Method for determination of residual acids in TATP." *Forensic Science International* 2013; 228(1–3): 170–173.

Matyas, R., Selesovsky, J., and Musil, T., "Decreasing the Friction Sensitivity of TATP, DADP and HMTD." *Central European journal of Energetic Materials* 2013; 10(2): 263–275.

Peterson, G. R., Bassett, W. P., Weeks, B. L., and Hope-Weeks, L. J., "Phase pure triacetone triperoxide: The influence of ionic strength, oxidant source, and acid catalyst." *Crystal Growth and Design* 2013; 13(6): 2307–2311.

Romolo, F. S., Cassioli, L., Grossi, S., Cinelli, G., and Russo, M. V., "Surface-sampling and analysis of TATP by swabbing and gas chromatography/mass spectrometry." *Forensic Science International* 2013; 224(1–3): 96–100.

Bali, M.S., Armitt, D., Wallace, L., and Day, A.I., "Cyclic pentanone peroxide: Sensitiveness and suitability as a model for triacetone triperoxide." *Journal of Forensic Sciences* 2014; 59(4): 936–942.

Gerber, M., Walsh, G., and Hopmeier, M., "Sensitivity of TATP to a TASER electrical output." *Journal of Forensic Sciences* 2014; 59(6): 1638–1641.

Giordano, B. C., Lubrano, A. L., Field, C. R., and Collins, G. E., "Dynamic headspace generation and quantitation of triacetone triperoxide vapor." *J. Chromatogr. A.* 2014; 1331: 38–43.

Klapotke, T. M., and Wloka, T., "Peroxide explosives." *PATAI'S Chemistry of Functional Groups*, Online 2009–2014, John Wiley & Sons, Ltd. doi: 10.1002/9780470682531.pat0879

Kotrly, M., and Turkova, I., "Analysis of nonstandard and home-made explosives and post-blast residues in forensic practice." *Proceedings of SPIE* 2014; 9073: 90730U.

Kunzel, M., Yan, Q.-L., Selesovsky, J., Zemen, S., and Matyas, R., "Thermal behavior and decomposition kinetics of ETN and its mixtures with PETN and RDX." *J Therm Anal Calorim* 2014; 115: 289–299.

Manner, V. W., Tappan, B. C., Scott, B. L., Preston, D. N., and Brown, G. W., "Crystal structure, packing analysis, and structural-sensitivity correlations of erythritol tetranitrate." *Cryst Growth Des* 2014; 14: 6154–6160.

Nazarian, A., and Presser, C., "Forensic analysis methodology for thermal and chemical characterization of homemade explosives." *Thermochimica Acta* 2014; 576: 60–70.

Manner, V. W., Preston, D. N., Tappan, B. C., Sander, V. E., Brown, G. W., Hartline, E., and Jensen, B., "Explosive performance properties of erythritol tetranitrate (ETN)." *Propellants, Explos, Pyrotech* 2015; 40: 460–462.

Marsh, C. M., Mothershead, R. F., and Miller, M. L., "Post-blast analysis of Hexamethylene Triperoxide Diamine using liquid chromatography-Atmospheric pressure chemical Ionization-Mass Spectrometry." *Science and Justice* 2015; 55(5): 299–306.

Matyas, R., Kunzel, M., Ruzicka, A., Knotek, P., and Vodochodsky, O., "Characterization of erythritol tetranitrate. Part 1: physical properties." *Propellants Explos, Pyrotech.* 2015; 40: 185–188.

Oxley, J. C., Smith, J. L., Porter, M., McLennan, L., Colizza, K., Zeiri, Y., Kosloff, R., and Dubnikova, F., "Synthesis and degradation of hexamethylene triperoxide diamine (HMTD)." *Propellants Explos, Pyrotech.* 2015; 40: 1–18.

Kotrly, M., Mares, B., Turkova, I., and Beroun, I., "Identification of improvised explosives residues using physical-chemical analytical methods under real conditions after an explosion." *Proceedings of SPIE* 2016; 9823: 98230S.

Matyas, R., Lycka, A., Jirasko, R., Jakovy, Z., Maixner, J., and Miskova, L., "Analytical characterization of Erythritol Tetranitrate, an improvised explosive." *Journal of Forensic Sciences* 2016; 61(3): 759–764.

Steinkamp, F. L., DeGreeff, L. E., Collins, G., and Rose-Pehrsson, S. L., "Factors affecting the intramolecular decomposition of hexamethylene triperoxide diamine and implications for detection." *J. Chromatogr A.* 2016; 1451: 83–90.

DeGreeff, L. E., Cerreta, M. M., and Katilie, C. J., "Variation in the headspace of bulk Hexamethylene Triperoxide Diamine (HMTD) with time, environment, and formulation." *Forensic Chemistry* 2017; 4: 41–50.

Fraga, C. G., Mitroshkov, A. V., Mirjankar, N. S., Dockendorff, B. P., and Melville, A. M., "Elemental source attribution signatures for Calcium Ammonium Nitrate (CAN) fertilizers used in homemade explosives." *Talanta* 2017; 174: 131–138.

Gamble, S. C., Campos, L. C., and Morgan, R. M., "Detection of trace peroxide explosives in environmental samples using solid phase extraction and liquid chromatography mass spectrometry." *Environmental Forensics* 2017; 18(1–2): 50–61.

Assis, A. C. A., Caetano, J., Florencio, M. H., and Cordeiro, C., "Triacetone triperoxide characterization by FT-ICR mass spectrometry: Uncovering multiple forensic evidence." *Forensic Science International* 2019; 301: 37–45. https://www.sciencedirect.com/science/article/pii/S0379073819301549?via%3Dihub

Balachandar, K. G., and Thangamani, A., "Studies on some of the Improvised Energetic Materials (IEMs): Detonation, blast impulse and TNT equivalence parameters." *Oriental Journal of Chemistry.* 2019; 35(6): 1813–1823. https://www.orientjchem.org/vol35no6/studies-on-some-of-the-improvised-energetic-materials-iems-detonation-blast-impulse-and-tnt-equivalence-parameters/

Bezemer, K., McLennan, L., van Duin, L., Kuijpers, C. J., Koeberg, M., van den Elshout, J., van der Heijden, A., Busby, T., Yevdokimov, A., Schoenmakers, P., and Smith, J., "Chemical attribution of the home-made explosive ETN–Part I: Liquid chromatography-mass spectrometry analysis of partially nitrated erythritol impurities." *Forensic Science International.* 2019 Dec 19: 307, 110102. https://www.sciencedirect.com/science/article/pii/S0379073819305146

Trace and Post Blast Residues Analysis

Crowson, A., and Beardah, M. S., "Development of an LC/MS method for the trace analysis of hexamethylenetriperoxidediamine (HMTD)." *Analyst* 2001; 126: 1689–1693.

Klapec, D. J., and Ng, D., "The use of capillary electrophoresis in the detection of Monomethylamine and Benzoate Ions in the forensic examination of explosives residues." *J Forensic Sci.* 2001 Sep; 46(5): 1168–1173.

Sigman, M. E., and Ma, C. Y., "Detection limits for GC/MS analysis of organic explosives." *Journal of Forensic Sciences* 2001; 46(1): 6–11.

Sigman, M. E., Ma, C. Y., and Ilgner, R. H., "Performance Evaluation of an In-injection Port Thermal Desorption /Gas Chromatographic/Negative Ion Chemical Ionization Mass Spectrometric Method for Trace Explosive Vapor Analysis." *Anal Chem.* 2001 Feb 15; 73(4): 792–798.

Kelleher, J. D., "Explosives residue: Origin and Q5 distribution." *Forensic Science Communications* 2002; 4(2).

Calderera, S., Gardebas, G., and Martinez, F., "Solid phase micro extraction coupled with on-column GC/ECD for the post-blast analysis of organic explosives." *Forensic Sci Int.* 2003 Oct 14; 137(1): 6–12.

Kamyshny, A., Magdassi, S., Avissar, Y., and Almog, J., "Water-soaked evidence: Detectability of explosive traces after immersion in water." *J Forensic Sci.* 2003 Mar; 48(2): 312–317.

Kosanke, K.L., Dujay, R.C., and Kosanke, B., "Characterization of pyrotechnic reaction residue particles by SEM/EDS." *J Forensic Sci.* 2003 May; 48(3): 531–537.

Oxley, J. C., Smith, J. L., Resende, E., and Pearce, E., "Quantification and aging of the post-blast residue of TNT landmines." *J Forensic Sci.* 2003 Jul; 48(4): 742–753.

Biswas, S., Chowdhury, B., and Ray, B. C., "A novel Spectrofluorometric method for the ultra-trace analysis of nitrite and nitrate in aqueous medium and its application to air, water, soil and forensic samples." *Talanta.* 2004 Oct 8; 64(2): 308–312. doi: 10.1016/j.talanta.2004.02.018.

Cullum, H. E., McGavigan, C., Uttley, C. Z., Stroud, M. A., and Warren, D. C., "A second survey of background levels of explosives and related compounds in the environment." *J. Forensic Sci* 2004. 49(4): 684–690.

Muller, D., Levy, A., Shelef, R., Abramovich-Bar, S., Sonenfeld, D., and Tamiri T. "Improved method for the detection of TATP after explosion." *J Forensic Sci.* 2004 Sep; 49(5): 935–938.

Xu, X., van de Craats, A. M., Kok, E. M., and de Bruyn, P. C., "Trace analysis of peroxide explosives by high performance liquid chromatography-atmospheric pressure chemical ionization-tandem mass spectrometry (HPLC-APCI-MS/MS) for forensic applications." *J Forensic Sci.* 2004 Nov; 49(6): 1230–1236.

Xu, X., van de Craats, A. M., and de Bruyn, P. C., "Highly sensitive screening method for nitroaromatic, nitramine and nitrate ester explosives by high performance liquid chromatography-atmospheric pressure ionization-mass spectrometry (HPLC-API-MS) in forensic applications." *J Forensic Sci.* 2004 Nov; 49(6): 1171–1180.

Bradley, K. S., "Determination of Elemental Sulfur in explosives and explosive residues by gas chromatography-mass spectrometry." *J Forensic Sci.* 2005 Jan; 50(1): 96–103.

Hopper, K. G., LeClair, H., and McCord, B. R., "A novel method for analysis of explosives residue by simultaneous detection of Anions and Cations via Capillary Zone Electrophoresis." *Talanta.* 2005 Aug 15; 67(2): 304–312. doi: 10.1016/j.talanta.2005.01.037. Epub 2005 Feb 24.

Kinghorn, R., and Miller, C., "Analysis of Trace Residues of Explosive Materials by Time-of-Flight LC/MS", Agilent Technologies, Wilmington, DE, 2005.

Oxley, J. C., Smith, J. L., Kirschenbaum, L., Shinde, K. P., and Marimganti, S., "Accumulation of explosives in hair." *Journal of Forensic Sciences* 2005; 50(4): 826–831.

Borton, C., and Olson, L., "Trace Level Analysis of Explosives in Ground Water and Soil", Applied Biosystems/MDS Sciex, Foster City, CA, 2006.

Almog, J., Burda, G., Shloosh, Y., Abramovich-Bar, S., Wolf, E., and Tamiri, T., "Recovery and detection of Urea Nitrate in traces." *J Forensic Sci.* 2007 Sep; 52(6): 1284–1290. Available at: https://doi.org/10.1111/j.1556-4029.2007.00551.x

Martin, A. N., Farqar, G. R., Gard, E. E., Frank, M., and Fergensen, D. P., "Identification of high explosives using single-particle Aerosol Mass Spectrometry." *Anal Chem.* 2007 Mar 1; 79(5): 1918–1925.

Oxley, J. C., Smith, J. L., Kirschenbaum, L., and Marimganti, S., "Accumulation of explosives in hair: part II: factors affecting sorption." *Journal of Forensic Sciences* 2007; 52(6): 1291–1296.

Sanchez, J.C., Toal, S. J., Wang, Z., Dugan, R. E., and Trogler, W. C., "Selective detection of trace nitroaromatic, nitramine, and nitrate ester explosive residues using a three-step fluorimetric sensing process: a tandem turn-off, turn-on sensor." *J Forensic Sci.* 2007 Nov; 52(6): 1308–1313. Epub 2007 Oct 17.

The Technical and Scientific Working Group on Fire and Explosion Analysis (T/SWGFEX). "Recommended Guidelines for Forensic Identification of Post Blast Explosive Residues." 2007. Available at https://www.swgfex.com/publications.

Verkouteren, J. R., "Particle Characteristics of Trace High Explosives: RDX and PETN." *J Forensic Sci.* 2007 Mar; 52(2): 335–340.

Hutchinson, J. P., Johns, C., Breadmore, M. C., Hilder, E. F., Guijt, R. M., Lennard, C., Dicinoski, G., and Haddad, P. R., "Identification of inorganic ions in postblast explosive residues using portable CE instrumentation and capacitively coupled contactless conductivity detection." *Electrophoresis.* 2008 Nov; 29(22): 4593–4602. doi: 10.1002/elps.200800226.

Johns, C., Shellie, R. A., Potter, O. G., O'Reilly, J. W., Hutchinson, J. P., Guijt, R. M., Breadmore, M. C., Hilder, E. F., Dicinoski, G. W., and Haddad, P. R., "Identification of homemade inorganic explosives by ion chromatographic analysis of post-blast residues." *Journal of Chromatography A* 2008; 1182: 205–214.

Lahoda, K. G., Collin, O. L., Mathis, H. E., LeClair, W. S. H., and McCord, B. R., "A survey of background levels of explosives and related compounds in the environment." *J. Forensic Sci.* 2008; 53(4): 802–806.

Meng, H.-B., Wang, T.-R., Guo, B.-Y., Hashi, Y., Guo, C.-X., and Lin, J.-M., "Simultaneous determination of inorganic anions and cations in explosive residues by ion chromatography." *Talanta* 2008; 76: 241–245.

Oxley, J. C., Smith, J. L., Kirschenbaum, L. J., Marimganti, S., and Vadlamannati, S., "Detection of explosives in hair using ion mobility spectrometry." *Journal of Forensic Sciences* 2008; 53: 690–693.

Perret, D., Marchese, S., Gentili, A., Terracciano, A., Bafile, E., and Romolo, F. S., "LC-MS-MS determination of stabilizers and explosives residues in handswabs." *Chromatographia.* 2008; 68: 517–524.

Prest, J. E., Beardah, M. S., Baldock, S. J., Doyle, S. P., Fielden, P. R., Goddard, N. J., Treves Brown, B. J., "Determination of chlorine containing species in explosive residues using chip-based isotachophoresis" *Journal of Chromatography A* 2008; 1195: 157–163.

Sharma, S. P., and Lahiri, S. C., "Preparation of a database for the determination and estimation of high explosives and a comparative study of different methods of estimation of RDX." *Journal of the Indian Chemical Society* March 2008; 85(3): 273–279.

Talaty, M., Mulligan, C. C., Justes, D. R., Jackson, A. U., Noll, R. J., and Cooks, R. G., "Fabric Analysis by Ambient Mass Spectrometry for Explosives and Drugs." *Analyst* 2008 Nov; 133(11): 1532–1540.

Banas, A., Banas, K., Bahou, M., Moser, H. O., Wen, L., Yang, P., Li, Z. J., Cholewa, M., Lim, S. K., and Lim, Ch. H., "Post-blast detection of traces of explosives by means of Fourier transform infrared spectroscopy." *Vibrational Spectroscopy* 10 November 2009; 51(2): 168–176.

Gottfried, J. L., De Lucia, F. C., and Mizolek, A. W., "Discrimination of explosive residues on organic and inorganic substrates using laser-induced breakdown spectroscopy." *Journal of Analytical Atomic Spectrometry* March 2009; 24(3): 249–356.

Jander, P., and Noll, R., "Automated Detection of Fingerprint Traces of High Explosives Using Ultraviolet Raman Spectroscopy." *Appl Spectrosc.* 2009 May; 63(5): 559–563.

Lancaster, S. L., Marshall, M., and Oxley, J. C., "Laboratory Analysis of Explosion Debris." *Wiley Encyclopedia of Forensic Science*, Sept 2009. Available at: https://doi.org/10.1002/9780470061589.fsa213.

Lancaster, S. L., Marshall, M., and Oxley, J. C., "Explosion debris: Laboratory analysis of." In *Wiley Encyclopedia of Forensic Science*, Jamieson, A., and Moenssens, A., (Eds). Wiley, Chichester, UK 2009.

Lazic, V., Palucci, A., Poggi, C., and Buono, E., "Analysis of explosive and other organic residues by laser induced breakdown spectroscopy." *Spectrochimica Acta Part B* 2009; 64: 1028–1039.

MacCrehan, W. A., "A NIST Standard Reference Material (SRM) to Support the Detection of Trace Explosives." *Anal Chem.* 2009 Sep 1; 81(17): 7189–7196.

Martinez-Lozano, P., Rus, J., Fernandez de la Mora, G., Hernandez, M., and Fernandez de la Mora, J., "Secondary Electrospray Ionization (SESI) of Ambient Vapors for Explosive Detection at Concentrations Below Parts Per Trillion." *J Am Soc Mass Spectrom* 2009; 20: 287–294.

Oxley, J. C., Smith, J. L., Bernier, E., Moran, J. S., and Luongo, J., "Hair as forensic evidence of explosive handling." *Propellants, Explosives, Pyrotechnics* 2009; 34(4): 307–314.

Quirk, A. T., Bellerby, J. M., Carter, J. F., Thomas, F. A., and Hill, J. C., "An initial evaluation of stable isotopic characterisation of post-blast plastic debris from improvised explosive devices." *Science and Justice* 2009; 49: 87–93.

Szakal, C., and Brewer, T. M., "Analysis and Mechanisms of Cyclotrimethylenetrinitramine Ion Formation in Desorption Electrospray Ionization." *Analytical Chemistry* July 2009; 81(13): 5257–5266.

Banas, A., Banas, K., Bahou, M., Moser, H. O., Yang, P., Li, Z. J., Cholewa, M., and Lim, S. K., "Multivariate analysis techniques in the forensic investigation of the post blast residues by means of Fourier transform infrared spectroscopy." *Anal. Chem.* 2010; 82(7): 3038–3044.

Prest, J. E., Beardah, M. S., Baldock, S. J., Doyle, S. P., Fielden, P. R., Goddard, N. J., Treves Brown, B. J., "Determination of the potassium content of explosive residues using miniaturised Isotachophoresis" *Electrophoresis* 2010; 31: 3775–3782.

Sarazin, C., Delaunay, N., Varenne, A., Costanza, C., Eudes, V., and Gareil, P., "Simultaneous capillary electrophoretic analysis of inorganic Ions and Cations in post-blast extracts of Acid-Aluminum Mixtures." *J Sep Sci.* 2010 Oct; 33(20): 3177–3183.

Sarazin, C., Delaunay, N., Varenne, A., Vial, J., Costanza, C., Eudes, V., Minet, J. J., and Gareil, P., "Identification and determination of inorganic Anions in real extracts from pre - and post - blast residues by capillary electrophoresis." *J Chromatogr A.* 2010 Oct 29; 1217(44): 6971–6978.

Turillazzi, E., Monaci, F., Neri, M., Pomara, C., Riezzo, I., Baroni, D., and Fineschi, V., "Collection of trace evidence of explosive residues from the skin in a death due to a disguised letter bomb. The synergy between confocal laser scanning microscope and inductively coupled plasma atomic emission spectrometer analyses." *Forensic Science International* 2010; 197(1–3): e7–e12.

Ahmad, U. K., Tze, O. S., Ghazali, M. F., Hooi, Y. C., and Abdullah, M. K., "Analysis of anionic post-blast residues of low explosives from soil samples of forensic interest." *Malaysian Journal of Analytical Sciences* 2011; 15: 213–226.

Bowen, A. M., "A method for isolating very small particles from plastic explosive samples." *The Microscope* 2011; 59(3): 117–128.

Cummins, J., Hull, J., Kits, K., and Goodpaster, J. V., "Separation and identification of anions using porous graphite carbon and electrospray ionization mass spectrometry: Application to inorganic explosives and their post blast residues. *Anal. Methods* 2011; 3: 1682.

de Perre, C., and McCord, B., "Trace analysis of urea nitrate by liquid chromatography-UV/fluorescence." *Forensic Science International* 2011; 211: 76–82.

Douglas, T. A., Walsh, M. E., McGrath, C. J., Weiss, C. A., Jaramillo, A. M., and Trainor, T. P., "Desorption of nitramine and nitroaromatic explosive residues from soils detonated under controlled conditions." *Environ Toxicol Chem.* 2011 Feb; 30(2): 345–353. Available at: doi: 10.1002/etc.383.

Gregory, K. E., Kunz, R. R., Hardy, D. E., Fountain III, A. W., and Ostazeski, S. A., "Quantitative Comparison of Trace Organonitrate Explosives Detection by GC-MS and GC-ECD Methods with Emphasis on Sensitivity." *Journal of Chromatographic Science* 2011; 49(1): 1–7.

Saiz, J., Ferrando, J. L., Atoche, J. C., Torre, M., and Garcia-Ruiz, C., "Study of losses of volatile compounds from dynamites. Investigation of cross-contamination between dynamites stored in polyethylene bags." *Forensic Science International* 2011; 211(1–3): 27–33.

Doyle, S., "Quality and trace detection and identification of organic high explosives." In *Forensic Investigation of Explosives*, 2nd Edition, ed. A. Beveridge. 2012 Boca Raton, FL. CRC Press (Taylor and Francis Group).

Kozole, J., Tomlinson-Philips, J., Stairs, J. R., Harpe, J. D., Lukow, S., Lareau, R., Boudries, H., Lai, H., and Brauer, C. S., "Characterizing the gas phase ion chemistry of an ion trap mobility spectrometry based explosive trace detector using a tandem mass spectrometer." *Talanta* 2012; 99: 799–810.

Oxley, J. C., Smith, J. L., Kirschenbaum, L. J., Marimiganti, S., Efremenko, I., Zach, R., and Zeiri, Y., "Accumulation of Explosives in Hair-Part 3: Binding Site Study." *Journal of Forensic Sciences* 2012; 57(3): 623–635.

Song-im, N., Benson, S., and Lennard, C., "Evaluation of different sampling media for their potential use as a combined swab for the collection of both organic and inorganic explosive residues." *Forensic Science International* 2012; 222 (1–3): 102–110.

Abdul-Karim, N., Morgan, R., Binions, R., Temple, T., and Harrison, K., "The spatial distribution of post blast rdx residue: Forensic implications." *Journal of Forensic Sciences* 2013; 58(2): 365–371.

Ahmad, U. K., Liew, C. Y., Huri, M. A. M., and Abdullah, S. A. A. S., "Forensic analysis of inorganic anions from post-blast pyrotechnic residues." *Journal Teknologi* 2013; 62(3): 77–82.

Almog, J., Espino, D., Tamiri, T., and Sonenfeld, D., "Trace analysis of urea nitrate in post-blast debris by GC/MS." *Forensic Science International* 2013; 224(1–3): 80–83.

Borusiewicz, R., Zadora, G., and Zieba-Palus, J., "Chemical analysis of post explosion samples obtained as a result of model field experiments." *Talanta* 2013; 116: 630–636.

Brust, H., van Asten, A., Koeberg, M., van der Heijden, A., Kuijpers, C-J., and Schoenmakers, P., "Pentaerythritol tetranitrate (PETN) profiling in post-explosion residues to constitute evidence of crime-scene presence." *Forensic Science International* 2013; 230: 37–45.

Clemons, K., Dake, J., Sisco, E., and Verbeck, G.F., "Trace analysis of energetic materials via direct analyte-probed nanoextraction coupled to direct analysis in real time mass spectrometry." *Forensic Science International* 2013; 231(1–3): 98–101.

DeTata, D., Collins, P., and McKinley, A., "An investigation into the fate of organic explosives in soil." *Australian Journal of Forensic Sciences* 2013; 45(1): 71–84.

Forbes, T. P., and Sisco, E., "Recent advances in ambient mass spectrometry of trace explosives." *Analyst* 2018; 143(9): 1948–1969.

King, S., Benson, S., Kelly, T., and Lennard, C., "Determining the effects of routine fingermark detection techniques on the subsequent recovery and analysis of explosive residues on various substrates." *Forensic Science International* 2013; 233(1–3): 257–264.

Lordel-Madeleine, S., Eudes, V., and Pichon, V., "Identification of the nitroaromatic explosives in post-blast samples by online solid phase extraction using molecularly imprinted silica sorbent coupled with reversed-phase chromatography." *Anal Bioanal Chem* 2013; 405: 5237–5247.

Nuntawong, N., Eiamchai, P., Limwichean, S., Wong, B., Horprathum, M., Patthanasettakul, V., Leelapojanaporn, A., Nakngoenthong, S., and Chindaudom, P., "Trace detection of perchlorate in industrial-grade emulsion explosive with portable surface-enhanced Raman spectroscopy." *Forensic Science International* 2013; 233(1–3): 174–178.

Saha, S., Mandal, M. K., Chen, L. C., Nimoniya, S., Shida, Y., and Hiraoka, K., "Trace level detection of explosives in solution using Leidenfrost phenomenon assisted thermal desorption ambient mass spectrometry." *Mass Spectrometry* 2013; 2: S0008.

Sisco, E., Dake, J., and Bridge, C., "Screening for trace explosives by AccuTOF™-DART®: An in-depth validation study." *Forensic Science International* 2013; 232(1–3): 160–168.

Swider, J.R., "Optimizing Accu time-of-flight/direct analysis in real time for explosive residue analysis." *Journal of Forensic Sciences* 2013; 58(6): 1601–1606.

Bors, D., Cummins, J., and Goodpaster, J., "The anatomy of a pipe bomb explosion: Measuring the mass and velocity distributions of container fragments." *Journal of Forensic Sciences* 2014; 59(1): 42–51.

Ceco, E., Onnerud, H., Menning, D., Gilljam, J. L., Baath, P., and Ostmark, H., "Stand-off imaging Raman spectroscopy for forensic analysis of post-blast scenes: Trace detection of ammonium nitrate and 2, 4, 6-trinitrotoluene." *Proceedings of SPIE* 2014; 9073: 90730G.

Huri, M. A. M., and Ahmad, U. K., "Forensic analysis of high explosive residues from selected cloth." *Malaysian Journal of Analytical Sciences* 2014; 18(1): 68–77.

Kotrly, M., and Turkova, I., "Analysis of nonstandard and home-made explosives and post-blast residues in forensic practice." *Proceedings of SPIE* 2014; 9073: 90730U.

Martin-Alberca, C., de la Ossa, M. A. F., Saiz, J., Ferrando, J.L., and Garcia-Ruiz, C., "Anions in pre- and post-blast consumer fireworks by capillary electrophoresis." *Electrophoresis* 2014; 35(21–22): 3273–3280.

Nguyen, V., Furstenberg, R., Carr, N., McGill, R., Mott, D.R., Papantonakis, M., Kendziora, C. A., and McGill, R. A., "Fate and effects of trace particulate explosives." *Proceedings of SPIE* 2014; 9073: 90730R.

Wright, S., and Gillen, G., "Cathodoluminescence and DART mass spectrometry for the forensic identification of explosive and narcotic particle residues on surfaces." *Microscopy and Microanalysis* 2014; 20(3): 920–921.

Bianchi, F., Gregori, A., Braun, G., Crescenzi, C., and Careri, M., "Micro-solid-phase extraction coupled to desorption electrospray ionization-high-resolution mass spectrometry for the analysis of explosives in soil." *Analytical and Bioanalytical Chemistry* 2015; 407(3): 931–938.

Brust, H., Koeberg, M., Van Der Heijden, A., Wiarda, W., Mugler, I., Schrader, M., Vivo-Truyols, G., Schoenmakers, P., and Van Asten, A., "Isotopic and elemental profiling of ammonium nitrate in forensic explosives investigations." *Forensic Science International* 2015; 248: 101–112.

Chakrabortty, A., Bagchi, S., and Chandra Lahiri, S., "Studies of fire debris from bomb blasts using ion chromatography, gas chromatography-mass spectrometry and fluorescence measurements-evidence of ammonium nitrate, wax-based explosives and identification of a biomarker." *Australian Journal of Forensic Sciences* 2015; 47(1): 83–94.

ENFSI. "Best practice manual for the forensic recovery, identification and analysis of explosives traces." ENFSI-BPM-EXP-01, Version 01, November 2015.

Furstenberg, R., Nguyen, V., Fischer, T., Abrishami, T., Papantonakis, M., Kendziora, C., Mott, D.R., and McGill, R. A., "Advances in sublimation studies for particles of explosives." *Proceedings of SPIE* 2015; 9455: 94550R.

Kotrly, M., and Turkova, I., "New possibilities to analyse non-standard explosives and post blast residues in forensic practice." *Proceedings of SPI* 2015; 9486: 948614.

Marsh, C. M., Mothershead, R. F., and Miller, M. L., "Post-blast analysis of Hexamethylene Triperoxide Diamine using liquid chromatography-atmospheric pressure chemical ionization-mass spectrometry." *Science and Justice* 2015; 55(5): 299–306.

van der Voort, M. M., van Wees, R. M. M., Brouwer, S. D., van der Jagt-Deutekom, M. J., and Verreault, J., "Forensic analysis of explosions: Inverse calculation of the charge mass." *Forensic Science International* 2015; 252: 11–21.

Zapata, F., de la Ossa, M. A. F., Gilchrist, E., Barron, L., and Garcia-Ruiz, C., "Progressing the analysis of improvised explosive devices: Comparative study for trace detection of explosive residues in handprints by Raman spectroscopy and liquid chromatography." *Talanta* 2016; 161: 219–227.

Yu, H.A., Lewis, S. W., Beardah, M. S., and NicDaeid, N., "Assessing a novel contact heater as a new method of recovering explosives traces from porous surfaces." *Talanta* 1 February 2016; 148: 721–728.

Degnan, J. R., Böhlke, J. K. Pelham, K., Langlais, D. M., and Walsh, G. J., "Identification of groundwater Nitrate contamination from explosives used in road construction: Isotopic, Chemical, and Hydrologic evidence." *Environ. Sci. Technol.* 2016; 50(2): 593–603.

Fraga, C. G., Mitroshkov, A. V., Mirjankar, N. S., Dockendorff, B. P., and Melville, A. M., "Elemental source attribution signatures for calcium ammonium nitrate (CAN) fertilizers used in homemade explosives." *Talanta* 2017; 174: 131–138.

Gamble, S. C., Campos, L. C., and Morgan, R. M., "Detection of trace peroxide explosives in environmental samples using solid phase extraction and liquid chromatography mass spectrometry." *Environmental Forensics* 2017; 18(1–2): 50–61.

Kotrly, M., Mares, B., Turkova, I., and Beroun, I., "Identification of improvised explosives residues using physical-chemical analytical methods under real conditions after an explosion." *Proceedings of SPIE* 2016; 9823: 98230S.

Mauricio, F. G. M., Pralon, A. Z., Talhavini, M., Rodrigues, M. O., and Weber, I. T., "Identification of ANFO: Use of luminescent taggants in post-blast residues." *Forensic Science International* 2017; 275: 8–13.

Nic Daeid, N., Yu, H. A., and Beardah, M. S., "Investigating TNT loss between sample collection and analysis" *Science and Justice* 2017; 57: 95–100.

Rapp-Wright, H., McEneff, G., Murphy, B., Gamble, S., Morgan, R., Beardah, M., and Barron, L., "Suspect screening and quantification of trace organic explosives in wastewater using solid phase extraction and liquid chromatography-high resolution accurate mass spectrometry." *Journal of Hazardous Materials* 5 May 2017; 329: 11–21.

Yenel Avci, G. F., Anilanmert, B., and Cengiz, S., "Rapid and simple analysis of trace levels of three explosives in soil by liquid chromatography-tandem mass spectrometry." *Acta Chromatographica* 2017; 29(1): 45–56.

Yu, H. A., DeTata, D. A., Lewis, S. W., and Daeid, N. N., "The stability of TNT, RDX and PETN in simulated post-explosion soils: Implications of sample preparation for analysis." *Talanta* 2017; 164: 716–726.

Wen, P., Amin, M., Herzog, W. D., and Kunz, R. R., "Key challenges and prospects for optical standoff trace detection of explosives." *TrAC* 2018; 100: 136–144.

Black, C., D'Souza, T., Smith, J. C., and Hearns, N. G. R. "Identification of post-blast explosive residue using Direct-Analysis-in-Real-Time and Mass Spectrometry (DART-MS)." *Forensic Chemistry* 2019; 16: 100185. https://doi.org/10.1016/j.forc.2019.100185

Gaiffe, G., Cole, R. B., Sonnette, A., Floch, N., and Bridoux, M. C. "Identification of post-blast residues by DART-High resolution mass spectrometry combined with multivariate statistical analysis of the Kendrick mass defect." *Analytical Chemistry* 2019. https://pubs.acs.org/doi/abs/10.1021/acs.analchem.9b00137

Irlam, R. C., Parkin, M. C., Brabazon, D. P., Beardah, M. S., O'Donnell, M., and Barron, L. P. "Improved determination of femtogram-level organic explosives in multiple matrices using dual-sorbent solid phase extraction and liquid chromatography-high resolution accurate mass spectrometry." *Talanta* 2019; 203: 65–76. https://www.sciencedirect.com/science/article/pii/S0039914019305314

Kober, S. L., Hollert, H., and Frohme, M., "Quantification of nitroaromatic explosives in contaminated soil using MALDI-TOF mass spectrometry." *Analytical and Bioanalytical Chemistry* 2019; 411(23): 5993–6003. https://link.springer.com/article/10.1007/s00216-019-01976-y

Kotrly, M., Eisner, A., Beroun, I., Ventura, K., and Turkova, I. "New possibilities of post-blast residues analysis in forensic science." *Proceedings SPIE 11012, Detection and Sensing of Mines, Explosive Objects, and Obscured Targets XXIV* 2019: 1101212. https://www.spiedigitallibrary.org/conference-proceedings-of-spie/11012/1101212/New-possibilities-of-post-blast-residues-analysis-in-forensic-science/10.1117/12.2519118.short

Mauricio, F. G. M., Abritta, V. R. M., Aquino, R. L., Ambrosio, J. C. L., Logrado, L. P. L., and Weber, I. T., "Evaluation of interferers in sampling materials used in explosive residue analysis by Ion Chromatography." *Forensic Science International Volume 307, Article* 2019; 109908. https://www.sciencedirect.com/science/article/pii/S0379073819303202

Murphy, B. R. "Suspect screening and identification of energetic materials using liquid chromatography coupled to high-resolution mass spectrometry (Doctoral dissertation, King's College London). https://ethos.bl.uk/OrderDetails.do?uin=uk.bl.ethos.789232

Romolo, F. S., and Palucci, A., "Advances in the analysis of explosives," *Emerging technologies for the analysis of forensic traces.* 2019; 207–240. https://link.springer.com/chapter/10.1007/978-3-030-20542-3_15

Taranto, V., Ueland, M., Forbes, S. L., and Blanes, L. "The analysis of nitrate explosive vapour samples using Lab-on-a-chip instrumentation." *Journal of Chromatography A* 2019. https://www.sciencedirect.com/science/article/pii/S0021967319306004

Thangadurai, S., "Ion mobility spectrometry: A tool in the forensic science for the post detonation residue analysis." *Journal of Forensic Science and Criminology* 2019; 7(2): 201. http://www.annexpublishers.com/articles/JFSC/7201-Ion-Mobility-Spectrometry-A-Tool-in-the-Forensic-Science-for-the-Post-Detonation-Residue-Analysis.pdf

Smokeless Powder

MacCrehan, W. A., and Reardon, M. R., "Developing a quantitative extraction technique for determining the organic additives in smokeless handgun powder," *Journal of Forensic Sciences* 2001; 46(4): 802–807.

Heramb, R. M., and McCord, B. R., "The manufacture of smokeless powders and their forensic analysis: A brief review." *Forensic Science Communications* April 2002; 4: 1–7.

MacCrehan, W. A., and Reardon, M. R., "A qualitative comparison of smokeless powder measurements." *Journal of Forensic Sciences* 2002; 47(5): 996–1000.

MacCrehan, W. A., and Reardon, M. R., "A quantitative comparison of smokeless powder measurements." *Journal of Forensic Sciences* 2002; 47(6): 1283–1287.

Wissinger, C. E., and McCord, B. R., "A gradient reversed phase HPLC procedure for smokeless powder comparison." *Journal of Forensic Sciences* 2002; 47(1): 168–174.

Mathis, J. A., and McCord, B. R., "Gradient reversed-phase liquid chromatography-electrospray mass spectrometric method for the comparison of smokeless powders." *Journal of Chromatography A* 2003; 988: 107–116.

Hernandez, N. M., Rosario, S. V., Hernandez, S. P., and Mina, N., "Detection and Characterization of Smokeless Powders with Ion mobility Spectrometry." *Proceedings of the SPIE* 2005; 5778: 607–616.

Hopper, K. G., and McCord, B. R., "A comparison of smokeless powders and mixtures by capillary zone electrophoresis." *Journal of Forensic Sciences* March 2005; 50(2): 307–315.

Mathis, J. A., and McCord, B. R., "Mobile phase influence on electrospray ionization of the analysis of smokeless powders by gradient reversed phase high performance liquid chromatography-ESIMS." *Forensic Science International* 2005; 154: 159–166.

Mahoney, C. M., Gillen, G., and Fahey, A. J., "Characterization of gunpowder samples using Time-of-Flight Secondary Ion Mass Spectrometry (TOF-SIMS)." *Forensic Sci Int.* 2006 Apr 20; 158(1): 39–51.

Muller, D., et al., "A novel method for the analysis of discharged smokeless powder residues." *Journal of Forensic Sciences* January 2007; 52(1): 75–78.

West, C., Baron, G., and Minet, J. J., "Detection of gunpowder stabilizers with Ion Mobility Spectrometry." *Forensic Sci Int.* 2007 Mar 2; 166(2–3): 91–101.

Burleson, G. L., Gonzalez, B., Simons, K., and Yu, J. C., "Forensic analysis of a single particle of partially burnt gunpowder by solid phase micro-extraction-gas chromatography-nitrogen phosphorus detector." *J Chromatogr A.* 2009 May 29; 1216(22): 4679–4683. doi: 10.1016/j.chroma.2009.03.074. Epub 2009 Apr 1.

Joshi, M., Delgado, Y., Guerra, P., Lai, H., and Almirall, J. R., "Detection of odor signatures of smokeless powders using solid phase microextraction coupled to an ion mobility spectrometer." *Forensic Science International* 2009; 188: 112–118.

Scherperel, G., Reid, G. E., and Waddell Smith, R., Characterization of smokeless powders using nanoelectrospray ionization mass spectrometry (nESI-MS). *Anal Bioanal Chem.* 2009 Aug; 394(8):2019–28.

Joshi, M., Rigsby, K., and Almirall, J. R., "Analysis of the headspace composition of smokeless powders using GC-MS, GC-μECD and ion mobility spectrometry." *Forensic Sci Int.* 2011 May 20; 208(1–3): 29–36. doi: 10.1016/j.forsciint.2010.10.024. Epub 2010 Nov 24.

López-López, M., Ferrandoa, J. L., and García-Ruiz, C., "Comparative analysis of smokeless gunpowders by Fourier transform infrared and Raman spectroscopy." *Analytica Chimica Acta* 2012; 717: 92– 99.

Fernandez de la Ossa, M. A., Ortega-Ojeda, F., and Garcia-Ruiz, C., "Discrimination of non-explosive and explosive samples through nitrocellulose fingerprints obtained by capillary electrophoresis." *Journal of Chromatography A* 2013; 1302: 197– 204.

Thomas, J. L., Lincoln, D., and McCord, B. R., "Separation and detection of smokeless powder additives by ultra performance liquid chromatography with Tandem Mass Spectrometry (UPLC/MS/MS)." *Journal of Forensic Sciences* May 2013; 58(3): 609–615.

Chang, K. H., Yew, C. H., and Abdullah, A. F. L., "Optimization of headspace solid-phase microextraction technique for extraction of volatile smokeless powder compounds in forensic applications." *Journal of Forensic Sciences* 2014; 59(4): 1100–1108.

Roberts, M., Petraco, N., and Gittings, M., "Novel method for the detection of nitroglycerin in smokeless powders." *Science and Justice* 2015; 55(6): 467–471.

ASTM E2998-16 "Standard Practice for Characterization and Classification of Smokeless Powder." ASTM International, West Conshohocken, PA, 2016, Available at www.astm.org

Dennis, D. M. K., Williams, M. R., and Sigman, M. E., "Assessing the evidentiary value of smokeless powder comparisons." *Forensic Science International* 2016; 259: 179–187.

Chajistamatiou, A. S., and Bakeas, E. B., "A rapid method for the identification of nitrocellulose in high explosives and smokeless powders using GC-EI-MS." *Talanta* 2016; 151: 192–201.

Li, F., Tice, J., Musselman, B. D., and Hall, A. B., "A method for rapid sampling and characterization of smokeless powder using sorbent-coated wire mesh and direct analysis in real time - mass spectrometry (DART-MS)." *Science and Justice* 2016; 56(5): 321–328.

Lopez-Lopez, M., Merk, V., Garcia-Ruiz, C., and Kneipp, J., "Surface-enhanced Raman spectroscopy for the analysis of smokeless gunpowders and macroscopic gunshot residues." *Analytical and Bioanalytical Chemistry* 2016; 408: 4965–4973.

Alvarez, A., Yanez, J., Contreras, D., Saavedra, R., Saez, P., and Amarasiriwardena, D., "Propellant's differentiation using FTIR-photoacoustic detection for forensic studies of improvised explosive devices." *Forensic Science International* 2017; 280: 169–175.

ASTM E2999-17. *Standard Test Method for Analysis of Organic Compounds in Smokeless Powder by Gas Chromatography-Mass Spectrometry and Fourier Transform Infrared Spectroscopy.* ASTM International, West Conshohocken, PA, 2017, Available at www.astm.org

Bors, D., and Goodpaster, J., "Mapping smokeless powder residue on PVC pipe bomb fragments using total vaporization solid phase microextraction." *Forensic Science International* 2017; 276: 71–76.

Reese, K. L., Jones, A. D., and Smith R. W., "Characterization of smokeless powders using multiplexed collision-induced dissociation mass spectrometry and Chemometric procedures." *Forensic Science International* 2017; 272: 16–27.

Taudte, R. V., Roux, C., and Beavis, A., "Stability of smokeless powder compounds on collection devices." *Forensic Science International* 2017; 270: 55–60.

Scene Response, Investigation and Sampling

National Institute of Justice, *A Guide for Explosion and Bombing Scene Investigation*. 2000. Available at https://www.swgfex.com/publications

Cullum, H., Lowe, A., Marshall, M. and Hubbard, P. "Physical and chemical evidence remaining after the explosion of large improvised bombs. Part 2: firings of calcium ammonium nitrate/sugar mixtures." *Journal of Forensic Sciences* 2000; 45(2): 333–348.

Phillips, S. A., Lowe, A., Marshall, M., Hubbard, P., Burmeister, S. G., and Williams, D. R., "Physical and chemical evidence remaining after the explosion of large improvised bombs. Part 1: firings of ammonium nitrate/sugar and urea nitrate." *Journal of Forensic Sciences* 2000; 45(2): 324–332.

Gowadia, H. A., and Settles, G. S., "The natural sampling of airborne trace signals from explosives concealed upon the human body." *J Forensic Sci.* 2001 Nov; 46(6): 1324–1331.

Lowe, A. M., Marshall, M., Walker, C. L., and Hubbard, P. "Physical and chemical evidence remaining after the explosion of large improvised bombs. Part 3: firings of calcium carbonate ammonium nitrate/sugar." *Journal of Forensic Sciences* 2001; 46(3): 535–548.

Oxley, J. C., Smith, J. L, Resende E., Rogers, E., Strobel, R. A., and Bender, E. C., "Improvised explosive devices: pipe bombs." *Journal of Forensic Sciences* 2001; 46(3): 510–534.

Kelleher, J. D., "Explosives residue: Origin and Q5 distribution." *Forensic Science Communications* 2002; 4(2)

NIJ A Guide for Explosion and Bombing Scene Investigations. 2002 NCJ 181869, U.S. Department of Justice.

Crowl, D. A., *Understanding Explosions*. Wiley, Hoboken, NJ, 2003.

Christian, D. R., "The multifaceted demands and dangers you face when investigating clandestine laboratories." *Evidence Technol, Mag* 2005 May-June; 18–21.

Royds, D., Lewis, S. W., and Taylor, A. M., "A case study in forensic chemistry: The Bali bombings." *Talanta* 2005 Aug 15; 67(2): 262–268.

Barakey, M., "Complacency can kill at haz-mat incidents." *Fire Engineering*, 2006. http://www.fireengineering.com/display_article/251456/25/none/none/Feat/Complacency-can-kill-at-haz-matincidents.

Kuila, D. K., Sharma, S. P., Chakrabortty, A., and Lahiri, S. C., "Composition Profile of Low Explosives Cases in India." *Forensic Science International* June 2006; 159(2): 127–131.

Reynolds, J., Nunes, P., Whipple, R., and Alcaraz, A., (2006). *On-Site Analysis of Explosives in Various Matrices*, NATO Security through Science Series, Springer, the Netherlands.

Hargather, M. J., and Settles, G. S., "Optical measurement and scaling of blasts from gram-range explosive charges." *Shock Waves* Dec 2007; 17(4): 215–223.

Kuzmin, V., Solovev, M., and Tuzkov, Y., "Forensic investigation of some peroxides explosives." *Central European Journal of Energetic Materials* 2008; 5: 77–85.

Kunz, R. R., Gregory, K. C., Hardy, D., Oyler, J., Ostazeski, A., and Fountain III, A. W., "Measurement of trace explosive residues in a surrogate operational environment: implications for tactical use of chemical sensing in C-IED operations." *Anal Bioanal Chem* 2009; 395: 357–369.

Laska, P. R., "Post Blast Crime Scenes." *Evidence Technology Magazine*, July-August 2009.

Oxley, J. C., Smith, J. L., Bernier, E., Moran, J. S., and Luongo, J., "Hair as forensic evidence of explosive handling." *Propellants, Explosives, Pyrotechnics* 2009; 34(4): 307–314.

Oxley, J. C., Smith, J. L., Yue, J., and Moran, J., "Hypergolic reactions of TNT." *Propellants, Explosives, Pyrotechnics* 2009; 34(5): 421–426.

Widory, D., Minet, J. J., Barbe-Leborgne, B., "Sourcing explosives: A Multi-Isotope approach." *Sci Justice* 2009 Jun; 49(2): 62–72.

Zeichner, A., Abramovich-Bar, S., Tamiri, T., and Almog, J., "A feasibility study on the use of double-sided adhesive coated stubs for sampling of explosive traces from hands." *Forensic Sci Int.* 2009 Jan 30; 184(1–3): 42–46.

Turillazzi, E., Monaci, F., Neri, M., Pomara, C., Riezzo, I., Baroni, D., and Fineschi, V., "Collection of trace evidence of explosive residues from the skin in a death due to a disguised letter bomb. The synergy between confocal laser scanning microscope and inductively coupled plasma atomic emission spectrometer analyses." *Forensic Science International* 2010; 197(1–3): e7–e12.

Ramasamy, A., Hill, A. M., Masouros, S., Gibb, I., Bull, A. M. J., and Clasper, J. C., "Blast-related fracture patterns: A forensic biomechanical approach." *Journal of the Royal Society Interface* 2011; 8(58): 689–698.

Sorensen, A., and McGill, W. L., "What to look for in the aftermath of an explosion? A review of blast scene damage observables." *Engineering Failure Analysis* 2011; 18(3): 836–845.

Tarvin, M., McCord, B., Mount, K., and Miller, M. L., "Analysis of hydrogen peroxide field samples by HPLC/FD and HPLC/ED in DC mode." *Forensic Science International* 2011; 209(1–3): 166–172.

Beveridge, A. D., editor. *Forensic Investigation of Explosions.* CRC Press Taylor & Francis, Boca Raton, FL, 2012.

Bjelovuk, I. D., Jaramaz, S., and Mickovic, D., "Estimation of explosive charge mass used for explosions on concrete surface for the forensic purpose." *Science and Justice* 2012; 52(1): 20–24.

Jardine, J. W., "Presentation of Explosive Casework Evidence." 2012. In: *Investigation of Explosives*, 2nd Edition, ed. A. Beveridge. 2012 Boca Raton, FL. CRC Press (Taylor and Francis Group).

Matyas, R., Selesovsky, J., and Musil, T., "Sensitivity to friction for primary explosives." *J. Hazard. Mater.* 2012: 213–214: 236–241.

Miller, C. J., and Cespedes, E. R., "Methodologies for removing/desorbing and transporting particles from surfaces to instrumentation." *Sensing and Imaging* 2012; 13(3–4): 101–117.

Sachleben, D. J., *Vehicle-borne Improvised Explosives Devices: Collection Analysis and Presentation of Evidence in Forensic Investigation of Explosives.* 2nd Edition, ed. A. Beveridge, Boca Raton, FL. CRC Press (Taylor and Francis Group), 2012.

Sorensen, A., and McGill, W. L., "Utilization of existing blast analysis software packages for the back-calculation of blast loads." *Journal of Performance of Constructed Facilities* 2012; 26(4): 544–546.

Strobel, R. A., "Recovery of material from the scene of an explosion and its subsequent forensic laboratory examination – A team approach." In *Forensic Investigation of Explosives*, 2nd Edition, ed. A. Beveridge. CRC Press (Taylor and Francis Group), Boca Raton, FL, 2012.

Vermette, J. Y., *General Protocols at the Scene of an Explosion in Forensic Investigation of Explosives*. 2nd Edition, ed. A. Beveridge. CRC Press (Taylor and Francis Group), Boca Raton, FL, 2012.

Bjelovuk, I., Jaramaz, S., Elek, P., Mickovic, D., and Kricak, L., "Preliminary estimation of the explosive mass based on the crater resulting from the surface explosion on asphalt." *Strojarstvo* 2013; 55(3): 203–210.

Breeze, J., Hunt, N., Gibb, I., James, G., Hepper, A., and Clasper, J., "Experimental penetration of fragment simulating projectiles into porcine tissues compared with simulants." *Journal of Forensic and Legal Medicine* 2013; 20(4): 296–299.

DeTata, D. A., Collins, P. A., and McKinley, A. J., "A comparison of common swabbing materials for the recovery of organic and inorganic explosive residues." *Journal of Forensic Sciences* 2013; 58(3): 757–763.

Matyas, R., Selesovsky, J., and Musil, T., "Decreasing the friction sensitivity of TATP, DADP and HMTD." *Central European journal of Energetic Materials* 2013; 10(2): 263–275.

Bainazarova, G. Z., and Dilbarkanova, Z. R., "Complex counter explosive examination in investigations of the acts of terrorism accomplished with the use of explosive devices." *Life Science Journal* 2014; 11(11): 367–371.

Bors, D., Cummins, J., and Goodpaster, J., "The anatomy of a pipe bomb explosion: Measuring the mass and velocity distributions of container fragments." *Journal of Forensic Sciences* 2014; 59(1): 42–51.

Bors, D., Cummins, J., and Goodpaster J., "The anatomy of a pipe bomb explosion: The effect of explosive filler, container material and ambient temperature on device fragmentation." *Forensic Science International* 2014; 234: 95–102.

Gonzalez-Nicieza, C., Alvarez-Fernandez, R., Alvarez-Fernandez, M.I., Lopez-Gayarre, F., and Fabian-Alvarez, V., "Forensic analysis of a methane gas explosion in a block of apartments." *Engineering Failure Analysis* 2014; 36: 243–252.

Bjelovuk, I. D., Jaramaz, S., Elek, P., Mickovic, D., and Kricak, L., "Estimation of the explosive mass based on the surface explosion crater on asphalt." *Tehnicki Vjesnik* 2015; 22(1): 227–232.

Smith, J., *Brodie's Bombs and Bombings - A Handbook to Protection, Security, Detection, Disposal and Investigation for Industry, Police and Fire Departments.* 4th Edition. Charles C Thomas Pub Ltd, 2015.

Chetti, S. O., Malve, M. K., and Daundkar, B. B., "Crime scene management of low intensity explosion near Air India Colony." *Journal of Forensic Medicine and Toxicology* 2016; 33(2): 110–112.

Diaz, E., and Thiboutot, S., "Deposition Rates from Blow-in-Place of Different Donor Charges: Comparison of Composition C-4 and Shaped Charges." *Propellants, Explosives, Pyrotechnics* 2017; 42(1): 90–97.

Kotrly, M., Mares, B., Turkova, I., and Beroun, I., "Identification of improvised explosives residues using physical-chemical analytical methods under real conditions after an explosion." *Proceedings of SPIE* 2016; 9823: 98230S.

Thurman, J. T., *Practical Bomb Scene Investigation*, 3rd edition. CRC Press, July 2017.

Wilks, B., Morgan, R. M., and Rose, N. L., "An experimental study addressing the use of geoforensic analysis for the exploitation of improvised explosive devices (IEDs)." *Forensic Science International* 2017; 278: 52–67.

Verolme, E. K., Van der Voort, M. M., Smits, R., Weerheijim, J., Koh, Y. H., and Kang, K. W., "A method for backward calculation of debris in a post blast scene." *Journal of Loss Prevention in the Process Industries* 2018; 51: 54–64.

Czetina, A., Hofstätter, M., Schraml, S., Hubner, M., Sulzer, P., Rothbacher, D., Wurglitsch, R., Riedl, E., Sonntag, M., Moser, S., and Fey, J. "Robot assisted analysis of suspicious objects in public spaces using CBRN sensors in combination with high-resolution LIDAR." IEEE International Symposium on Safety, Security, and Rescue Robotics (SSRR) 2019 Sep 2 (pp. 256–262). IEEE. https://ieeexplore.ieee.org/abstract/document/8848950

Appendix Q
Homemade Explosive Manufacturing Schemes

1 Generic Manufacturing

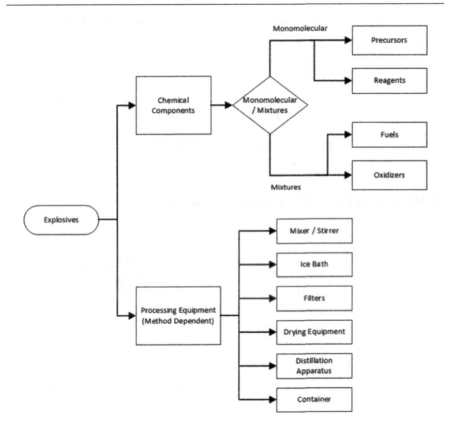

2 Azide Primary Explosives

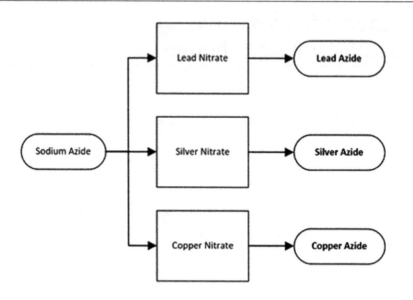

3 Hydrogen Peroxide-Based Explosives

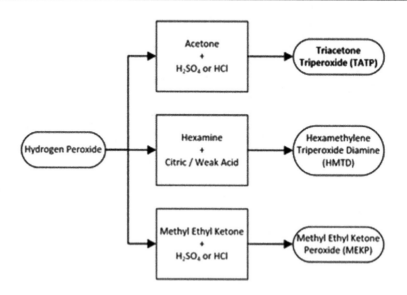

4 Nitrated Organic Explosives

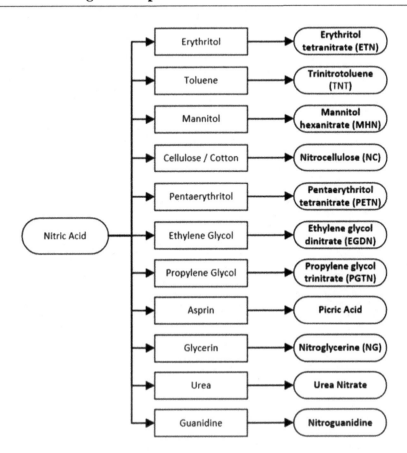

Appendix R
Drug
Chemical Table 1

	Amphetamine	Cocaine	Heroin	Methylenedioxy Amphetamine / Methylenedioxy Methamphetamine (MDA/MDMA)	Methamphetamine	Phencyclidine (PCP) analogs	P-2-P + Phenylacetone
Acetic Acid	*		*	*			*
Acetic Anhydride			*				*
Acetaldehyde					*		
Acetonitrile	*						
a-Acetylphenylacetonitrile							*
Allyl Benzene	*				*		
Allyl Chloride	*				*		
4-Allyl-1,2-Methylenedioxy benzene				*			
Aluminum foil	*				*		
Aluminum Chloride				*			*
Ammonia gas	*		*	*	*		
Ammonium Acetate	*			*			
Ammonium Chloride			*		*		
Ammonium Formate	*			*		*	
Ammonium Hydroxide	*					*	
Benzaldehyde					*		
Benzene	*						
Benzyl Chloride							*
Benzyl Cyanide							*
Bromobenzene						*	*
Bromothiophene						*	

(Continued)

	Amphetamine	Cocaine	Heroin	Methylenedioxy Amphetamine / Methylenedioxy Methamphetamine (MDA/MDMA)	Methamphetamine	Phencyclidine (PCP) analogs	P-2-P + Phenylacetone
Butylamine	*						
Calcium Oxide (anhydrous lime)		*	*				
Chloroacetone	*				*		*
3-Chloropropene					*		
Copper Sulfate				*			
Cuprous Oxide							
Cyclohexanone						*	
Dibromomethane				*			
Ephedrine/Pseudoephedrine					*		*
Ethyl Acetate	*						
Formamide	*			*	*		
Formic Acid	*			*	*		
Hydriodic Acid (HI)	*			*	*		
Hydrobromic Acid (HBr)				*	*		
Hydrochloric Acid (HCl)	*	*	*	*	*	*	
Hydrogen	*				*		
Hydrogen Peroxide				*			
Hydroxylamine HCl					*		
Iodine							*
Iron Filings							

(Continued)

	Amphetamine	Cocaine	Heroin	Methylenedioxy Amphetamine / Methylenedioxy Methamphetamine (MDA/MDMA)	Methamphetamine	Phencyclidine (PCP) analogs	P-2-P + Phenylacetone
Isosafrole				★			
Lithium					★		
Lithium Aluminum Hydride	★						
Magnesium Turnings				★	★	★	
Manganous Carbonate							★
Manganous Chloride							★
Mercuric Chloride	★			★	★		
Mercury	★						
Methylamine				★	★		★
Methylformamide				★	★		★
Nitroethane				★			★
Norpseudoephedrine	★				★		
Palladium Sulfate					★		
Perchloric Acid							
Phenylacetic Acid							
Phenylacetone	★				★		★
Phenylacetonitrile							
Phenylmagnesium Bromide	★					★	
Phosphorus					★		★
Phosphorus Pentachloride					★		

(Continued)

	Amphetamine	Cocaine	Heroin	Methylenedioxy Amphetamine / Methylenedioxy Methamphetamine (MDA/MDMA)	Methamphetamine	Phencyclidine (PCP) analogs	P-2-P + Phenylacetone
Piperidine						⋆	
Piperonal				⋆			
Platinum					⋆		
Platinum Chloride					⋆		
Platinum Oxide					⋆		
Potassium Carbonate						⋆	
Potassium Cyanide						⋆	
Potassium Hydroxide	⋆			⋆	⋆	⋆	⋆
Pumice							⋆
Pyridine	⋆					⋆	
Raney Nickel					⋆		⋆
Sodium					⋆		
Sodium Acetate	⋆						⋆
Sodium Amalgam	⋆						
Sodium Bisulfite						⋆	⋆
Sodium Carbonate			⋆				
Sodium Hydroxide	⋆	⋆		⋆	⋆	⋆	⋆
Sodium Sulfate	⋆						⋆
Sulfuric Acid	⋆	⋆		⋆	⋆		⋆
Thionyl Chloride					⋆		

Appendix S
Drug
Chemical Table 2

AMPHETAMINE	Precursors	Reagents
A1 : Amphetamine via Leuckart reaction	Ammonia Ammonium formate Formamide Phenylacetone	Formic acid Hydrochloric acid Sulfuric acid
A2 : Amphetamine via Benzaldehyde/Nitroethane	Benzaldehyde Nitroethane	Acetic acid Butylamine Hydrogen Lithium aluminum hydride Sodium amalgam Raney nickel
A3: Amphetamine via P-2-P/ Hydroxylamine	Hydroxylamine HCl Phenylacetone	Acetic acid Hydrogen Lithium aluminum hydride Raney nickel Palladium black Sodium acetate Sodium amalgam
A4: Amphetamine via P-2-P/ Ammonia	Ammonia Phenylacetone	Aluminum Mercuric chloride
A5 : Amphetamine via Benzene/ Allyl Chloride/Ammonia	Benzene Allyl chloride Ammonia	Ferric chloride
A6: Amphetamine via Allylbenzene/Acetonitrile	Allylbenzene Acetonitrile	Hydrochloric acid
A7: Amphetamine via Phenylpropanolamine/HI	Phenylpropanolamine	Hydriodic acid Red phosphorus

COCAINE	Precursors	Reagents
	Coca Leaves	Calcium Oxide (Lime) Sulfuric Acid Hydrochloric Acid

FENTANYL Analogs	Precursors	Reagents
Fentanyl/Acetylfentanyl	4-piperidone monohydrate hydrochloride (2-bromoethyl) benzene Aniline Diisopropylethylamine (DIPEA) Acetic anhydride	Acetic acid Cesium carbonate (CsCO) Sodium triacetoxyborohydride Propionyl chloride Hydrochloric acid Citric acid
Thiofentanyl/ Acetylthiofentanyl	2-Thiophene Triethylamine (TEA) 4-piperidone monohydrate hydrochloride Diisopropylethylamine (DIPEA) Acetic anhydride	Acetic acid Mesyl chloride (MsCl) Cesium carbonate (CsCO) Sodium triacetoxyborohydride Propionyl chloride Hydrochloric acid Citric acid

HEROIN	Precursors	Reagents
White Powder	Opium Acetic anhydride	Calcium oxide Ammonium chloride Sodium carbonate Hydrochloric acid
Black Tar	Opium Acetic acid	Calcium carbonate

MDA/MDMA	Precursors	Reagents
MD1: MDA via Isosafrole to 3,4-Methylenedioxy P-2-P using the Leuckart reaction	Ammonia Ammonium formate Formamide Isosafrole	Acetic acid Formic acid Hydrochloric acid Hydrogen peroxide Sulfuric acid
MD2: MDA via Piperonal/Nitroethane	Nitroethane Piperonal	Acetic acid Ammonium acetate Lithium aluminum hydride
MD3: MDMA via 3,4-Methylenedioxy P-2-P/Methylamine	Isosafrole Methylamine	Aluminum foil Mercuric chloride
MD4: MDMA via 3,4-Methylenedioxy P-2-P using the Leuckart reaction	Isosafrole Methylamine Methylformamide	Acetic acid Formic acid Hydrochloric acid Hydrogen peroxide Sulfuric acid
MD5a: MDA via Safrole/HBr/Ammonia	Safrole Ammonia (MDA)	Cuprous oxide Hydrobromic acid
MD5b: MDMA via Safrole/HBr/ Methylamine	Safrole Methylamine (MDMA)	Mercuric chloride Sodium carbonate Sodium hydroxide

METHAMPHETAMINE	Precursors	Reagents
M1: Methamphetamine via P-2-P/ Methylamine	Phenylacetone Methylamine	Aluminum Mercuric chloride
M2: Methamphetamine via Ephedrine/H_2	Ephedrine	Hydrogen Palladium black Palladium sulfate Perchloric acid Phosphorus pentachloride Platinum Platinum chloride Sodium acetate Sulfuric acid Thionyl chloride
M3: Methamphetamine via Ephedrine/HI	Ephedrine	Hydriodic acid Iodine Red phosphorus
M4: Methamphetamine via Benzyl Chloride/Acetaldehyde/Methylamine	Acetaldehyde Benzyl chloride Methylamine	Iodine Magnesium
M5: Methamphetamine via Leuckart reaction	Phenylacetone Methylamine Methylformamide	Formic acid Hydrochloric acid
M6: Methamphetamine via Benzene/ Allyl Chloride/Methylamine	Allyl chloride Benzene Methylamine	Ferric chloride
M7: Methamphetamine via Birch Reduction	Ephedrine	Ammonia Ammonium chloride Lithium or Sodium Sodium chloride Tetrahydrofuran

PHENYLACETONE	Precursors	Reagents
P1: P-2-P via Phenylacetic acid/Acetic anhydride	Phenylacetic acid Acetic anhydride	Sodium acetate
P2: P-2-P via Benzyl cyanide/Ethyl acetate	Benzyl cyanide Ethyl acetate	Acetic acid Phosphoric acid Sodium Sulfuric acid
P3: P-2-P via Phenylacetic acid/Lead acetate	Phenylacetic acid	Lead acetate
P4: P-2-P via Phenylacetic acid/Acetone	Phenylacetic acid Acetic acid	Hydrochloric acid Manganous carbonate Manganous chloride Nitric acid Pumice Sodium carbonate Thorium nitrate

(Continued)

PHENYLACETONE	Precursors	Reagents
P5: P-2-P via Benzaldehyde/Nitroethane	Benzaldehyde Nitroethane	Butylamine Ferric chloride Iron filings
P6: P-2-P via Phenyl-2-propanol/Dichromate	Phenyl-2-propanol	Potassium dichromate
P7: P-2-P via Benzene/Chloroacetone	Benzene Chloroacetone	Aluminum chloride Sodium bisulfate

PHENCYCLIDINE analogs	Precursors	Reagents
PC1: Piperidine/Cyclohexane intermediate **PC1a:** Phenyl addition **PC1b:** Thiophene addition	Cyclohexanone Piperidine Bromobenzene (a) Phenylmagnesium bromide (a) Bromothiophene (b)	Magnesium Potassium cyanide Sodium cyanide
PC2: Pyridine/Cyclohexane intermediate **PC2a:** Phenyl addition **PC2b:** Thiophene addition	Cyclohexanone Pyridine Bromobenzene (a) Phenylmagnesium bromide (a) Bromothiophene (b)	Magnesium Potassium cyanide Sodium cyanide
PC3: Morpholine/Cyclohexane intermediate **PC3a:** Phenyl addition **PC3b:** Thiophene addition	Cyclohexanone Morpholine Bromobenzene (a) Phenylmagnesium bromide (a) Bromothiophene (b)	Magnesium Potassium cyanide Sodium cyanide

Appendix T
Explosives
Chemical Table 1

Chemical	Primary Explosive	Secondary Explosive	Black Powder	Flash Powder	Pyrotechnics
Acetic anhydride		*			
Acetone	*				*
Aluminum, dark pyro				*	*
Aluminum, powder		*		*	*
Ammonia	*				
Ammonium dichromate			*		*
Ammonium nitrate	*	*	*	*	*
Ammonium perchlorate					*
Ammonium picrate			*		*
Antimony, powder				*	*
Antimony trisulfide				*	*
Barium chlorate				*	*
Barium nitrate				*	*
Barium peroxide				*	*
Barium sulfate				*	*
Benzene		*		*	
Benzene hexachloride					*
Black copper oxide				*	*
Calcium hydride				*	*
Calcium/magnesium 75/25				*	*
Calcium metal				*	*
Charcoal			*	*	
Copper oxychloride					*
Cotton (cellulose)		*			

(Continued)

Chemical	Primary Explosive	Secondary Explosive	Black Powder	Flash Powder	Pyrotechnics
Dimethylaniline		★			
Glycerin		★			
Hexachlorobenzene					★
Hexachloroethane					★
Hexamethylene-tetramine		★			
Hexamine	★	★			★
Hydrochloric acid	★				
Hydrogen peroxide	★				
Hydroxyethyl cellulose				★	
Iodine crystals	★				
Lead acetate	★				
Lead nitrate	★				
Lead peroxide					★
Lithium perchlorate				★	
Magnalium				★	★
Magnesium, powder	★	★		★	★
Mercury	★				
Methylaniline		★			
Nitric acid	★	★			
Petroleum jelly		★			★
Phenol		★			
Potassium chlorate		★		★	★
Potassium dichromate				★	★
Potassium iodide	★				

(Continued)

Chemical	Primary Explosive	Secondary Explosive	Black Powder	Flash Powder	Pyrotechnics
Potassium nitrate			*	*	*
Potassium perchlorate		*		*	
Potassium permanganate	*			*	*
Potassium picrate	*		*	*	
Red phosphorus	*				
Sodium azide	*				
Sodium chlorate			*		*
Sodium nitrate				*	*
Sodium perchlorate			*	*	
Starch		*			
Strontium nitrate				*	*
Strontium perchlorate				*	*
Strontium peroxide					*
Sucrose				*	*
Sulfur			*	*	
Sulfuric acid		*			
Titanium powder		*		*	*
Toluene		*			
Trichlorophenol	*			*	
Trinitroresorcinate		*			
Zinc, powder					*

Appendix U
Explosives
Chemical Table 2

Primary/Initiating Explosives	Precursors	Reagents
Ammonium triiodide ($NH_4)I_3$	Iodine crystals Ammonia	Potassium iodide
Armstrong Mixture	Potassium chlorate Red phosphorus	
HMDT (Hexamethylene Triperoxide Diamine)	Hexamine Hydrogen peroxide	Citric acid Hydrochloric acid Sodium bicarbonate Sodium carbonate
Lead Azide ($Pb(N_3)_2$)	Lead acetate Sodium azide	
Lead Styphnate (Lead Trinitroresorcinate)	Lead nitrate Trinitroresorcinate	Magnesium oxide
Mercury Fulminate	Mercury Nitric acid	
Nitrogen Trichloride (NCl_3)	Ammonium nitrate	Hydrochloric acid Potassium permanganate
TATP (Triacetone Triperoxide)	Acetone Hydrogen peroxide	Hydrochloric acid Sodium bicarbonate Sodium carbonate

Secondary Explosives	Precursors	Reagents
Nitrocellulose	Cotton (cellulose) Nitric acid	Sulfuric acid
Nitroglycerin	Glycerin Nitric acid	Sulfuric acid
Nitrostarch	Starch Nitric acid	Sulfuric acid
PETN (Pentaerythritol tetranitrate)	Pentaerythritol Nitric acid	Sulfuric acid
Picrates	Picric acid Sodium hydroxide Potassium hydroxide Ammonia Ammonium hydroxide	

(Continued)

Secondary Explosives	Precursors	Reagents
Picric Acid (2,4,6-trinitrophenol)	Phenol Aspirin Nitric acid	Sulfuric acid
RDX (Cyclonite)	Hexamine Methenamine Nitric acid Paraformaldehyde Hexamethylenetetramine	Sulfuric acid Ammonium nitrate Acetic anhydride
Tetryl (trinitro-2,4,6-phenylmethylnitramine)	Methylaminedimethylaniline Nitric acid	Sulfuric acid
TNT (Trinitrotoluene)	Toluene Nitric acid	Sulfuric acid

FOX Mixtures	Oxidizer	Fuel
ANAL	Ammonium nitrate	Aluminum powder Zinc powder
ANFO	Ammonium nitrate	Motor oil Kerosene Nitromethane
Chlorate Mixtures	Potassium chlorate Sodium chlorate	Petroleum jelly Aluminum powder Magnesium powder Sugar Zinc powder
Perchlorate Mixtures	Potassium perchlorate Sodium perchlorate	Petroleum jelly Aluminum powder Magnesium powder Sugar Zinc powder
Flash Powders	Potassium chlorate Potassium perchlorate Ammonium perchlorate Potassium nitrate Barium nitrate	Metal fuels: Al, Mg, Ti, Fe Sulfur Phosphorus Antimony sulfide Arsenic sulfide Sugar
Pyrotechnic Mixtures	Potassium chlorate Potassium perchlorate Ammonium perchlorate Potassium nitrate Barium nitrate Strontium nitrate	Sulfur Phosphorus Sawdust Wax Oil Sugar Al, Mg, Ti, Fe, Sr, Ba

Black Powders	Oxidizer	Fuel
Black Powder	Potassium nitrate	Charcoal
	Sodium nitrate	Sulfur
Black Powder	Potassium nitrate	Charcoal
Substitutes	Sodium nitrate	Sulfur
		Sugar
		Borax
		Dextrin
		Ascorbic acid
Golden Powder®	Potassium nitrate	Ascorbic acid
Black Canyon®		
Pyrodex® and	Potassium nitrate	Sodium benzoate
Triple Seven®	Potassium perchlorate	Dicyandiamide
		Charcoal
		Borax
		Dextrin
		Sulfur (Pyrodex)
		Sodium 3 nitrobenzoate (Triple Seven)

Index

Printed in the United States
by Baker & Taylor Publisher Services